Radio Propagation Measurements and Channel Modeling: Best Practices for Millimeter-Wave and Sub-Terahertz Frequencies

This book offers comprehensive, practical guidance on RF propagation channel characterization at mmWave and sub-terahertz frequencies, with an overview of both measurement systems and current and future channel models. It introduces the key concepts required for performing accurate mmWave channel measurements, including channel sounder architectures, calibration methods, channel sounder performance metrics and their relationship to propagation channel characteristics. With a comprehensive introduction to mmWave channel models, the book allows readers to carefully review and select the most appropriate channel model for their application. The book provides fundamental system theory accessible in a step by step way with clear examples throughout. With inter- and multi-disciplinary perspectives, the reader will observe the tight interaction between measurements and modeling for these frequency bands and how different disciplines, of necessity, interact in this area. This book gathers insights from the NextG Channel Model Alliance, with over over 180 participants from more than 80 organizations, arguably the world's most active and knowledgeable researchers on the subject. This is an excellent reference for researchers, including graduate students, working on mmWave and sub-THz wireless communications, and for engineers developing communication systems.

Theodore S. Rappaport founded NYU WIRELESS, a multi-disciplinary research center, and the wireless research centers at the University of Texas at Austin and Virginia Tech. Professor Rappaport is a member of the National Academy of Engineering and the Wireless Hall of Fame, and a fellow of the National Academy of Inventors.

Kate A. Remley is the leader of the Metrology for Wireless Systems Project at the National Institute of Standards and Technology. Here, Dr. Remley specializes in the development of calibrated measurements for microwave and millimeter-wave wireless systems and standardized over-the-air test methods for the wireless industry.

Camillo Gentile is leading the Radio Access and Propagation Metrology Group and the NextG Channel Measurement and Modeling Project at the National Institute of Standards and Technology. Here, Dr. Gentile prototypes novel radio-frequency channel measurement and modeling techniques for the millimeter-wave and sub-terahertz bands.

Andreas F. Molisch is *Solomon Golomb–Andrew and Erna Viterbi* Chair Professor at the University of Southern California. He is a Fellow of the National Academy of Inventors, AAAS, IEEE, IET, a member of the Austrian Academy of Sciences and the recipient of numerous awards.

Alenka Zajić is Ken Byres Professor in the School of ECE at Georgia Tech. She is the recipient of many awards, including NSF CAREER Award, multiple Best Paper Awards and the Neal Shepherd Memorial Best Propagation Paper Award. Her research interests span applied electromagnetics, wireless communications and compute systems.

Radio Propagation Measurements and Channel Modeling: Best Practices for Millimeter-Wave and Sub-Terahertz Frequencies

Edited by

THEODORE S. RAPPAPORT
New York University

KATE A. REMLEY
National Institute of Standards and Technology, Boulder

CAMILLO GENTILE
National Institute of Standards and Technology, Gaithersburg

ANDREAS F. MOLISCH
University of Southern California

ALENKA ZAJIĆ
Georgia Institute of Technology

CAMBRIDGE
UNIVERSITY PRESS

CAMBRIDGE
UNIVERSITY PRESS

University Printing House, Cambridge CB2 8BS, United Kingdom

One Liberty Plaza, 20th Floor, New York, NY 10006, USA

477 Williamstown Road, Port Melbourne, VIC 3207, Australia

314–321, 3rd Floor, Plot 3, Splendor Forum, Jasola District Centre, New Delhi – 110025, India

103 Penang Road, #05–06/07, Visioncrest Commercial, Singapore 238467

Cambridge University Press is part of the University of Cambridge.

It furthers the University's mission by disseminating knowledge in the pursuit of education, learning, and research at the highest international levels of excellence.

www.cambridge.org
Information on this title: www.cambridge.org/9781009100717
DOI: 10.1017/9781009122740

© Theodore S. Rappaport, Kate A. Remley, Camillo Gentile,
Andreas F. Molisch and Alenka Zajić 2022

First published 2022

A catalogue record for this publication is available from the British Library.

ISBN 978-1-009-10071-7 Hardback

TSR: To my brilliant, industrious and creative graduate students.

KAR: To my father, Frederick M. Remley, whose love and support enabled my lifelong journey in radio and wireless engineering, even when this was an unconventional choice for females.

CG: To my father who passed away last year, for his unwavering love, strength and support. You will always be my role model, my hero.

AFM: To the memory of my dear friend, mentor and role model Larry Greenstein.

AZ: To my family, for all their patience and support over the years.

Contents

Contributors

Mustapha Bennai
Communications Research Centre, Ottawa, Canada

Anmol Bhardwaj
NIST Communications Technology Laboratory, Gaithersburg, USA

Hyun-Kyu Chung
ETRI, Daejeon, Korea

Yvo de Jong
Communications Research Centre, Ottawa, Canada

Diego Dupleich
Electronic Measurements and Signal Processing Group, Technische Universität Ilmenau, Ilmenau, Germany

Camillo Gentile
NIST Communications Technology Laboratory, Gaithersburg, USA

Ismail Guvenc
North Carolina State University, Raleigh, USA

Katsuyuki Haneda
Department of Electronics and Nanoengineering, University of Aalto, Espoo, Finland

Ruisi He
State Key Lab of Rail Traffic Control and Safety, Beijing Jiaotong University, Beijing, China

Robert W. Heath, Jr.
Department of Electrical and Computer Engineering, North Carolina State University, Raleigh, USA

Aki Hekkala
Keysight Technologies, Oulu, Finland

Martin Käske
Electronic Measurements and Signal Processing Group, Technische Universität Ilmenau, Ilmenau, Germany

Myung-Don Kim
ETRI, Daejeon, Korea

Ozge Hizir Koymen
Qualcomm Technologies, Inc., Bridgewater, USA

Russell W. Krueger
Waterborne Environmental, Champaign, USA

Pekka Kyosti
Centre for Wireless Communications, University of Oulu, Oulu, Finland; Keysight Technologies, Oulu, Finland

Juyul Lee
ETRI, Daejeon, Korea

George MacCartney, Jr.
NYU WIRELESS, New York University, New York, USA

Alexander Maltsev
University of Nizhny Novgorod, LG Electronics, Nizhny Novgorod, Russia

David Matolak
University of South Carolina, Columbia, USA

David Michelson
University of British Columbia, Vancouver, Canada

Andreas F. Molisch
University of Southern California, Los Angeles, USA

Robert Müller
Electronic Measurements and Signal Processing Group,Technische Universität Ilmenau, Ilmenau, Germany

Ozgur Ozdemir
North Carolina State University, Raleigh, USA

Peter Papazian
retired, formerly NIST Communications Technology Laboratory,
Boulder, USA

Jae-Joon Park
ETRI, Daejeon, Korea

Andrey Pudeyev
LG Electronics, Nizhny Novgorod, Russia

Jeanne T. Quimby
NIST Communications Technology Laboratory, Boulder, USA

Theodore S. Rappaport
NYU WIRELESS, New York University, New York, USA

Kate A. Remley
NIST Communications Technology Laboratory, Boulder, USA

Sana Salous
Department of Engineering, Durham University, Durham, UK

Akbar Sayeed
Independent Researcher, Madison, USA

Christian Schneider
Electronic Measurements and Signal Processing Group, Technische Universität
Ilmenau, Ilmenau, Germany

Jelena Senic
NIST Communications Technology Laboratory, Boulder, USA

Ruoyu Sun
CableLabs, Louisville, Kentucky, USA

Shu Sun
NYU WIRELESS, New York University, New York, USA

Reiner Thomä
Electronic Measurements and Signal Processing Group, Technische Universität
Ilmenau, Ilmenau, Germany

Vutha Va
Department of Electrical and Computer Engineering, University of Texas at Austin,
Austin, USA

Usman Virk
Keysight Technologies, Oulu, Finland

Yunchou Xing
NYU WIRELESS, New York University, New York, USA

Alenka Zajić
Georgia Institute of Technology, Atlanta, USA

Preface

In 2015 something special and unusual occurred in the field of wireless communications research. Several researchers in academia and industry received an email from Dr. Thyaga Nandagopal, a highly regarded program manager at the National Science Foundation in the USA. The email from Thyaga (as he is known to his friends and colleagues) contained a persuasive invitation to join a budding organization called the *5G mmWave Channel Model Alliance*. Dr. Nandagopal's invitation further explained that this new "Alliance" was being organized by Drs. Nada Golmie and Kate Remley of the National Institute of Standards and Technology (NIST), and that this collaborative effort could be an excellent venue to share knowledge and information, and to compare notes on the new area of millimeter-wave (mmWave) channel measurement and modeling that was just beginning to spawn within industry, academia and federal agencies across the world. Thyaga even suggested that the effort could be complementary to the channel modeling activities that were launching in various standards bodies for the global mobile communications industry, and could be used to promote the academic creed of the open sharing of data, measurement best practices, models and ideas.

The idea of sharing data with others to enable common metrics and models of measurement is something that NIST has historically advocated, and now it seemed that this worthy goal was being pursued through the 5G mmWave Channel Model Alliance. In fact, Drs. Golmie and Remley made clear when establishing the 5G mmWave Channel Model Alliance in the summer of 2015 that NIST wanted it to serve as a neutral, open and transparent forum for fostering collaboration and tackling long-term channel propagation measurement and modeling challenges for 5G. Indeed, as the national metrology laboratory for the United States, NIST has always had a strong interest in measurement science, with a passion and mission for developing sound techniques for measurement and modeling that can be shared with organizations, industry, academia and the public throughout the world.

From those early beginnings, at a time when the mobile communications world had very little understanding of mmWave radio propagation, not to mention channel measurement and modeling at those frequencies, a remarkable collaboration began to take hold. NIST contracted with Marc Leh, a consultant from Corner Alliance, Inc., who began to coordinate what would soon become one of the most active and productive global collaborations in the fields of wireless communications propagation measurements and modeling. Today, the 5G mmWave Channel Model Alliance has expanded its mission and scope, and in 2021 was renamed as the NextG Channel Model Alliance, involving over 183 participants from 84 organizations worldwide (see: https://sites.google.com/a/corneralliance.com/5g-mmwave-channel-model-alliance-wiki/home).

Over the course of five years and nearly 100 teleconference meetings, through thousands of shared documents and discussions and countless hours of collaboration, coordination, debate, editing and benchmarking, a massive body of knowledge has been compiled through the voluntary efforts of hundreds of engineers, scientists, researchers and students from around the world. This manuscript is an earnest attempt by the authors to distill and compile this vast body of knowledge, and is the culmination of the massive collaborative effort undertaken by dozens of dedicated volunteers and "spectrum explorers" who so kindly and willingly shared their knowledge and best practices with the engineering world as it embarks on the deployment of 5G mmWave networks globally.

It is rare to find a publication that brings together such a vast collection of experiences and understanding from the world's most active and knowledgeable researchers who have worked in the relatively unexplored mmWave and sub-terahertz spectrum bands. The authors have worked tirelessly to organize the vast and valuable material to allow a newcomer to the fields of radio-propagation measurement and channel modeling to rapidly learn the fundamentals and key findings that helped launch the mmWave wireless revolution.

This book is roughly divided into two parts, with the first part focusing on measurement and the latter part focusing on channel modeling. Chapter 1 provides context and background for the work presented here, with a focus on the essential link between measurement and modeling for mmWave and terahertz (THz) wireless systems. Chapter 2 cements this link by describing important channel metrics to be estimated from measurement. In Chapter 3, an overview of some common channel sounder architectures is provided, including calibration and timing, which are critical for mmWave and higher frequencies, and methods to ascertain key channel sounder system parameters. The characteristics of the participants' state-of-the-art in channel sounders are also provided as a useful benchmark in Chapter 3. In Chapter 4, various methods for verifying the performance of channel sounders are presented. Understanding the nonidealities of a channel sounder is key to providing meaningful and accurate measurement-derived channel characteristics. Verification techniques are illustrated with examples from the contributors' own systems. Chapter 5 introduces important concepts in mmWave channel modeling, including the types of models that are described in subsequent chapters. Chapter 5 provides background on parameters used in deterministic models that focus on a specific time and/or location for the measurements. Chapter 6 describes current approaches to modeling path loss and shadowing, with models that are especially relevant to mmWave and sub-THz bands. Chapter 7 provides recent progress in clustering and tracking algorithms, which are widely used in the analysis of field data and the creation of channel models for new communication systems. In Chapter 8, time dispersion characteristics are described. These are the backbone of efficient, lower-complexity, geometry-based stochastic channel models that can quickly generate a representative impulse response of the multipath channel. Chapter 9 briefly describes the growing field of peer-to-peer networking, which includes device-to-device and vehicular applications. With the increased focus on handheld wireless devices, human blockage models are becoming increasingly important, which is the focus of Chapter 10, including both blockage but also Doppler effects. Finally, Chapter 11 introduces important concepts in the next generation of THz channel models, covering frequencies from 100 GHz to 3 THz. THz wireless systems promise ultrahigh data rates and super-fast download speeds, as well as imaging, sensing and spectroscopy applications, motivating a growing body of research in this area. Chapters 12 and 13 conclude the entire body of work with a cohesive discussion on the connection between measurements and models. Looking to the future, the synergy between these two disciplines is made clear.

The mobile communications industry made a fundamental and historic shift when it added mmWave operating carrier frequencies that moved everything higher in the spectrum by an order of magnitude for the first time. The international cellular telephone standards body, 3GPP, ratified Release 15 in 2018, and the world's first 5G mmWave cellular network began operation soon after that. With 5G mmWave now a reality, the engineering advances in semiconductors, computing, materials, device integration and packaging, antenna arrays and signal processing will surely continue to move up in frequency, where even greater bandwidths and temporal/spatial resolution will unleash new applications that we can only imagine today. It is our hope that this book will help those future explorers understand the fundamental principles and practice that will help open up these exciting new frequency bands in the decades to come.

1 Introduction

Theodore S. Rappaport, Kate A. Remley, Camillo Gentile,
Andreas F. Molisch and Alenka Zajić

The demand for the next generation of wireless networks has been spurred by smartphones and other data-intensive mobile devices. The popularity of these devices has led to an exponential increase in wireless data transmission, and hence the need to provide massive capacity increases (upwards of 1,000 times) and connectivity (billions of users and machines) in addition to supporting an increasingly diverse set of services and applications. It is anticipated that future generations of mobile wireless networks will be a combination of technologies, such as those used in wireless local-area networks (e.g., Wi-Fi) and those used in cellular communications (e.g., fifth-generation, or 5G) operating in both licensed and unlicensed bands between 300 MHz and 300 GHz. In particular, the use of millimeter-wave (mmWave) frequencies, between 28 and 300 GHz, is expected to increase bandwidth by as much as 100 times over what is currently available for cellular networks today.

To meet this challenge, several international fora have been established to explore issues related to hardware, networks and standards [1–12]. The goal of many of these groups is to address short-term needs that solve specific problems, as opposed to longer-duration research efforts. For example, in ITU-R there are several groups studying relevant issues in 5G general and mmWave channel modeling. As another example, the cellular industry standards group Third Generation Partnership Project (3GPP) has recently published its technical specifications for mmWave channel models in Release 14 [2], and the Working Party 5D in Study Group 5 (ITU-R WP5D) is studying not only the vision and requirements for 5G, but also 5G channel models for evaluation purposes [13]. The mmWave channel models coming from this group are almost identical to the 3GPP channel models. Study Group 3 in ITU-R [14] also studies channel models at mmWave frequencies for outdoor short-range propagation [15], indoor propagation [16], outdoor-to-indoor propagation [17], outdoor-to-indoor propagation model definitions and measurement guides [18].

The 5G mmWave Channel Model Alliance was formed to take a longer view by addressing issues related to measurement and modeling that impede progress in standards development and hardware optimization [19].[1] The kick-off meeting was held in July 2015 at the US National Institute of Standards and Technology in Boulder, CO, USA. More than 30 participants discussed a vision for cooperative engagement intended to fill the need for longer-term research into propagation effects specific

---

[1] Here, "5G" refers to the next generation of mobile wireless communication systems.

to mobile wireless applications at mmWave frequencies and how those effects can be translated into robust channel models. In May 2021 the 5G mmWave Channel Model Alliance was renamed the NextG Channel Model Alliance and included 183 participants from 84 organizations worldwide. The group primarily meets through teleconferences and currently has subgroups dedicated to measurement and channel-modeling activities.

One key Alliance strength is that participants utilize a wide range of channel sounders with various architectures. This allows the group to study representative propagation environments with several different channel-sounder technologies that will ultimately result in more robust channel models. For example, vector-network-analyzer (VNA)-based channel sounders provide a high dynamic range, allowing detailed insight into the fading characteristics of a specific environment, but primarily for slowly varying or static channels. On the other hand, sampler-based channel sounders are often fast, providing instantaneous channel information. As another example, some sounders have active antenna arrays capable of resolving the angle of arrival of multipath components (MPCs) in the plane of the antenna to within a few degrees, while others have antenna coverage over a hemisphere with nominally lower angular resolution, and yet others use lens antenna arrays for analog multi-beamforming that can provide spatial resolution comparable to or finer than existing phased arrays at the cost of increased physical size.

In order for Alliance members to combine measured data from sounders having different architectures, it is essential to have confidence that each channel sounder is performing as expected. This includes verifying that the resulting measured data and the postprocessing routines provide results in agreement with the theory. Verified data can then be used to extract statistically representative metrics that feed into channel models, such as path loss and power delay profile.

The propagation channel is independent of the measurement system that samples it. That is, the channel sounder will provide only an estimate of the characteristics of the true channel. The extent to which the sounder captures the channel's features will depend on the hardware employed and how well that hardware works, which is why verification is so important. However, there are many channel sounder architectures, each of which may utilize various types of hardware. Such hardware is typically selected to capture the channel effects that are of most relevance to the channel models to be employed, which are in turn chosen to be relevant to the wireless system that will operate in the channel. That is, the channel sounder must capture the channel features that are required for parameterizations of the channel model and this will dictate the type of hardware that is selected. Thus, there is a strong link between the modeling and measurement communities.

The second part of this book describes the main mmWave models that have been presented in the literature, including tapped delay line stochastic models, geometry-based stochastic channel models and quasi-deterministic models. Various path loss and shadowing models have also been included. Because the clustering of MPCs into groups with similar behavior is important for all of these types of models, a discussion on this has been included.

Each of these types of models has been considered for mmWave channel modeling by various academic and industrial groups, including members of the 5G mmWave Channel Model Alliance. Having a compendium of these models and model types in a single volume will allow users to compare and select the model that may be of the most utility to their projects.

References

[1] Institute of Electrical and Electronics Engineers (IEEE), "Future Networks Initiative," online: https://futurenetworks.ieee.org.

[2] 3GPP, "Technical specification group radio access network: Study on channel model for frequencies from 0.5 to 100 GHz (Release 14)," 3rd Generation Partnership Project (3GPP), TR 38.901 V14.2.0, Sept. 2017, online: www.3gpp.org/DynaReport/38901.htm.

[3] COST IC1004, "White paper on channel measurements and modeling for 5G networks in the frequency bands above 6 GHz," online: www.ic1004.org.

[4] European Cooperation in Science and Technology (COST), "Inclusive radio communications (IRACON)," online: www.iracon.org.

[5] National Science Foundation (NSF), "Millimeter-wave Research Coordination Network (RCN)," online: http://mmwrcn.ece.wisc.edu.

[6] European Telecommunications Standards Institute (ETSI), "Millimetre wave transmission," online: http://www.etsi.org/technologies-clusters/technologies/millimetre-wave-transmission.

[7] ETSI, "5th generation," online: www.etsi.org/technologies-clusters/technologies/5g.

[8] The 5G Infrastructure Public Private Partnership (5G-PPP), "Homepage," online: https://5g-ppp.eu.

[9] METIS-II, "Homepage," online: www.metis2020.com.

[10] 5G Forum, "Homepage," online: 5gforum.org/english.

[11] Fifth Generation Mobile Communications Promotion Forum (5GMF), "Homepage," online: http://5gmf.jp/en.

[12] Global Information and Communications Technology Standardisation Forum for India (GISFI), "Homepage," online: www.gisfi.org.

[13] International Telecommunication Union (ITU-R) Working Party 5D, "International mobile telecommunications (IMT) systems," online: www.itu.int/en/ITU-R/study-groups/rsg5/rwp5d/Pages/default.aspx.

[14] International Telecommunication Union (ITU-R) Study Group 3, "Radiowave propagation," online: www.itu.int/en/ITU-R/study-groups/rsg3/Pages/default.aspx.

[15] International Telecommunication Union, "Recommendation ITU-R P.1411-9: Propagation data and prediction methods for the planning of short-range outdoor radiocommunication systems and radio local area networks in the frequency range 300 MHz to 100 GHz," online: www.itu.int/rec/R-REC-P.1411-9-201706-I/en.

[16] International Telecommunication Union, "Recommendation ITU-R P.1238-9: Propagation data and prediction methods for the planning of indoor radiocommunication systems and radio local area networks in the frequency range 300 MHz to 100 GHz," online: www.itu.int/rec/R-REC-P.1238-9-201706-I/en.

[17] International Telecommunication Union, "Recommendation ITU-R P.2109: Prediction of building entry loss," online: www.itu.int/rec/R-REC-P.2109/en.

[18] International Telecommunication Union, "Recommendation ITU-R P.2040-1: Effects of building materials and structures on radiowave propagation above about 100 MHz," online: www.itu.int/rec/R-REC-P.2040/en.

[19] 5G mmWave Channel Model Alliance, "Homepage," online: https://sites.google.com/a/corneralliance.com/5g-mmwave-channel-model-alliance-wiki/home.

2　Estimating Channel Characteristics from Measurements

Kate A. Remley, Sana Salous, Theodore S. Rappaport, Ruoyu Sun, Camillo Gentile, Andreas F. Molisch and Alenka Zajić

Many metrics used to characterize propagation-channel conditions (such as path loss and time delay) are derived from measured data collected by channel sounders. Parameters extracted from these metrics are often used in channel models. Channel model selection can have a strong influence on deployment and system performance metrics, such as spectral efficiency and coverage, as well as hardware/signal processing requirements. Typically, averaged metrics and parameters are calculated from hundreds or thousands of measurements made in a specific category of environment, such as an urban canyon or an indoor office. The large number of measurements is needed because each city and office is different. However, no measurement instrument is ideal, so the metrics and parameters derived from each single measurement will only approximate the true channel conditions. To emphasize this fact, we refer to channel metrics derived from measurement as "estimated" quantities.

Estimated quantities will be affected both by known system limitations and by unknown measurement errors. For example, all channel sounders use electronic circuits that restrict operation to a certain frequency range, frequency step, amplitude range and time step. They also use antennas with limited angular resolution and specific polarization characteristics to capture channel characteristics. These are known limitations arising from system design. In addition, unintentional nonidealities in the electronic hardware (for example, amplitude compression of a power amplifier, excessive noise floor of a receiver and timing errors between transmitters and receivers) and mechanical hardware (for example, system and antenna positioning errors or nonideal antenna machining) will also impact the accuracy of the measurement. To a large extent, it is the goal of channel sounder verification to reveal system limitations and unintentional errors inherent in the measurement system and procedures. This knowledge allows uses of channel sounders having different architectures to anticipate whether or not the metrics extracted from their data should be directly comparable.

In the next section we provide a commonly used impulse-response representation of the propagation channel to illustrate the effects of the nonideal hardware on estimates of channel metrics. In the subsequent chapters we define metrics of interest that are often derived from channel sounder measurements.

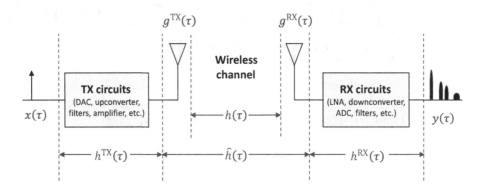

Figure 2.1 Impulse response model for characterizing a wireless propagation channel. The measured output $y(\tau)$ is a function of the input modified by the channel as well as the nonideal responses of the hardware and antennas.

2.1   Measuring Propagation-Channel Characteristics

We define the propagation channel as the environment between the transmitter (TX) and receiver (RX) of the channel sounder. At a minimum, most channel sounders try to estimate the path loss or attenuation between the transmitter and receiver, and many do this by measuring the "impulse response" of the channel. In the impulse response model (see Figure 2.1), propagation-channel characteristics that can impact the propagation of a signal between the transmitter and receiver, such as diffraction, reflection and scattering are considered. Additionally, shadowing may occur due to movement of the transmitter, receiver and/or objects in the environment. In this model, the received voltages, corresponding to the electromagnetic (EM) waves incident on the RX antenna, are a superposition of the waves coming from the line-of-sight (LoS) path and waves coming from different directions due to the scatterers in the channel. This superposition of multiple reflections is known as *multipath propagation*.

An impulse-response-based model for a wireless propagation channel measurement is given in Figure 2.1, where the goal is to estimate the complex channel impulse response (CIR), mathematically denoted as $h(\tau)$, from a measurement $y(\tau)$. Here, $x(\tau)$ is the input signal, $g^{\mathrm{TX}}(\tau)$ and $g^{\mathrm{RX}}(\tau)$ are the nonideal antenna responses and $h^{\mathrm{TX}}(\tau)$ and $h^{\mathrm{RX}}(\tau)$ are the nonideal TX and RX responses, respectively. Note that the variable τ, or "excess time delay," is often used to denote the time dependence of the system, as opposed to the absolute time. The excess time delay is the time difference between a signal that arrives along an LoS path and the various multipath components (MPCs) that arrive later. This variable, as opposed to the absolute time delay, t, is often more useful to system designers who must compensate for multipath with equalization.

It is important to note that, for this model, the antenna responses are part of the measured channel response, so the impulse response, or power delay profile, measured by a channel sounder may differ depending on the type of antenna used. For example, a directional transmit antenna facing the receive antenna may not excite many reflections off of surfaces within the environment, resulting in a measurement of fewer MPCs than an omnidirectional antenna. The channel plus antenna response is denoted as $\hat{h}(\tau)$ in Figure 2.1. In some cases, the gains of the antennas are de-embedded in postprocessing. Often, $g^{TX}(\tau)$ and $g^{RX}(\tau)$ are assumed constant and their values are simply subtracted (in dB) from the final path loss value.

However, to characterize directional effects, techniques for de-embedding directional antennas are used in postprocessing of channel measurements. Characterizing directional effects is especially important at millimeter-wave (mmWave) frequencies because mmWave applications often require that antennas are directional to overcome path loss and minimize spatial interference between users. As a result, many mmWave channel models require information on the antenna response as a function of the transmitted signal's angle of departure (AoD) and the received signal's angle of arrival (AoA). Methods for verifying AoA and AoD are discussed in subsequent chapters, and channel models depending on these parameters are discussed in Chapter 7.

The model in Figure 2.1 can be used to provide an estimate of the channel response $h(\tau)$, which we call $\tilde{h}(\tau)$ to denote that it is derived from measurement. Assuming the system is linear, this can be written as

$$y^{(\text{meas, uncorr})}(\tau) = x(\tau) \times h^{TX}(\tau) \times g^{TX}(\tau) \times \tilde{h}(\tau) \times g^{RX}(\tau) \times h^{RX}(\tau), \quad (2.1)$$

or

$$y^{(\text{meas, uncorr})}(\tau) = x(\tau) \times h^{TX}(\tau) \times \hat{\tilde{h}}(\tau) \times h^{RX}(\tau), \quad (2.2)$$

where "uncorr" indicates that calibrations have not been applied to correct for the nonideal response of the system hardware. In the frequency domain, the convolutions given in eq. (2.2) correspond to multiplications and may be written as

$$Y^{(\text{meas, uncorr})}(f) = X(f)H^{TX}(f)\hat{\tilde{H}}(f)H^{RX}(f). \quad (2.3)$$

The frequency domain representation often provides a straightforward way for mathematically applying calibrations, mismatch corrections or other corrections. The impulse-response model described here can inform designers what conditions to expect in deployment or it may be convolved with a signal to simulate performance. Calibration methods and issues for single-antenna systems are discussed in Chapter 3.

2.2   Path Loss

Path loss is commonly expressed as the ratio between the transmitted power P_t and received power P_r, both in watts. For transmit and receive antennas having gain G_t and G_r, respectively, separated by a distance d in meters, path loss PL for an LoS link in free space (without reflections) may be expressed as [1–5]

$$PL = \frac{P_t}{P_r} = \frac{1}{G_t G_r} \frac{4\pi d^2}{\lambda}, \quad (2.4)$$

where λ is the electrical wavelength at the frequency of operation in meters and d satisfies the conditions

$$d \gg \lambda \qquad (2.5)$$

and

$$d > D_f = \frac{2D^2}{\lambda}, \qquad (2.6)$$

where D_f is referred to as the Fraunhofer distance and D is the size of the antenna aperture. Under the condition $d > D_f$, the propagating electromagnetic wave starts to approximate a plane wave.

Equation (2.4) is sometimes referred to as the Friis formula (although in [1] Friis based his derivation on the ratio of the received power to the transmitted power). However, eq. (2.4) represents the path loss as a positive quantity corresponding to the value of attenuation. Millimeter-wave bands are the first frequencies where antenna gains can be made sufficiently large within a small form factor such that path loss can be dramatically reduced on a link to a mobile handset [6].

The Friis equation shows that the path loss increases as the square of the distance d. This factor of two in the exponent of the numerator of eq. (2.4) is an important parameter called the path loss exponent. Often, verification methods check whether measured results produce a free-space path-loss exponent equal to 2 (or very close to 2). In practice, the path-loss exponent is often used to evaluate the degree to which an environment has propagation conditions similar to free space. For example, in highly reflective environments, multipath reflections change the slope of a path-loss-versus-distance curve, typically resulting in higher values of path-loss exponent. To form a statistically accurate estimate of path loss, measurements should be repeated over several separation distances between the transmitter and the receiver, and the results compared with theory.

Often, the path-loss exponent is determined by applying linear regression to calculate the slope of a measured path-loss-versus-distance curve. This "curve" will be a straight line when plotted on a log–log scale. The standard deviation of the difference between the measured points and the points on the line provides an indication of the confidence in the estimate of the path-loss exponent. In addition to environmental factors that may affect the path-loss exponent, the measurement process itself can affect the estimate, including factors such as the number of points collected in the path-loss-versus-distance curve, the positioning accuracy of those points and the channel sounder's amplitude resolution, as well as its repeatability. The verification process tries to separate the measurement-process-induced factors from the environmental factors.

In addition to the analytic expression for free-space path loss in eq. (2.4), there are also several statistical models that can be used to describe the path-loss exponent for other environments, such as the single-frequency, close-in (CI) free-space reference distance model [3]; the single-frequency floating-intercept (FI) model [2]; the multi-frequency CI model with a frequency-dependent term (CIF) [7]; the CI model with a height-dependent term (CIH) [8]; and the multi-frequency alpha–beta–gamma (ABG) model [9–12]. Sensitivity analysis over several measurement campaigns suggests a 1 m free-space path-loss reference could hold over a wide range of mmWave frequencies [12–14].

2.3   Delay Spread

Typically, multipath propagation is evaluated using a metric termed the root-mean-square (RMS) delay spread, which is derived from the power delay profile (PDP). The PDP is the magnitude squared of the channel's complex impulse response, and may be given as

$$PDP(\tau) = |h(\tau)|^2. \tag{2.7}$$

Often, to model an environment, many measurements, typically on the order of thousands, are collected and the PDPs are averaged. The averaged PDP is often computed as the spatial average of the channel's magnitude-squared baseband impulse response over a local area. It is important to distinguish between the instantaneous PDP and the average PDP. The former is defined as the squared magnitude of the instantaneous impulse response, while the latter is defined as the ensemble average over multiple realizations of the small-scale fading. Multiple measurements for averaging are generated by moving the TX and/or RX of the sounding system over different locations. The size of the region over which the spatial average is done is also important. The size of the region should be large enough to span multiple wavelengths (at least as many as the number of resolvable MPCs within the delay spread; see Chapter 4). The larger the region, the better from the viewpoint of statistical averaging. However, it cannot be too large. In particular, the channel parameters – delay and AoD/AoA associated with each MPC – should not vary beyond the corresponding resolution limits of the channel sounder (see also Chapter 4). Nor should the amplitudes of the underlying MPCs change (although the phases can change), a description that is formalized by the stationarity region [15, 16]. Averaging becomes particularly sensitive in ultrawideband channels [17]. Satisfying these conditions will yield an estimate of the "local" channel statistics, and the averaging of multiple PDP realizations is primarily averaging over the changes in the complex path amplitude (that is, path phases). However, we can also average the "local" PDPs over a larger area (over which the delays, and AoAs/AoDs vary substantially) to yield an estimate of an aggregate (global) PDP for the entire area. The local and global PDPs can be used for different aspects of wireless network design.

We can also view the multiple spatial measurements as a long sequence of temporal measurements in which the TX and/or RX moves over the region in which the multiple spatial measurements are made. Then spatial averaging is equivalent to temporal averaging and the local versus global aspect refers to which temporal segments are combined for computing the average. The spatial region/time span over which the averaging is done defines the ensemble of channel realizations over which the channel statistics (local versus global) are computed [18, 19]. In essence, the "local" PDP statistics can be computed by averaging the PDPs for multiple spatial measurements over a smaller local region, or by averaging the PDPs over multiple temporal segments in which the TX/RX move over the same local region. Global PDP statistics can be obtained by averaging the local PDPs corresponding to different smaller local regions, partitioning the global coverage region. The global statistics are equivalent to averaging multiple spatial PDP measurements over the entire global spatial region, or by averaging the PDPs over multiple temporal segments in which the TX/RX move over the global spatial region. Note that other kinds of statistics can be derived from the ensemble of measured PDPs as well.

2.4   Directional Channel Impulse Response

For channel sounders with directional antenna arrays at the receiver, the spatial–temporal channel impulse response, $h(\tau, \theta, \phi)$, where θ denotes elevation and ϕ denotes azimuth angles, can be ideally modeled as

$$h(\tau, \theta, \phi) = \sum_{n=1}^{N_p} \alpha_n \delta(\tau - \tau_n(\theta, \phi)), \qquad (2.8)$$

where α_n is the complex path amplitude and $\tau_n(\theta, \phi)$ is the relative path delay for the *nth* propagation path (out of N_p total paths) coming from the direction θ and ϕ. In order to develop this model, the different paths coming from different directions have to be determined using measurements collected by the sounder antenna for different transmitter and sounder locations.

2.5   Doppler

The Doppler power spectrum $|H_p(\tau, \Delta f)|^2$ for a given MPC resolved along the sounder path may be calculated from the discrete Fourier transform of the transfer function

$$H_p(\tau, \Delta f) = \frac{1}{M} \sum_{m=1}^{M} h_p(\tau, m\Delta t) e^{(-j2\pi \Delta f m \Delta t / T)}, \qquad (2.9)$$

where, as illustrated in Figure 2.2, m denotes the PDP measurement number made along the channel sounder's path in time, p is the pth MPC, and Δf denotes the Doppler frequency shift associated with the channel sounder's path [9]. The observable range of the shift depends on the sampling rate as $-1/2\Delta t \leq \Delta f \leq 1/2\Delta t$,

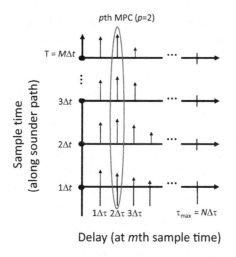

Delay (at mth sample time)

Figure 2.2 The Doppler power spectrum for the pth MPC may be calculated from $m = 1 \ldots M$ power delay profiles measured along the sounder's path.

where Δt is the sampling-time increment, as shown in Figure 2.2. In the above relation, τ denotes the relative delay for the pth MPC.

2.6   Final Comments

The variability and uncertainty in the quantities measured by channel sounders is important if we wish to know whether our measured data support the conclusion that "the channel sounder is operating within specification" or "the channel sounder is not performing within specification." Thus, one key goal of the measurement verification process is to provide standardized techniques in controlled or characterized environments that allow users to determine their sounder's variability.

However, the absolute accuracy with which channel parameters must be estimated in order to yield useful results in wireless network design and simulation is a topic of current research in the channel measurement community (see, for example, [12, 20, 21]). Ultimately, the sensitivity of the final quantity of interest (such as throughput, capacity or reliability) to the changes in metrics such as path loss, RMS delay spread, angle of arrival or Doppler will determine the corresponding accuracy requirements. For example, as mentioned above, in LoS measurements there are many factors that may cause the free-space path-loss exponent to differ from 2, only some of which are due to the accuracy of the channel sounder. As such, obtaining a path-loss exponent of 1.9 may be acceptable in some cases, whereas a path-loss exponent of 1.98 may be required for other applications. Thus, in the chapters that follow, we simply provide the values of the metrics obtained in order to illustrate best practices for verifying sounder performance, rather than judging the utility of these values.

References

[1] H. T. Friis, "A note on a simple transmission formula," *Proceedings of the IRE*, vol. 34, no. 5, pp. 254–256, May 1946.

[2] A. Zajić, *Mobile-to-Mobile Wireless Channels,* Artech House: Boston, MA, 2013.

[3] T. S. Rappaport, *Wireless Communications: Principles and Practice*, 2nd ed., Prentice Hall: Upper Saddle River, NJ, 2002.

[4] A. Molisch, *Wireless Communications*, 2nd ed., IEEE Press/Wiley: Piscataway, NJ, 2011.

[5] S. Salous, *Radio Propagation Measurement and Channel Modeling*, Wiley: Chichester, 2013.

[6] T. S. Rappaport, R. W. Heath, Jr., R. C. Daniels and J. N. Murdock, *Millimeter Wave Wireless Communications*, Prentice Hall: Upper Saddle River, NJ, 2015.

[7] G. R. MacCartney, Jr., T. S. Rappaport, S. Sun and S. Deng, "Indoor office wideband millimeter-wave propagation measurements and channel models at 28 and 73 GHz for ultra-dense 5G wireless networks," *IEEE Access*, vol. 3, pp. 2388–2424, Dec. 2015.

[8] G. R. MacCartney and T. S. Rappaport "Rural macrocell path loss models for millimeter wave wireless communications," *IEEE Journal on Selected Areas in Communications*, vol. 35, no. 7, pp. 1663–1677, July 2017.

[9] P. B. Papazian, C. Gentile, K. A. Remley, J. Senic and N. Golmie, "A radio channel sounder for mobile millimeter-wave communications: System implementation and measurement assessment," *IEEE Transactions on Microwave Theory and Techniques*, vol. 64, no. 9, pp. 2924–2932, Sept. 2016.

[10] S. Piersanti, L. A. Annoni and D. Cassioli, "Millimeter waves channel measurements and path loss models," *2012 IEEE International Conference on Communications (ICC)*, June 2012, pp. 4552–4556.

[11] G. R. MacCartney, Jr., J. Zhang, S. Nie and T. S. Rappaport, "Path loss models for 5G millimeter wave propagation channels in urban microcells," *2013 IEEE Global Communications Conference (GLOBECOM)*, Dec. 2013, pp. 3948–3953.

[12] S. Sun, T. S. Rappaport, T. Thomas, A. Ghosh, H. Nguyen, I. Kovacs, I. Rodriguez, O. Koymen and A. Partyka, "Investigation of prediction accuracy, sensitivity, and parameter stability of large-scale propagation path loss models for 5G wireless communications," *IEEE Transactions on Vehicular Technology*, vol. 65, no. 5, pp. 2843–2860, May 2016.

[13] Y. Xing and T. S. Rappaport, "Propagation measurements and path loss models for sub-THz in urban microcells," *2021 IEEE International Conference on Communications*, June 2021.

[14] Y. Xing and T. S. Rappaport, "Terahertz wireless communications: Co-sharing for terrestrial and satellite systems above 100 GHz (invited)," *IEEE Communications Letters*, vol. 25, no. 10, pp. 3156–3160, Oct. 2021.

[15] A. Gehring, M. Steinbauer, I. Gaspard and M. Grigat, "Empirical channel stationarity in urban environments," *Proceedings of the 4th European Personal Mobile Communications Conference (EPMCC'01)*, Feb. 2001.

[16] R. Wang, C. U. Bas, S. Sangodoyin, S. Hur, J. Park, J. Zhang and A. F. Molisch, "Stationarity region of mm-wave channel based on outdoor microcellular measurements at 28 GHz," *IEEE Military Communications Conference (MILCOM 2017)*, Oct. 2017, pp. 782–787.

[17] A. Meijerink and A. F. Molisch, "On the physical interpretation of the Saleh–Valenzuela model and the definition of its power delay profiles," *IEEE Transactions on Antennas and Propagation*, vol. 62, no. 9, pp. 4780–4793, Sept. 2014.

[18] P. Bello, "Characterization of randomly time-variant linear channels," *IEEE Transactions on Communications Systems*, vol. 11, no. 4, pp. 360–393, Apr. 1963.

[19] R. Kattenbach, "Characterization of time-variant indoor radio channels by means of their system and correlation functions." PhD dissertation *(in German)*, University of Kassel, Shaker Verlag, Aachen, Germany, 1997.

[20] M. Landmann, M. Käske and R. S. Thomä, "Impact of incomplete and inaccurate data models on high resolution parameter estimation in multidimensional channel sounding," *IEEE Transactions on Antennas and Propagation*, vol. 60, no. 2, pp. 557–573, Feb. 2012.

[21] J. Quimby, A. E. Curtin, D. R. Novotny, K. A. Remley, C.-M. Wang and R. Candell, "Variability of sounder measurements in manufacturing facilities," *IEEE Antennas and Propagation Symposium Conf. Dig.*, pp. 1689–1690, Aug. 2016.

3 Channel Sounders

Reiner Thomä, Robert Müller, Christian Schneider, Diego Dupleich,
Kate A. Remley, Jelena Senic, Ruoyu Sun, Peter Papazian,
Jeanne T. Quimby, Camillo Gentile, Juyul Lee, Myung-Don Kim,
Jae-Joon Park, Hyun-Kyu Chung, David Michelson,
Theodore S. Rappaport, George MacCartney, Jr., Yunchou Xing, Shu Sun,
Sana Salous, Akbar Sayeed, Robert W. Heath, Jr., Vutha Va,
Alenka Zajić and Andreas F. Molisch

Channel sounders in the millimeter-wave (mmWave) bands need to meet stringent requirements in order to provide meaningful, accurate and repeatable measurements. Requirements include wide bandwidth to cover the frequency range of interest, fine time delay resolution to distinguish the multipath components (MPCs), high waveform repetition rate for Doppler coverage and multiple-antenna arrays for dual polarization applications and to distinguish AoD and AoA. The Channel Alliance participants use a wide variety of channel sounder architectures. Most fall into three major categories: VNA-based, correlation-based and FMCW (chirp). Each is described here followed by a general discussion on sounder calibration, timing synchronization and system parameters.

3.1 Channel Sounder Architectures

3.1.1 Vector Network Analyzer

The vector network analyzer (VNA) is a ready-to-use commercial off-the-shelf channel-measurement solution capable of providing wide bandwidths and high dynamic range. Because VNAs acquire data in a series of narrowband, frequency-by-frequency measurements over the desired band of frequencies, they typically have very high dynamic range compared to instruments that sample over a wider bandwidth. However, the slow acquisition time requires that static channels are measured. The complex frequency response measured by the VNA can be converted to the impulse response via the inverse Fourier transform (often implemented via the inverse fast Fourier transform, or IFFT). The TX (i.e., the source) and the receivers used to measure the channel's scattering parameters are both physically located within the VNA. They share the same internal sources, local oscillators and other timing circuits, giving it very high accuracy.

Cables that may connect the VNA ports to the antennas at mmWave frequencies suffer from phase instability and phase nonlinearity over a wide bandwidth that can lead to distortion of the frequency sweep, which reduces the time delay resolution and may provide a time-variant response that is characteristic of the cables rather than the radio channel. Consequently, VNAs are often employed over short links and indoors, where the channel characteristics are assumed to be time-invariant for the duration of the frequency sweep. To extend the range, an RF-over-fiber solution may be used to reduce cable loss and timing errors [1]. For mmWave applications, VNAs that operate in the lower frequency bands sometimes include up- and down-converters. Such frequency converters may also be used for the other architecture types as well [2].

Vector network analyzers may be used as reference instruments to which other channel sounders can be compared. There are several reasons for this. One reason is that they have well-established calibrations and uncertainty analyses (through the manufacturer or from external calibration packages such as the NIST Microwave Uncertainty Framework [3]). Additionally, well-established techniques have been developed for shifting the reference plane of the VNA from one circuit location to another, which is useful for comparing VNA and channel sounder measurements of, effectively, the same channel. Finally, they can acquire data over a bandwidth that is wider than real-time sampling channel sounders. These data can then be filtered to match the response of another sounder type. These three features allow comparison of channel characteristics between VNAs and other sounders.

3.1.2   Correlation-Based Sounders

An alternative to frequency stepping of sinusoids is to utilize a broadband, periodic excitation signal. Correlation processing, where the received signal is convolved with a time-reversed copy of the transmitted one, is carried out to obtain the complex impulse response of the channel and to improve the dynamic range. The excitation signal may be periodic pseudo-random binary sequences (PRBS) [4–14] or another type of periodic multicarrier signal often called a "multisine" [15–18].

At the receiver, two architectures are commonly employed for correlation-based sounders. For the "sliding correlator," the same wideband signal is generated with slightly different clock rates at the transmitter and receiver. At the receiver, the replica of the transmitted signal is multiplied, for example, with a mixer, with the incoming received signal and low-pass filtered to enable bandwidth compression. Bandwidth compression has the advantage that a narrowband receiver may be used. The resulting time dilation provides near real-time channel measurements, although the time dilation of the sliding correlator creates a slight delay in the channel measurement as the two signals slide past each other in the correlator. The correlated signals are sampled at or above the Nyquist rate to obtain the channel impulse response (CIR). Because multiple correlations must be sampled, the measurement time may be extended for sequential correlation processing. Alternatively, a hardware configuration designed for parallel correlation may be used.

An alternative architecture known as "direct correlation" or "real-time correlation" is based on Nyquist sampling of the received signal before correlation [7–12, 19–23]. The received signal is down-converted and directly digitized at a high sampling rate in real time, and correlation of the received signal with the ideal transmitted signal is performed in postprocessing. For direct correlation, an FFT is performed on the raw, periodic, complex channel samples. The frequency domain signal representation is then multiplied with the conjugated frequency response of the transmitted waveform with or without filtering. Filtering is sometimes applied to limit the occupied band or to correct for sounder hardware nonidealities. The latter is discussed in Section 3.2. An IFFT is then performed on the filtered response, resulting in the time-domain CIR. The direct correlation FFT method allows for circular convolution to be performed via multiplication in the frequency domain. Linear convolution can also be implemented using zero-padded FFTs.

Direct correlation channel sounders support very fast acquisition of channel data. Sampling of one PRBS/multisine signal period at the Nyquist rate or above often supports very fast acquisition of channel data and offers fine temporal resolution for Doppler measurements while capturing very rapid fading events [10–12, 19–22]. To this end, the period of the transmit signal should be at least as long, but not too much longer, than the relevant CIR excess time delay.

3.1.3   FMCW Sounders

A third architecture is the frequency-modulated, continuous-wave (FMCW) or "chirp" sounder, which transmits a linear frequency sweep with time [24]. The spectrum of the transmitted signal is flat over the frequency sweep for bandwidth (B, in hertz) × time (T, in seconds) products over 100. (As noted in [25], the amount of energy outside B is a function of the dispersion factor, where, for $BT = 10$ and 100, 95% and 98–99% of the signal energy is contained within B, respectively.) The time delay resolution of an FMCW sounder is inversely proportional to its bandwidth. The receiver either mixes the incoming signal with a synchronized replica of the transmitted signal for bandwidth compression or uses a quadrature down-converter as in the correlation-based sounders. Due to the high bandwidth requirements for mmWave channel characterization, quadrature or real-time sampling can be limited in terms of data transfer to storage and, in this case, bandwidth compression enables logging of long records of data if desirable.

3.2   Calibration Techniques and Timing Synchronization

3.2.1   Calibration Techniques

Sounder calibration is crucial for the interpretation of the data and the estimation of the channel parameters. Calibration implies precise characterization of the system through measurement and/or models [6]. For a channel sounder, this includes, for example, knowledge of the frequency response function, antenna characteristics, attenuation, noise level, and more. In general, knowledge of these (nonideal) system characteristics allows for compensation of their influence on measured results. For example, precise knowledge of the frequency response function and the complex array

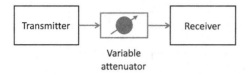

Figure 3.1 Setup for back-to-back tests.

radiation patterns of antenna arrays allows high-resolution estimation of delay and direction of arrival. While both linear and nonlinear behavior have to be considered, calibration of nonlinear distortion may be much more complicated than calibration of linear response.

3.2.2   Back-to-Back Calibrations and Predistortion Filters

To correct for system hardware nonidealities and to characterize the linearity of a channel sounder, back-to-back tests are performed which connect the transmitter and receiver via a calibrated attenuator, as illustrated in Figure 3.1. The measured signal is divided by the system response in postprocessing, reducing or eliminating measurement artifacts such as spurs and other distortion. The back-to-back test also allows the user to measure the linearity of the sounder and its dynamic range.

Starting from eq. (2.3), we can correct $Y^{(\text{meas, uncorr})}$ when the antennas are removed and a characterized variable attenuator is placed between the transmitter and receiver. In this case, the back-to-back response is (see, e.g., [26])

$$Y^{\text{B2B}}(f) = X(f)H^{\text{TX}}(f)H^{\text{B2B}}(f)H^{\text{RX}}(f), \tag{3.1}$$

where $H^{\text{B2B}}(f)$ is the frequency response of the attenuator and connecting cables. Correcting a channel measurement by the back-to-back measurement gives

$$Y^{\text{corr}}(f) = \frac{X(f) \cdot H^{\text{TX}}(f) \cdot \hat{H}(f) \cdot H^{\text{RX}}(f)}{X(f) \cdot H^{\text{TX}}(f) \cdot H^{\text{B2B}}(f) \cdot H^{\text{RX}}(f)} = \frac{Y^{(\text{meas, uncorr})}(f)}{Y^{\text{B2B}}(f)/H^{\text{B2B}}(f)}. \tag{3.2}$$

Multiplying by the known back-to-back response yields the desired result:

$$Y^{\text{corr}}(f) = \hat{H}(f). \tag{3.3}$$

Figure 3.2 shows an example of the back-to-back test in the E band (here, 72–81 GHz), which gives the received power for each attenuator setting. The figure shows the linearity of the sounder, where the output received signal decreases in approximately 10 dB steps corresponding to the applied attenuation.

As described above, the back-to-back test also gives the impulse response of the sounder, which determines its time delay resolution and any spurious signals generated within the system that need to be either corrected for or taken into account when interpreting the measured channel data (e.g., by restricting the multipath threshold). A typical back-to-back measurement is illustrated in Figure 3.3, which shows the system response over the entire time delay window, where the noise floor of the system

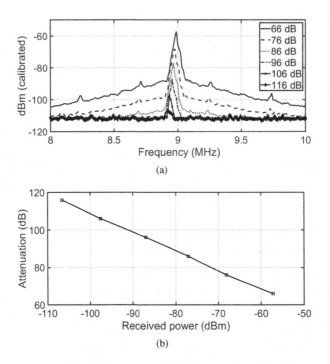

Figure 3.2 (a) Back-to-back test in E band (72–81 GHz) for the Durham channel sounder for attenuator settings from 66 dB to 116 dB; (b) received power versus attenuation setting, demonstrating a linear response.

is −110 dBm, with spurious components appearing at up to 10 dB higher than the noise floor of the sounder.

As an alternative, or in addition to dividing out the nonideal system response as in eq. (3.2), another method for reducing the level of distortion of the impulse response for wideband signals is to reduce the processed bandwidth to a width over which the system's transfer function is flat, giving the ideal impulse response. Figure 3.4 illustrates a conducted measurement of the back-to-back impulse response for an FMCW signal in the ISM band (58–64 GHz) with a 4.4 GHz bandwidth. The figure shows that, before compensation, the impulse response is not the desired ideal response with a single peak, but the response is dispersed in time delay, which limits the time delay resolution of the sounder as well as its dynamic response. This effect can be reduced by the method of eq. (3.2) or by selecting a section of the sweep for ideal channel response. After compensation, the impulse response of the sounder hardware is closer to an ideal delta function.

Finally, a free-field alternative to the conducted back-to-back test is to perform the correction from a measurement in an anechoic environment, which includes the antennas. In the case where the antennas cannot be detached from the rest of the RF unit (which commonly occurs, since mmWave RF units often have integrated antennas), this is actually the only viable method. Again, channel measurements are

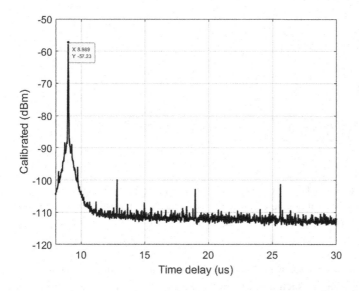

Figure 3.3 Back-to-back test of the Durham sounder illustrating spurious components. In this figure, distortion was reduced by limiting the measurement to a 1 GHz bandwidth.

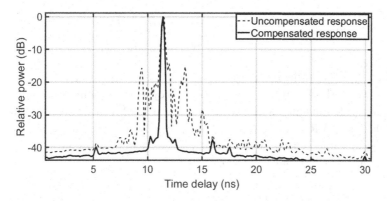

Figure 3.4 Durham sounder's impulse response over a 4.4 GHz bandwidth measured in a conducted back-to-back test.

divided by the measured system response in postprocessing, but this time the antenna responses are included. This is shown in Figure 3.5, which displays the frequency-limited, free-field back-to-back-corrected impulse response of that shown in Figure 3.4, which is now closer to the desired ideal response.

Some channel sounders utilize arbitrary waveform generators to create the transmitted signal. An example is the correlation-based sounder that transmits a BPSK (binary phase-shift keying) modulated signal. If the back-to-back system response is measured prior to system deployment, a predistortion filter can be applied. This approach can have the advantage of improving the dynamic range of the measurement since the transmitted signal has been optimized (not to be confused with the channel's noise

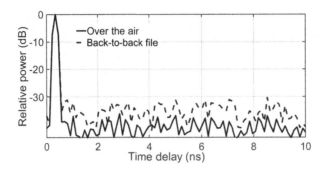

Figure 3.5 Corrected system response from both back-to-back test and from over-the-air test showing that the calibration eliminates system-induced multipath artifacts below a 30-dB threshold [6].

characteristics, which do not change because they are measurement-system indepen-dent). Calculating such a predistortion filter for a correlation-based channel sounder can be carried out as follows: The frequency response of the B2B channel $H^{B2B}(f)$ is measured by a VNA. The predistortion filter $W(f)$ is then given by [27]

$$W(f) = \frac{H^{B2B}(f) \cdot P(f)}{Y^{B2B}(f)} = \frac{H^{B2B}(f) \cdot P(f)}{P(f) \cdot H^{TX}(f) \cdot H^{B2B}(f) \cdot H^{RX}(f)}, \qquad (3.4)$$

where $Y^{B2B}(f)$ is the received signal through the B2B channel; $P(f)$ is a bandpass filter whose shape corresponds to the frequency response of the autocorrelation of the ideal pseudo-random sequence; and $H^{TX}(f)$ and $H^{RX}(f)$ are the nonideal responses of the TX and RX system hardware, respectively. The predistortion filter is applied to the transmitted waveform, so that $H^{TX}(f)$ and $H^{RX}(f)$ are removed from the calibrated received signal $Y^{Cal}(f)$:

$$Y^{Cal}(f) = [P(f) \cdot W(f)]H^{TX}(f) \cdot \hat{H}(f) \cdot H^{RX}(f) = P(f) \cdot \hat{H}(f), \qquad (3.5)$$

where $\hat{H}(f)$ is the measured estimate of $H(f)$, the response of the radio propagation channel.

For the example below, the back-to-back measurement and predistortion filter cal-culation were repeated three times to remove both linear and nonlinear distortion. The calibrated power delay profiles (PDPs) and power spectra for one TX/RX antenna pair in the 60 GHz sounder are presented in Figure 3.6. In Figure 3.6(a) the nonideal sidelobes in the PDP caused by nonideal system hardware are decreased from approx-imately 25 dB (dashed line) to over 50 dB (solid line) below the peak. The calibrated spectrum is smooth and matches the ideal one very well, as shown by the zoomed-in power spectrum in Figure 3.6(b). As long as the channel sounder hardware does not change, the predistortion filter may be applied prior to a measurement in a complex environment, just as the standard back-to-back calibration is applied afterward.

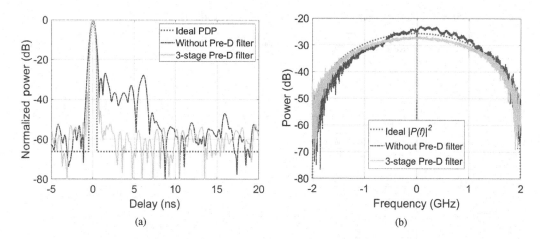

Figure 3.6 Effects of pre-distortion filter (Pre-D filter) on (a) PDP and (b) power spectrum. © 2017 IEEE. Reprinted, with permission, from [27].

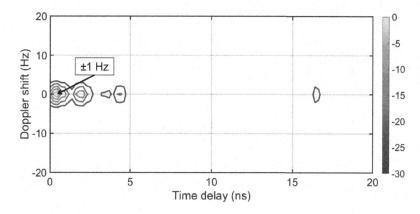

Figure 3.7 Delay Doppler from back-to-back test.

3.2.3 Phase Noise

Another factor to consider is the phase noise of the sounder, which impacts the delay Doppler function and the estimation of MIMO (multiple-input, multiple-output) capacity. Figure 3.7 displays the delay Doppler function where the Doppler spectrum is seen to be centered at zero frequency, which indicates the stability of the measured response over the acquisition time [24]. The figure shows that the inner ring is between ±1 Hz, which corresponds to the acquisition time of 1 s. Other sounder calibrations can be obtained from tests in an anechoic chamber, such as coupling between MIMO channels, and phase noise for MIMO capacity estimation, as illustrated in Figure 3.8 [24].

Finally, Table 3.1 summarizes channel sounder calibration techniques used by Alliance participants.

Table 3.1 Channel sounder calibrations used by Alliance participants.

Group, location, contact	Frequency response	Received power/path loss	Antennas and antenna arrays	TX/RX LO frequency synchronization
Communications Research Centre, Canada, Yvo de Jong, Mustapha Bennai, Jeff Pugh	Standard VNA two-port calibration	N/A	Anechoic chamber (complex radiation patterns, azimuth and elevation, full polarization)	Remote oscillators locked to 10 MHz reference. (Residual phase drift recorded and removed in post-processing)
Durham University, Durham, UK, Prof. Sana Salous	Back-to-back with up to 110 dB attenuation Anechoic chamber with antennas	Wideband	Azimuth calibration	Rubidium and 1 PPS GPS
ETRI, Daejeon, S. Korea, Juyul Lee, Myung-Don Kim	Correlation-based, PRBS	Wideband, 80 dB range power level (w/o antenna)	Complex radiation patterns (anechoic chamber, vertical polarization, azimuth)	Training of Rb. Clock
Georgia Tech, Atlanta, GA, Prof. Alenka Zajic	Standard VNA two-port calibration	N/A		
TU Ilmenau, Ilmenau, Germany, Prof. Reiner Thomä, Robert Müller	Frequency response of the transmission system (S_{21}), is de-embedded in post-processing. Antennas are considered part of the CIR. 60 GHz BW, 801 frequency points	Wideband, antennas included	Gain provided by manufacturer	LO freq. = 11~17 GHz
Keysight, Santa Rosa, CA, Robin Wang, Sheri Detomasi	Free-field back-to-back system response (antenna is included) Bandwidth: 1 GHz Number of frequency points: 5,000 Sa	Wideband. Number of power levels: one or more Antenna is included	Complex radiation patterns and gain provided by antenna vendor	GPS outdoors or separate operation after 1 PPS input/output sync indoors
NIST, Boulder, CO, Peter Papazian, Camillo Gentile, Jeanne Quimby, Kate Remley	Back-to-back conducted. 2 GHz bandwidth, 81,880 frequency points, 13 power levels. No AGC.	Wideband. 13 power levels No antennas	In-house near-field scan	PPS synchronization + GPS
North Carolina State University, Raleigh, NC, Prof. Ismail Guvenc, Ozgur Ozdemir	Conducted back-to-back measurement. Antennas are part of CIR.	First TX power is calibrated using power sensor. Then the attenuation due to calibration cable measured. This is used to calibrate the received power and path loss	Gain provided by antenna vendor	Two Rb clocks for TX/RX after training or single Rb clock used at the TX/RX
NYU WIRELESS, New York, NY, Prof. Ted Rappaport, Hangsong Yan, George MacCartney, Yunchou Xing	Antennas were included.	Wideband and narrowband calibration procedures. Calibration attenuation setting at the receiver was from 0 dB to 70 dB with 10 dB increments for each step. Received power/path loss calculated with antenna gain removed Gain of converter flat across 800 MHz frequency bandwidth. Free-space path loss calibration conducted using a 1, 2, 3, 4 and 5 meter close-in free-space power measurement, and at 33 m for narrowband	Antenna gain determined by the "three antenna method" and verified by antenna specification sheet. The antenna polarization configuration used in the calibration was VP/VP. Azimuth and elevation angles were included.	Frequency was tuned at the beginning of each day and synchronized.
University of British Columbia, Vancouver, Canada, Prof. Dave Michelson	Standard VNA two-port calibration	Wideband, antennas included	Manufacturer's data sheet	10 MHz reference distributed by fiber/coax
University of Southern California and Samsung, Prof. Andy Molisch	Anechoic chamber calibration at different power levels	Wideband, antennas included	Anechoic chamber (complex radiation pattern per beam, at different power levels)	Rb clock and GPS (1 PPS)
University of Wisconsin–Madison, Madison, WI, Akbar Sayeed	Frequency response with arbitrary resolution includes all RF elements and DAC/ADCs. Alternatively, determine frequency response of only antennas using VNAs (up to 50GHz) for short links	Measurements of antenna gains done in a particular direction with or without the lens (particular feed used for excitation). Confirmed Friis law for path loss. RF BW of 1 GHz, but lower baseband bandwidths as noted above	Both azimuth and elevation (2D arrays), but single polarization at this time. Full I/Q complex waveforms	Developed training signals and RX processing for remote time–frequency synch.

University of Wisconsin–Madison: Plan to develop algorithms to isolate different components of the channel response. We also have a digital oscilloscope with 13 GHz bandwidth with four channels that can provide a higher-quality alternative to ADC-based RX processing.

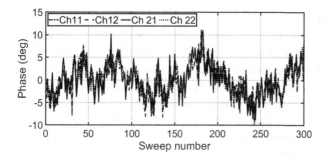

Figure 3.8 Phase noise measured over the 2 × 2 MIMO channel in the anechoic chamber at Durham.

3.2.4   Timing Synchronization

There are many areas where time synchronization is important in channel sounders. For example, it is often desirable to synchronize the initial starting point of the transmitted PRBS signal or chirp/frequency sweep with the receiver to measure the absolute time of flight. This is significant for establishing an "absolute time reference" for the measurements. Training-signal-based schemes may also be used to determine the time of flight if bidirectional communication is possible between the TX and RX. (Essentially, the RX sends back a training signal following a fixed delay after receiving the first signal from the TX from which the roundtrip time of flight can be estimated.) These schemes are new, and are a current area of research.

Also, frequency synchronization between the transmitter and receiver oscillators is necessary to minimize time drift, which can impede accurate delay spread analysis or cause artificial Doppler [28]. It is crucial that the time drift during the acquisition should not erroneously shift the MPCs by a time-delay resolution bin. The relative delays can be established a number of ways as long as there is good symbol timing and frequency synchronization between the TX and the RX.

We should distinguish between long-term and short-term drift of the local oscillators in the channel sounder, both between the TX and RX, but also within each subsystem. This drift, or LO phase variance, should be considered over the period of time corresponding to the processing interval of interest. For example, if we intend to scan an antenna array for the purpose of high-resolution angle of arrival (AoA) estimation, phase coherency will be needed over the length of this entire scan acquisition. The resulting acquisition time can be very long for synthetic aperture scanning or very short if fast switched-array scanning is used. For averaging over long received-signal periods, we will need to minimize drift over that averaging time. Finally, if we intend to resolve Doppler, the drift should be minimized over the duration required to obtain an accurate Fourier transform. This can be estimated from the delay Doppler, where the acquisition time duration should ensure that the Doppler spectrum is centered at 0 Hz. The concept of Allan variance is the relevant measure for phase noise/drift characterization as it allows a proper choice of the observation time according to the application.

Synchronization can be achieved if the same reference or local oscillator is used for both ends of the link. However, this limits the mobility and, possibly, range of transmission as it requires a cable to connect both ends of the link, similar to the VNA solution. Long-range transmission of a local oscillator signal or clock reference signal can be achieved by radio-over-fiber transmission; this concept has been experimentally verified for ultra-wideband (UWB) signals below 10 GHz [29], but not yet for mmWave systems. A third, more common, technique uses calibrated rubidium standards at the transmitter and at the receiver, which can be synchronized to the "pulse-per-second" (PPS) reference signal that is output by either rubidium standards or GPS. GPS is used for outdoor applications. However, this method provides only a few nanoseconds of stability, which can lead to increased uncertainty due to drift for short-duration waveforms.

Temporary direct connection of two rubidium clocks (in a leader–follower configuration) can be used indoors or outdoors [10–12, 30]. For the case involving two clocks, after the clocks have synchronized they are disconnected, typically holding their synchronization anywhere from a few minutes to a few hours, depending on the type of timing reference standard used at both the transmitter and receiver.

In general, rubidium frequency standards are preferred over crystal oscillators for improved long-term timing accuracy, with orders of magnitude better phase noise and stability. However, short-term timing accuracy is still typically accomplished with quartz oscillators, which are often built into the rubidium clock assembly because quartz crystal oscillators typically have better phase noise characteristics than rubidium clocks. When rubidium frequency standards are locked to a GPS they may require up to 27 hours to stabilize.

Another approach for time and frequency synchronization is to use appropriate training signals, known at both the transmitter and the receiver, at the beginning of each measurement block, as in an actual communication system [31]. This approach is particularly suited for long-range measurements and when the local oscillators are stable over sufficiently long measurement blocks.

Synchronization Example

To illustrate synchronization with rubidium clocks, the following is a description of how ETRI conducts its sounder synchronization for long-distance outdoor measurements. Figure 3.9 illustrates a diagram for timing synchronization. As can be seen, both TX and RX parts of the sounder have rubidium clocks and auxiliary circuits for triggering and synchronization. The former is for the absolute time reference while the latter is to trigger or synchronize the start of the PN-code frame so that at the instant when the PN sequences are transmitted, the receiver starts to acquire data.

To synchronize the clocks, ETRI first connects $1PPS_{OUT}$ of the TX rubidium clock to $1PPS_{IN}$ of the RX clock and monitors with software until the timing difference between the 1 PPS signal of TX and that of RX is less than 1 ns. The connections are shown in Figure 3.9. Typically, this process takes several hours. When the timing difference is less than 1 ns, as shown by the first vertical line in Figure 3.10, they activate the EXT_{SYNC}. This trigger synchronizes the frames between TX and RX as shown by the second vertical line in Figure 3.10. Then, they disconnect the 1 PPS connection between TX and RX and the EXT_{SYNC}.

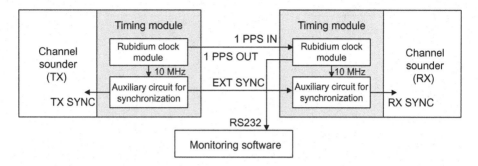

Figure 3.9 ETRI channel sounder timing synchronization.

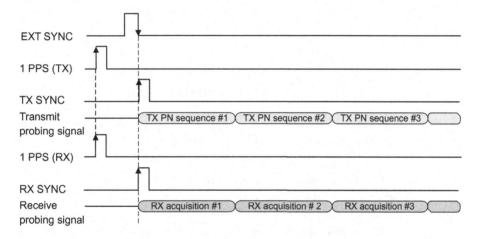

Figure 3.10 ETRI channel sounder timing diagram.

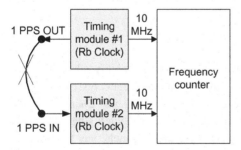

Figure 3.11 Time-dependent synchronization accuracy test. The X indicates that the cable between the TX and RX rubidium clocks disconnects once the two units are synchronized to within 1 ns.

The next step is to measure the time-dependent impacts on synchronization accuracy when the TX and RX rubidium clocks are disconnected. After synchronization between TX and RX is within 1 ns, they compare the 10 MHz reference signals with a frequency counter, as illustrated in Figure 3.11. The TX and RX reference signals are then monitored for many hours (about 17 hours in Figure 3.12).

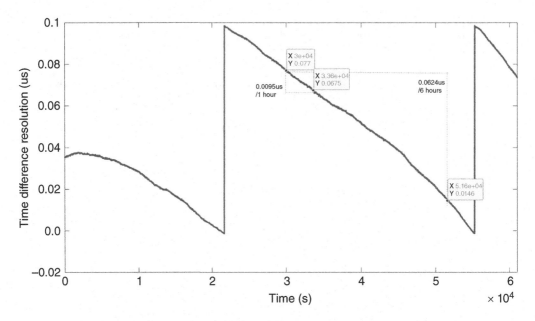

Figure 3.12 Timing synchronization accuracy measurements. Reproduced from [30] with permission by the Electronics and Telecommunications Research Institute.

Figure 3.12 shows typical monitoring results, where the discontinuities are simply due to a frequency-counter-overflow reset and do not affect the results. We observe that about 10 ns (49.6 ns per 5 hours) timing drift occurs every hour. This equates to about 0.16 ns timing error each minute. This would be acceptable for most delay-spread analyses.

3.3   Antenna Architectures

Radio systems in the mmWave bands often use highly adaptable and electrically steerable antennas that rely on multiple-antenna technology [8, 10, 32]. Many channel sounders use directional and/or steerable antennas. Channel sounders may use multiple antennas that can be enabled or selected via a single channel switched at the transmitter and at the receiver, or parallel channels may be used. Switching can be achieved either at IF, as in the Durham sounder [24], or at RF, as in the NIST sounder [10]. Switching at the IF enables higher channel isolation and avoids potential switching-related losses in the mmWave bands, while switching at RF avoids the cost of duplicating the RF heads.

Parallel reception allows for a higher channel sampling rate and synchronized acquisition of multiple-antenna systems, as in the University of Wisconsin's lens-array-based channel sounder [31]. Lens-array-based sounders also have the advantage of RF beamforming that is accomplished by the front-end lens. Yet another method is to have switchable phase shifters so that the sounder can switch between beams pointing into different directions; this concept is used in [18]. It has the advantage of providing beamforming gain in the RF domain. Finally, alternatively, absolute timing may be used, and a single rotatable, directional horn antenna may be used at the transmitter and/or receiver to capture channel characteristics, while the horn antennas are rotated to various angles. This procedure, used by NYU and others, takes more time and assumes the channel is relatively static during the rotation [7–9, 11, 12]. Synthetic aperture techniques may also be applied to these measurements to determine the angular characteristics of the channel [41].

The mmWave channel is much more directional and sparse than at UHF bands, such that ray-tracing may be used as a surrogate for accurate spatial prediction and propagation time synchronization, and, thus, absolute timing. The ray-tracing method for determining spatial characteristics of a channel and absolute propagation delays was performed in early mmWave channel modeling work where the measurements lacked time synchronization [12, 22, 28].

Participants in the 5G mmWave Channel Model Alliance use a wide range of channel sounders with various architectures and operating frequencies. These are summarized in Table 3.2.

3.3.1 Antenna Calibration

Another calibration is that of the antennas used in the measurements and for the estimation of the omnidirectional received signal from directional antennas. For this, the antenna's radiation pattern needs to be measured in an anechoic environment. If the channel sounder uses a rotating platform or a switched array, it is rotated with the same angular step as used in the measurements. For example, rotating the antenna in steps smaller than the 3 dB beamwidth (or even the 10 dB beamwidth) leads to overlap of the antenna beam, and the addition of the received power from the overlapped angles gives an apparent additional antenna gain that needs to be taken into account when estimating the received power/path loss [19].

If we intend to increase the angular resolution beyond that of the pattern corresponding to each antenna element, then we need to measure the complex radiation patterns of each antenna. We must also include the influence of the pivot point (thus, rotation on a circle may be advantageous). The omnidirectional PDP can then be reconstructed with a rotation step fine enough to reconstruct each individual antenna element pattern.

For other antenna array architectures, such as a discrete antenna array (with digital beamforming), a phased array, or a lens array, the antenna array characteristics for a "complete" set of array configurations would need to be measured in an anechoic chamber. The "complete set" in many cases may correspond to the finite set of (beamforming) array configurations that are used for channel measurements. For mmWave channel sounders, the influence of the channel sounding hardware on the antenna or antenna-array radiation pattern often cannot be ignored. It is instructive to measure the antenna pattern alone and with the channel sounder to understand the overall system effect on antenna performance. In the absence of an anechoic chamber, a wideband channel sounder can be used in an open environment to conduct antenna

Table 3.2 Overview of the various channel sounders used by the 5G mmWave Channel Model Alliance participants.

Group, location, contact	Sounder architecture, transmit signal	Signal recording hardware	TX/RX synch.	TX antenna architecture	RX antenna architecture	Antenna array architecture	Dual polarization capability
Communications Research Centre, Ottawa, Canada, Yvo de Jong, Mustapha Bennai, Jeff Pugh	VNA-based, sinusoid	VNA frequency sweep	Cable synch.	Either omni or steerable horn (18–30° typical HPBW)	Same as TX (several combinations possible)	N/A	Full dual pol. (2 × 2)
Durham University, Durham, UK, Prof. Sana Salous	Multi-band, chirp, FMCW	Heterodyne detector (BW compression)	Rb stds. TX and RX, with and without GPS	Either omni or directional (2 or 8 antennas)	Either omni or directional (2, 4, 8 antennas)	2 × 2 or 2 × 4 MIMO, in E and V bands and 8 × 8 MIMO in K band switched at TX/parallel at RX	With twists on the 2 × 2
ETRI, Daejeon, S.Korea, Juyul Lee, Myung-Don Kim	Correlation-based, PRBS	Sliding correlation (real-time in process)	Cable or Rb stds.	Either omni or directional (30°, 60°HPBW)	Either omni or directional (10° HPBW)	Single	1 × 1
Georgia Tech, Atlanta, GA, Prof. Alenka Zajic	VNA-based, sinusoid	VNA frequency sweep	Cable synch.	Directive, single	Directive, single	N/A	Vertical polarization
TU Ilmenau, Ilmenau, Germany, Prof. Reiner Thomä, Robert Müller	Correlation-based, PRBS	Periodic sub sampling	Tethered (coax or optical)	Pencil-beam radiation pattern. Main beam (2°, 15°, 30°, omni HPBW in az. and el.), low sidelobes	Same as TX (several combinations possible)	Rotated antenna (both sides)	Full dual pol (2 × 2)
Keysight, Santa Rosa, CA, Robin Wang, Sheri Detomasi	Correlation-based, Keysight proprietary sequence	Real-time sampling	Rb/GPS reference Function gen. for trigger source	Omni or antenna array: switched antenna, potential for synthetic aperture	Omni or antenna array: both parallel and switched antenna, potential for support synthetic aperture	2 × 2, 4 × 4, 8 × 8 and 16 × 16: TXside: switched RXside: true parallel or switched	HP/VP
NIST, Boulder, CO, Peter Papazian, Camillo Gentile, Jeanne Quimby, Kate Remley	Correlation-based, PRBS	Real-time sampling	Rb clocks, GPS outdoors	28 GHz and 83 GHz : omni or directive, single antenna 60 GHz: directive switched array	Directive, switched array	Octagonal, Switched antenna–RX	HP/VPwith twists
North Carolina State University, Raleigh, NC, Prof. Ismail Guvenc, Ozgur Ozdemir	Correlation-based, National Instruments (NI) proprietary sequence	Real time sampling Averaging on FPGA to reduce the streamingrate and improve SNR	Rb clocks Either use two Rb clocks by training clocks or use single Rb clock for TX and RX	Single directional horn antenna on a rotatable gimbal	Single directional horn antenna on a rotatable gimbal	N/A	Single Polarization
NYU WIRELESS, New York, NY, Prof. Ted Rappaport, Hangsong Yan, George MacCartney, Yunchou Xing	Correlation-based, PRBS	Sliding correlation (BW compression)/direct-correlation or real-time sampling	Free-running high-stability oscillator at TX and RX. Synch 1×/day or with cable or with cable and Rb clocks synched to 1#PPS via training	Single directive rotatable horn antenna	Single directive rotatable horn antenna	Single pyramidal horn antenna at both TX and RX: Gain 24.5 dBi, AZ. HPBW 10.9° and EL. HPBW 8.6°, and multiple other horn antenna options with various gain/HPBW	Vertical polarization and horizontal polarization
University of British Columbia, Vancouver, Canada, Prof. Dave Michelson	VNA-based, sinusoid	VNA frequency sweep	Tethered via optical fiber link	Either omni or horn (30 GHz). Horn (10 GHz)	Dual-polarized conical horn (30 GHz). Rectangular horn (10 GHz)	N/A	Full dual polarization: With TX twists (30 GHz). With TX and RX twists (10 GHz)
University of Southern California and Samsung, Prof. Andy Molisch	Correlation-based, proprietory multi-tone sequence	Real-time sampling, streaming to RAID array	Cable or Rb standards (with or without GPS)	90° patch antenna arrays	90° patch antenna arrays	16-element switched phased array at TX and RX (multiple panels at RX possible)	Single polarization
University of Wisconsin–Madison, Madison, WI, Akbar Sayeed	FPGA-based baseband processor: VNA, sinusoid; PRBS, Chirp FMCW, OFDM	Real-time sampling	Tethered for short links Untethered time/freq. synch. with training signal for longer links	Bidirectional CAP-MIMO transceiver for beamspace MIMO.	TX/RX inter-changeable	Lens antenna array: one sided hemisphere. Four multiple beams out of a max. of 16 (expandable).	Single polarization

University of Wisconsin–Madison: Lens antenna equivalent to 600–800 half wavelength-spaced antenna arrays (40 × 40 cm aperture at 10 GHz, 15.2 cm circular aperture at 28 GHz). In a given configuration, coverage area spanned by 4 × 4 array of feed antennas and measurements on four simultaneous beams can be made. Beamwidth: 4.2 degrees. The feed array can be enlarged or moved for larger angular coverage.

pattern measurements by temporally resolving individual MPCs, such that the LoS path is extracted and multipath with larger excess delays are eliminated [33].

While the calibration issues for single-antenna systems are fairly well understood, as discussed in this chapter, more work needs to be done to extend the methods to systems with multi-antenna arrays, including phased arrays or lens arrays. The calibration aspects of multi-antenna systems are also closely related to estimation of channel parameters from channel measurements.

3.4 Omnidirectional PDPs from Angular Data

Current channel measurements at mmWave frequencies are often carried out by rotating directional antennas, or electronically switching phased array or lens array beams. Such measurements give us insight into the directional structure of the channel. Adding the PDP from the different angles can be used to synthesize the PDP and, in some cases, the individual powers can be added as well [34].

However, the shape of the PDP and the calculated values of second-order statistics like delay spread and Doppler spread are influenced by the directivity/beamwidth of the measurements. In some cases, we might wish to calculate these statistics for an antenna with a different (larger) beamwidth. This is possible based upon (noncoherent) directive PDP measurements, as long as the target beamwidth is wider than the measured beamwidth. In the same way, we can synthesize channel statistics for the omnidirectional case [34–37].

Let $PDP(\phi, \tau)$ denote the underlying joint angle–delay PDP, from which the (azimuthally omnidirectional) PDP in delay can be obtained as $PDP_{\text{omni}}(\tau) = \int PDP(\phi, \tau)d\phi$. For the case illustrated here, without loss of generality, ϕ is defined as the azimuthal angle of rotation. Let us further define the measured angular resolved PDP as

$$PDP_{\text{meas}}(i\Delta\phi, \tau) = \int |g(\phi - i\Delta\phi)|^2 PDP(\phi, \tau)d\phi, \qquad (3.6)$$

where $|g(\phi)|^2$ is the magnitude squared radiation pattern of the directional antenna as a function of the angle ϕ; $i = 0, 1, \ldots, I - 1, I = 360/\Delta\phi$ is the number of directional scans; and $\Delta\phi$ is the angular step of the scans (which is on the order of the 3-dB beamwidth of the directional antenna). The estimated synthetic omnidirectional $PDP_{(\text{omni, est})}(\tau)$ is calculated as

$$PDP_{(\text{omni, est})}(\tau) = \sum_i PDP_{\text{meas}}(i\Delta\phi, \tau)$$

$$= \int \left[\sum_i |g(\phi - i\Delta\phi)|^2\right] PDP(\phi, \tau)d\phi. \qquad (3.7)$$

From the above relation, we can see that if the synthetic omnidirectional antenna pattern is constant over ϕ:

$$|g_{\text{omni}}(\phi)|^2 = \sum_i |g(\phi - i\Delta\phi)|^2 \approx c, \qquad (3.8)$$

then $PDP_{(\text{omni, est})}(\tau) = \int |g_{\text{omni}}(\phi)|^2 PDP(\phi, \tau)d\phi \approx cPDP_{\text{omni}}(\tau)$. The estimation error is given by

$$\epsilon_{\text{Ripple}}(\tau) = PDP_{\text{omni}}(\tau) - PDP_{\text{omni, est}}(\tau)$$

$$= \int \left[1 - |g_{\text{omni}}(\phi)|^2 PDP(\phi, \tau)d\phi \right]. \tag{3.9}$$

In order to minimize the error, we need to make $1 - |g_{\text{omni}}(\phi)|^2 \to 0$. First, we have to ensure that $|g_{\text{omni}}(\phi)|^2 \to c$ by choosing the right angular scan step, and then we have to compensate for the overlapping gain c.

Figure 3.13(a) shows the measured pattern of a horn antenna at 70 GHz and Figure 3.13(b) shows the synthetic omnidirectional pattern as the sum of shifted patterns for different angular steps. It can be observed that for steps smaller than the HPBW, the response is smoother than for steps equal to or larger than the theoretical HPBW. In general, a ripple can be observed and it depends on the angular step. For the HPBW step size the ripple is around 0.5 dB and for 20° it is about 1.46 dB. This ripple introduces an error and can be used as a metric to define a maximum tolerable ripple.

In the case of highly directional rotatable horn antennas, the overlap or ripple error leading to gain errors between a true azimuthally omnidirectional antenna pattern and a pattern that is synthesized through the superposition of received energy from different nonoverlapping 3-dB beamwidth pointing directions is remarkably small – typically much less than 1 dB – as was shown using both theory and measurement [34]. Since variability in connector losses and changes in signal strength due to the flexing of cabling at mmWave frequencies often result in losses of a few tenths of a decibel, [34] shows that well-characterized highly directional antennas may simply be pointed in successive orthogonal 3-dB beamwidth directions over the 2π azimuthal angular spread, and the received energy may be summed directly without the need for scaling or correction factors or extensive signal-processing techniques to determine azimuthally omnidirectional channel characteristics.

This technique may be extended to the full 4π steradians, as discussed in [34, 35], although in practical propagation measurements, where little energy is expected to be transmitted or received from directly overhead or directly below the antennas, it is possible to ignore many inconsequential antenna pointing directions (and thus save a great deal of time for measurements) in the summation of directional patterns to achieve a very close estimate (within one or two decibels) of the omnidirectional antenna power spectrum [35].

In summing the signals obtained from the orthogonal directional antenna patterns, the antenna gains must be removed to scale back the total received signal to what a unity gain omnidirectional antenna would receive [34, 35]. Note, however, that this technique requires an antenna with a pattern that fulfills the condition above; otherwise the ripple can increase significantly, and alternative methods have to be considered; for a discussion of such alternatives, see, for example, [36].

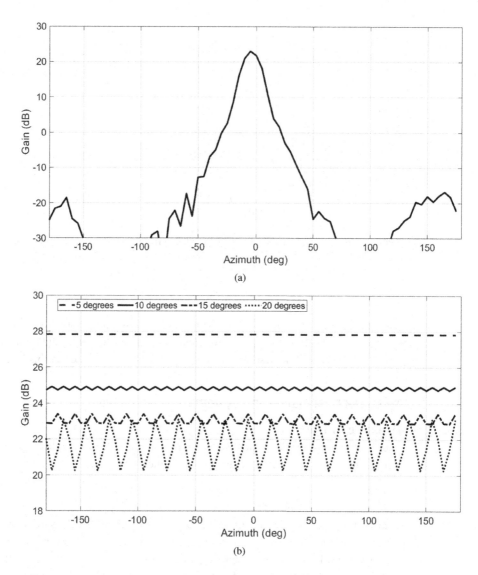

Figure 3.13 (a) Measured magnitude-squared radiation pattern plotted on a log scale for a horn antenna at 70 GHz with 15° HPBW in azimuth. (b) Synthesized omnidirectional response for different angular steps $\Delta\phi$.

3.5 Key Channel Sounder System Parameters

Here we describe the key technical parameters of the sounder equipment as it was used to collect the data used by the 5G mmWave Channel Model Alliance. We also describe common methods for estimating these parameters. In the Alliance, every set of measured channel data includes a version of this list. These parameters support the trustworthiness and usability of the measured propagation and channel parameter data

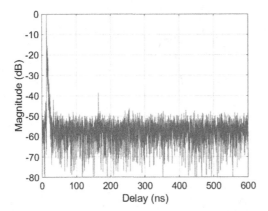

Figure 3.14 Example of an instantaneous PDP taken from a back-to-back measurement (see Section 3.2). The remaining artifacts (for example, near in time to the main peak and at around 150 ns) result from spurious reflections in the setup and from remaining nonlinear distortion.

set. They also indicate the potential strengths and limitations of both the data and the setup used. Procedures applied to calculate propagation-channel-related metrics and scenario metrics will be described elsewhere and are not included in this list.

3.5.1   Noise Floor Estimation and System Dynamic Range

3.5.1.1   Dynamic Range

There are various metrics related to dynamic range, each of which describes a parameter of importance in various channel measurement scenarios.

- The *instantaneous dynamic range* is defined with respect to the PDP and can be extracted from a line-of-sight (LoS)-only reference signal such as the one shown in Figure 3.14. It is specified as the ratio of the strongest PDP sample relative to the power of the noise floor, where noise floor estimation is described below. The single-path, LoS-only condition is required because, for multiple paths, the signal power is spread out, which reduces the maximum signal level between the strongest path and noise level. This and the below-defined quantities obviously depend on the bandwidth of the receiver. If the effective noise bandwidth can be varied (as happens, for example, with the IF bandwidth of a VNA), those settings should be described in the experimental setup.
- The *maximum instantaneous dynamic range* is defined for the best case of maximum input power for the single-path LoS-only case above with a spurious-free noise floor. The instantaneous dynamic range should not include further PDP averaging unless stated otherwise. Note that nonlinear distortion in a wideband signal may result in deterministic distortion that reduces the spurious-free noise floor. These spurs may depend on the characteristics of the transmitted signal and, therefore, it may be difficult to identify the origin of the spurious distortion, which may originate at RF, IF or baseband frequencies.

- The *system dynamic range* adds the receiver automatic gain control (AGC) sweep to the instantaneous dynamic range. It may be reasonable to utilize AGC at the receiver side along with transmit power control to extend the measurable path-loss range. The AGC can add attenuation for close-in measurements so that the receiver does not saturate for low values of path loss. However, in this case, the noise floor per acquisition may be variable, depending on the respective AGC setting. Likewise, if switched or rotating directional antennas are used, we may either adjust AGC per antenna or keep it constant. These conditions should be noted with respect to each measurement setup.
- The *maximum measurable path loss* consists of the system dynamic range, with the addition of the TX/RX antenna gains. This quantity represents the maximum measurable path loss that is the path loss of the smallest detectable multipath component. If AGC or averaging is used, these conditions should be noted. The maximum measurable path loss range is important for planning of a measurement campaign. It describes how the system can cope with LoS and NLoS situations. The maximum measurable path loss range relates to the receiver AGC and transmit power sweep. For example, a user may optimize a sounder for high path loss, such as by using high TX power. But in this case there may be problems with overloading the RX in LoS environments. Receiver AGC may help, but AGC range is, practically, rather limited. In this case, TX power control can extend the range. When reporting maximum measurable path loss range, it is helpful to note these conditions. Furthermore, multiple calibrations (for different transmit powers and different AGC settings) might have to be performed.

3.5.1.2 Noise Floor Estimation

Every measured CIR contains additive noise introduced by the measurement system's receiver hardware. The receiver's noise level limits the dynamic range of the measurement and, if not properly accounted for, may cause bias to channel characteristics that are related to received power, such as PDP, delay spread and delay window. Dominant reasons for receiver noise are low-noise amplifier (LNA) noise figure and analog-to-digital converter (ADC) quantization noise. Generally, the effective noise level varies with received power and AGC setting. Therefore, the effective noise floor depends on path loss and fading, and will also change when using directive antenna scans. The effective noise floor also depends on polarization match between the TX and RX antennas, since different channel noise and interference sources may be sensitive to antenna polarization (note this is a different effect than the channel loss caused by polarization mismatch between a transmitter and receiver). So, it is necessary to estimate the noise level for each measurement acquisition, especially when significant variations of received power between acquisitions and, therefore, noise level are observed.

If we have an estimate of the noise level P_N, we can apply a threshold $P_N \Delta_N$, where Δ_N is a relative threshold value applied to discard samples in the CIR that fall below this level. This can be written as

$$h(t_0, \tau) = \begin{cases} h(t_0, \tau) & |h(t_0, \tau)|^2 \geq \Delta_N P_N \\ 0 & \text{otherwise} \end{cases}. \tag{3.10}$$

We must first estimate the noise level P_N. There are various methods for accomplishing this. A popular, simple and effective method to estimate the noise level and its variation is to realize that the propagating signal, and thus the measured impulse response, will decay in energy at large excess delay times. By observing the received impulse response or PDP and noting the tail end where there is clearly no detectable signal, and where a simple physics calculation verifies that energy would not likely propagate to great delays, one may treat these late-arriving received samples as being purely noise after the detector. The average and standard deviation of these late tails may be used to estimate the average and standard deviation of the noise floor of the PDP. A signal-to-noise ratio (SNR) threshold (sometimes called "noise threshold") may then be set relative to the average noise floor, such that the SNR must be exceeded for a detectable MPC to be deemed to have arrived [12, 19]. A technique for finding this threshold is described in the following paragraphs.

Typically, the SNR threshold is set from one to a few standard deviations above the average noise level, as determined from the tail sample points. The SNR threshold, in relation to the average noise level, determines the false-alarm and missed-signal rate through the standard deviation of the noise, as discussed below. All samples below the signal threshold are treated as noise, and are ignored as signal, within the PDP. Work in [11, 19] shows how arriving multipath signals were determined by using a 5 dB SNR threshold, where the average noise floor was computed from the latest arriving 5–10% of all time samples captured in the measured PDPs.

An alternative approach is to measure the signal level at the delays preceding the LoS component, where physical considerations also dictate that no noise power can be present. However, great care must be taken that sidelobes (resulting, for example, from windowing in the frequency domain and transforming to the delay domain) do not impact the measured signal level. In situations where there is a large dynamic range with strong signals, however, it may be necessary to use a threshold below the strongest MPC to remove system artifacts and pulse sidelobes. This thresholding is typically performed simultaneously with the SNR noise floor threshold as a double thresholding technique [12].

A common method for finding the SNR threshold is derived from the instantaneous PDP domain (magnitude squared CIR) by applying a "zero-hypothesis test" (H0 hypothesis). Assuming only additive white Gaussian noise (AWGN) at the receiver, the PDP's noise power $|n(\tau)|^2 = |n_I(\tau) + jn_Q(\tau)|^2 = |n_I(\tau)|^2 + |n_Q(\tau)|^2$ follows a chi-squared distribution with two degrees of freedom (as the sum of two squared independent Gaussian variables n_I and $n_Q \approx \mathcal{N}(0, P_N)$). In most practical cases it is not possible to make a "noise-only" measurement that includes the antennas and the instantaneous AGC level when conditions are changing – for example, if the antennas are rotating or the receiver is moving. Therefore, we must estimate the noise level from the measured (in situ) PDP. This works well if the PDP has some regions where only noise is expected, as outlined in the preceding paragraph. This is the case if the CIR is sparse or if enough data have been collected in the delay domain so that the PDP decays well below the receiver's noise floor.

According to the principles of order statistics, we apply the following procedure: We sort the PDP samples in increasing order and calculate the mean-square summation

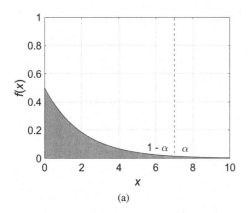
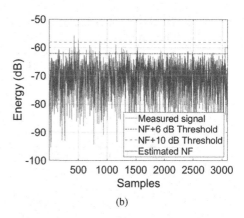

(a)   (b)

Figure 3.15 (a) Chi-square probability distribution function obtained from commonly available simulation software; (b) a noise measurement; the noise floor estimates thresholds.

of these ordered samples for different window lengths, with the window length going from 1 to, at most, the total number of samples. As soon as the next PDP sample is bigger than the current mean-squared value scaled by some auxiliary ΔN we stop. This mean-squared value is then taken as an estimate of the noise level. Finally, we discard the CIR values (set to zero according to the equation above) whose PDP values are smaller than the threshold. The advantage of order statistics is that we automatically collect all PDP values that fall under the zero-hypothesis test. So, we don't need to identify them by visual inspection.

The auxiliary scaling factor ΔN controls the probability of a false detection (taking a noise sample as a valid signal sample). It may be calculated with a chi-squared test [38] from:

$$P(x > P_N) = 1 - \int_0^{(P_N)} p(x)dx, \tag{3.11}$$

where $p(x)$ is the probability distribution function (PDF) $x \sim \chi_2^2$ and $\int_0^\infty p(x)dx = 1$. We first estimate the noise level P_N, normalize to variance 1, and then apply the threshold (quantile) which we take from tables of a normalized cumulative chi-squared distribution with two degrees of freedom. The reason for using this distribution is that we assume that the additive noise in the complex CIR is zero mean iid complex Gaussian (zero hypothesis). The probability of $x > P_N$ is 0.38 (for the normalized distribution) because we assume Gaussian distributed noise limits the receiver dynamic range. If we set a threshold of 3 dB ($P(x > 2P_N)$) the false alarm probability decreases to 0.13 and for 6 dB to 0.02. The false alarm probability of taking noise as signal is indicated with α in Figure 3.15(a) and Table 3.3.

For example, in a pure noise measurement with a threshold set to 6 dB, the false alarm probability is $\alpha = 0.02$. Depending on the application this may be considered too high. Increasing $\Delta N = 10$ dB would reduce the false alarm probability to almost 0. However, as shown in Figure 3.15(b), much of the measured signal would be discarded. To summarize, we can apply different values of ΔN for noise-level estimation and discarding of samples. Different threshold values may be needed for different

Table 3.3 False detection probability α for different threshold values Δ_N.

Threshold Δ_N (dB)	False alarm probability α^*
3 dB	0.13
5 dB	0.04
6 dB	0.02
10 dB	0

* Rounded to two significant digits.

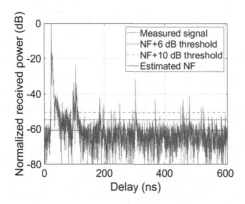

Figure 3.16 Different thresholds applied to an instantaneous PDP plotted on a log scale. The PDP was measured in a conference room at 70 GHz in an access point scenario, 15° half-power beamwidth (HPBW) TX antenna and 60° HPBW RX antenna.

applications: A low threshold increases probability that noise may be detected as multipath, whereas a high threshold may cause the user to miss relevant MPCs. An example in a real-world scenario is shown in Figure 3.16, where the effects of various thresholds can be seen.

A further improvement of the estimation might be obtained by additionally delay-gating the impulse response. In other words, if the environment ensures that MPCs with a delay larger than a threshold cannot occur, or would have a received power below the detection threshold, then any components with such large delay should be discarded – this prevents false alarms from showing up in this area [12, 18].

3.5.1.3   Delay Spread and Dynamic Range

As different channel sounders may have different dynamic ranges and because the noise level may change even during a measurement, we always have to indicate which dynamic range was applied for the delay spread calculation. Furthermore, for comparison to other systems or measurement environments, we may need to normalize the noise level. Because delay spread may be used as a design parameter for transceiver design (cyclic prefix, equalizer taps, predistortion), we may be forced to meet a certain target transmission system dynamic range, Δ_{DS}, where delay spread is calculated only over MPCs that exceed the threshold $P_N + \Delta_N$. Typical values of Δ_{DS} are 10 dB, 20 dB, etc.:

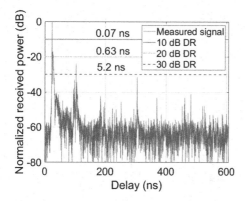

Figure 3.17 Delay spread calculated for different dynamic range target values Δ_{DS} (cf. Figure 3.16).

$$h(t_0, \tau) = \begin{cases} h(t_0, \tau) & |h(t_0, \tau)|^2 \geq \frac{\max(|h(t_0, \tau)|^2)}{\Delta_{DS}} \\ 0 & \text{otherwise} \end{cases}. \tag{3.12}$$

Of course, the channel sounder's dynamic range has to be better than the target transmission system dynamic range. Figure 3.17 shows the influence on the number of MPCs selected and the resulting value of delay spread by choosing different dynamic ranges.

The above discussion highlights the importance of sounder dynamic range in determining the "effective" delay spread as measured by the sounder. It is worth noting that this measurement/sounder-dependent definition of delay spread is different from some "textbook" definitions of the delay spread (which assume knowledge of noise-free CIR) and, thus, is best for practical system design only when the sounder noise floor is lower than that of the system under design (see, e.g., [19, 39]).

It is useful for channel modeling activities to adopt a common SNR and thresholding scheme as described above in order to allow repeatability and meaningful comparison of measurements taken with different channel sounding systems, architectures and frequencies, or in different environments, for example. Applying an absolute SNR threshold as described above also allows researchers to mimic the functionality of wideband receivers that will employ AGC in the receiver chain, which is a useful method for properly modeling and interpreting time dispersion characteristics of the channel [19].

Note that in some cases – for example, for evaluating a certain transceiver design – it may be desirable to mimic the power control of the transceiver chain. For instance, if the power control of the target transceiver tries to keep the total received power at a constant level, it may be better to define the delay spread for a dynamic range that follows the total power in the CIR (found by integrating the PDP) rather than to the power of the strongest path. There are different choices for different applications, with some common ones specified in [40].

Finally, Table 3.4 summarizes the key technical parameters of sounders used by Alliance participants.

Table 3.4 Technical parameters for sounders used by 5G mmWave Channel Model Alliance participants.

Group, location, contact	Center frequency (GHz)	Bandwidth (GHz)/ spectrum envelope shape	CIR period	TX power (dBm)	Maximum instantaneous dynamic range (dB)	Measurable channel attenuation range (dB)	CIR repetition rate f_{CIR} (Hz)	Averaging (number of averages)
Communications Research Centre, Canada, Yvo de Jong, Mustapha Bennai, Jeff Pugh	2.45 3.4375 5.8 13 25.875 38 61.25	0.1 0.075 0.15 0.5 1.25 1.00 0.5	Programmable. 1.3–2.7 µs typical	15 14 13 4 15 10 −1	70 70 70 70 60 50 50	130 130 130 120 140 140 120 (with SNR of 20 dB)	Programmable. ~5 channel responses per second max.	Programmable, typically 1–10 depending on SNR situation
Durham University, Durham, UK, Prof. Sana Salous	2.2–2.9 4.4–5.9 12–18 20–40 (K) 50–75 (V) 60–90 (E)	Programmable: 0.750 1.5 1.5 3 6 9 Rectangular	Programmable from ~16 µs to 1.6 ms	16 16 16 16 6–7 6–7	14-bit ADC: maximum dynamic range in principle corresponds to 14-bit but in practice this is limited by the phase noise and spurious signals	For 110 dB attenuation in back-to-back test: 10 dB SNR in E band and 15 dB in V band. For smaller attenuation: up to 70 dB SNR	Depends on the programmed sweep duration	Normally average 1 s of data
ETRI, Daejeon, S. Korea, Juyul Lee, Myung-Don Kim	28, 38	0.500, Sinc²	8.19 µs (122.1 kHz)	29 21 (at input port of antenna)	Max. 50 (peak to noise level of power delay profile)	60 dB (AGC range) with an SNR of 10 dB	Programmable in the range of 7.6–30.5 Hz	Programmable, typically 4–16 depending on SNR situation
Georgia Tech, Atlanta, GA, Prof. Alenka Zajic	33 140 310	14 (26–40 GHz) 60 (110–170 GHz) 20 (300–320 GHz) (with raised cosine filter with 0.2 roll off)	13 ns, 40 ns	0 dBm (power at IF input of the TX module)	55, noise level defined by thermal noise	90 dB	N/A	No sweep-to-sweep averaging is performed due to static channel environment
TU Ilmenau, Ilmenau, Germany, Prof. Reiner Thomä, Robert Müller	0.010–10 27–37 57–67 71–78 180–220	7 (null-to-null)	600 ns and 4.7 µs	38 27 24 30 −15	70–78	With 120 dB attenuation in back-to-back test: - 60 dB SNR @ 10 GHz - 45 dB SNR @ 30 GHz - 35 dB SNR @ 60 GHz - 35 dB SNR @ 70 GHz Depends on the antennas, with 15° HBPW = 21 dBi around 200 dB, AGC in the RF chain: 10 GHz up to 40 dB 30 GHz up to 35 dB 60 GHz up to 25 dB	100 Hz 13 kHz	Variable averaging
Keysight, Santa Rosa, CA, Robin Wang, Sheri Detomasi	MIMO and SISO: 0.010–44 SISO: 60–110 or above	Time domain: constant Envelope frequency domain: rectangular 1 GHz BW using multiple channel digitizer or 2 GHz BW using scope	10 µs or more	Power at input port of antenna: 10 dBm (w/o PA), 30 dBm or higher (with PA). Antenna gain: depends on antenna	89.9 (Thermal noise: −84 dBm @ 1 GHz bandwidth, 12-bit quantization, correlation gain 30 dB, no correlation noise due to Keysight designed waveform	86.9 dB @ 3dB SNR	50 kHz	No averaging. Time alignment based on calibrated system delay. Time trigger for TX and RX uses function generator

Institution	Frequency (GHz)	Null-to-null / waveform	Duration	Power	Dynamic range / gain	Measurable channel attenuation	Frequency resolution	Averaging
NIST, Boulder, CO. Peter Papazian, Camillo Gentile, Jeanne Quimby, Kate Remley	28.5	2 null-to-null, sinc²	2047 ns	33.5 dBm power at input port of antenna	66 dB correlation gain	140.7 dB, SNR = 20 dB	15 kHz	Variable averaging Synchronization circuit
			32.752 µs		90.3 dB correlation gain	152.7 dB, SNR = 20 dB	954 Hz	
	60.5	4 null-to-null, sinc²	2047 ns	20 dBm power at input port of antenna	66 dB correlation gain	149 dB, SNR = 20 dB	3.8 kHz	
	83.5	2 null-to-null, sinc²	2047 ns	15 dBm power at input port of antenna	66 dB correlation gain	120.2 dB, SNR = 20 dB	15 kHz	
North Carolina State University, Raleigh, NC, Prof. Ismail Guvenc, Ozgur Ozdemir	27.5–29.5	2 or 1	1.33 µs [2 GHz] or 2.67 µs [1 GHz]	Min: –30 dBm Max: +25 dBm	60 dB peak to noise ratio	170 dB (10 dB SNR, 32 averages, 17 dBi antennas at TX/RX)	100 kHz	Variable averaging
NYU WIRELESS, New York, NY, Prof. Ted Rappaport, Hansong Yan, George MacCartney, Yunchou Xing	28.0	0.800 null-to-null, sinc²	5,117.5 ns undilated or 40.9 ms dilated (slide factor of 8,000)	Power at input port of antenna: 30.1 dBm	Max. inst. dynamic range 30 dB, thermal noise and quantization noise (8-bit ADC digitizer)	Measurable channel attenuation range was 170–185 dB with an SNR of 5 dB. Linear response guaranteed with selective attenuator	24.45 Hz	Single PDP acquired by averaging 20 consecutive PDP samples, w/each sample 40.9 ms. Time alignment method based on peak voltage trigger
	73.5 142	1.000 null-to-null, sinc²	4,094 ns	30 dBm @ 28 GHz, 14.9 dBm @ 73.5 GHz, 0 dBm @ 142 GHz	Max. inst. dynamic range 40 dB, thermal noise and quantization noise (8-bit ADC digitizer)	Measurable channel attenuation range was up to 180 dB with an SNR of 5 dB. Linear response guaranteed with selective attenuator	Max. of 15.26 kHz	Single PDP, or variable PDP averaging in postprocessing
University of British Columbia, Vancouver, Canada, Prof. Dave Michelson	10 30	1	500 ns	30 dBm @ 10 GHz 46 dBm @ 30 GHz	50 dB @ 10 GHz 60 dB @ 30 GHz	90 dB, SNR = 20 dB	N/A	None
University of Southern California and Samsung, Prof. Andy Molisch	27.85	0.400 (can be increased to 1) OFDM – flat-top bandpass	Reconfigurable.	57 dBm EIRP	45 dB without spreading gain 74 dB with FFT processing gain (10*log10 (#fft points/2))	159 dB without averaging 198 dB with 10 averaging and FFT processing gain	For MIMO measurements, no averaging ~100 Hz; with 10 averages ~ 20 Hz; For SISO > 100 kHz	Reconfigurable: 1 for dynamic measurements; 10 for slower varying channels
University of Wisconsin-Madison, WI, Akbar Sayeed	10 28	RF bandwidth:1 GHz Baseband bandwidth: 125 MHz, 250 MHz, 370 MHz. Raised-cosine with 0.2 roll off (DAC-pulse-shaping)	Reconfigurable. Limited by the coherence of TX/RX oscillators. Significantly longer than the delay spreads to be measured.	22 dBm at the output of PA. Lens antenna gain ~33 dBi. Feed-aperture, other losses ~8 dB.	~65 dB w/o temporal averaging. Thermal, quantization (ADC) noise	Function of measurement distance. For max received SNR, difference ~ 55dB with min. SNR of 10 dB (or higher with temporal averaging)	Can change arbitrarily ~very long signaling durations and guard intervals possible	Possible to average an arbitrary number of blocks. Time alignment via training signals or cable-based for short links

References

[1] K. A. Remley, G. Koepke, C. L. Holloway, C. A. Grosvenor, D. Camell, J. Ladbury, R. T. Johnk and W. F. Young, "Radio-wave propagation into large building structures part 2: Characterization of multipath," *IEEE Transactions on Antennas and Propagation*, vol. 58, no. 4, pp. 1290–1301, Apr. 2010.

[2] S. Kim and A. Zajic, "Statistical characterization of 300-GHz propagation on a desktop," *IEEE Transactions on Vehicular Technology*, vol. 64, no. 8, pp. 3330–3338, Aug. 2015.

[3] D. F. Williams and A. Lewandowski, "NIST Microwave Uncertainty Framework," National Institute of Standards and Technology, online: www.nist.gov/ctl/rf-technology/relatedsoftware.cfm, 2011.

[4] W. G. Newhall, T. S. Rappaport and D. G. Sweeney, "A spread spectrum sliding correlator system for propagation measurements," *RF Design*, pp. 40–54, Apr. 1996.

[5] R. Zetik, M. Kmec, J. Sachs and R. S. Thomä, "Real-time MIMO channel sounder for emulation of distributed ultra-wideband systems," *International Journal of Antennas and Propagation*, vol. 2014, Article ID 317683.

[6] M. Landmann, M. Käske and R. S. Thomä, "Impact of incomplete and inaccurate data models on high resolution parameter estimation in multidimensional channel sounding," *IEEE Transactions on Antennas and Propagation*, vol. 60, no. 2, pp. 557–573, Feb. 2012.

[7] E. Ben-Dor, T. S. Rappaport, Y. Qiao and S. J. Lauffenburger, "Millimeter-wave 60 GHz outdoor and vehicle AOA propagation measurements using a broadband channel sounder," *2011 IEEE Global Telecommunications Conference (GLOBECOM 2011)*, 2011, pp. 1–6.

[8] T. S. Rappaport, S. Sun, R. Mayzus, H. Zhao, Y. Azar, K. Wang, G. N. Wong, J. K. Schulz, M. Samimi and F. Gutierrez, "Millimeter wave mobile communications for 5G cellular: It will work!" *IEEE Access*, vol. 1, pp. 335–349, May 2013.

[9] Sooyoung Hur, Y.-J. Cho, J. Lee, N.-G. Kang, J. Park and H. Benn, "Synchronous channel sounder using horn antenna and indoor measurements on 28 GHz," *2014 IEEE International Black Sea Conference on Communications and Networking (BlackSeaCom)*, May 2014.

[10] P. B. Papazian, C. Gentile, K. A. Remley, J. Senic and N. Golmie, "A radio channel sounder for mobile millimeter-wave communications: System implementation and measurement assessment," *IEEE Transactions on Microwave Theory and Techniques,* vol. 64, no. 9, pp. 2924–2932, Sept. 2016.

[11] G. R. MacCartney, Jr. and T. S. Rappaport, "A flexible wideband millimeter-wave channel sounder with local area and NLOS to LOS transition measurements," *2017 IEEE International Conference on Communications (ICC)*, May 2017, pp. 1–7.

[12] G. R. MacCartney, Jr. and T. S. Rappaport, "A flexible millimeter-wave channel sounder with absolute timing," *IEEE Journal on Selected Areas in Communications*, vol. 35, no. 6, pp. 1402–1418, June 2017.

[13] Y. Xing and T. S. Rappaport, "Propagation measurement system and approach at 140 GHz: Moving to 6G and above 100 GHz," *2018 IEEE Global Communications Conference (GLOBECOM)*, Dec. 2018, pp. 1–6.

[14] D. Shakya, T. Wu and T. S. Rappaport, "A wideband sliding correlator based channel sounder in 65 nm CMOS: An evaluation board design," *IEEE Global Communications Conference*, Dec. 2020.

[15] G. Simon and J. Schoukens, "Robust broadband periodic excitation design," *IEEE Transactions on Instrumentation and Measurement*, vol. 49, pp. 270–274, Apr. 2000.

[16] K. A. Remley, "Multisine excitation for ACPR measurements," *IEEE MTT-S International Microwave Symposium Digest,* June 2003, pp. 2141–2144.

[17] N. B. Carvalho, K. A. Remley, D. Schreurs and K. G. Gard, "Multisine signals for wireless system test and design," *IEEE Microwave Magazine*, pp. 122–138, June 2008.

[18] C. U. Bas, R. Wang, D. Psychoudakis, T. Henige, R. Monroe, J. Park, J. Zhang and A. F. Molisch, "A real-time millimeter-wave phased array MIMO channel sounder," *IEEE Vehicular Technology Conference*, Sept. 24–27, 2017, pp. 1–6.

[19] T. S. Rappaport, G. R. MacCartney, Jr., M. Samimi and S. Sun, "Wideband millimeter-wave propagation measurements and channel models for future wireless communication system design" *IEEE Transactions on Communications*, vol. 63, no. 9, pp. 3029–3056, Sept. 2015.

[20] G. R. MacCartney, Jr., S. Deng, S. Sun and T. S. Rappaport, "Millimeter-wave human blockage at 73 GHz with a simple double knife-edge diffraction model and extension for directional antennas," *2016 IEEE 84th Vehicular Technology Conference Spring (VTC2016-Fall)*, Sept. 2016, pp. 1–6.

[21] T. S. Rappaport, G. R. MacCartney, S. Sun, H. Yan and S. Deng, "Small-scale, local area, and transitional millimeter wave propagation for 5G cellular communications," *IEEE Transactions on Antennas and Propagation, Special Issue on 5G*, vol. 65, no. 12, pp. 6474–6490, Dec. 2017.

[22] G. R. MacCartney, Jr., T. S. Rappaport and S. Rangan, "Rapid fading due to human blockage in pedestrian crowds at 5G millimeter-wave frequencies," *2017 IEEE Global Communications Conference (GLOBECOM)*, Dec. 2017.

[23] T. S. Rappaport, *Wireless Communications: Principles and Practice*, 2nd ed., Prentice-Hall: Upper Saddle River, NJ, 2002.

[24] S. Salous, S. Feeney, X. Raimundo and A. Cheema, "Wideband MIMO channel sounder for radio measurements in the 60 GHz band," *IEEE Transactions on Wireless Communications*, vol. 15, no. 4, pp. 2825–2832, Apr. 2016.

[25] J. Klauder, A. Price, S. Darlington and W. Albersheim, "The theory and design of chirp radars," *The Bell System Technical Journal*, vol. 39, pp. 745–808.

[26] P. B. Papazian, J.-K. Choi, J. Senic, P. Jeavons, C. Gentile, N. Golmie, R. Sun, D. Novotny and K. A. Remley, "Calibration of millimeter-wave channel sounders for super-resolution multipath component extraction," *10th European Conference on Antennas and Propagation (EuCAP 2016)*, Apr. 2016, pp. 1–5.

[27] R. Sun, P. B. Papazian, J. Senic, Y. Lo, J. Choi, K. A. Remley and C. Gentile, "Design and calibration of a double-directional 60 GHz channel sounder for multipath component tracking," *11th European Conference on Antennas and Propagation (EuCAP 2017)*, Mar. 2017, pp. 1–5.

[28] R. Müller, R. Herrmann, D. A. Dupleich, C. Schneider and R. S. Thomä, "Ultrawideband multichannel sounding for mm-wave," *8th European Conference on Antennas and Propagation (EuCAP)*, Apr. 6–11, 2014, pp. 817–821.

[29] S. Sangodoyin, S. Niranjayan and A. F. Molisch, "A measurement-based model for outdoor near-ground ultrawideband channels," *IEEE Transactions on Antennas and Propagation*, vol. 64, no. 2, pp.740–751, Feb. 2016.

[30] J. Lee , M.-D. Kim, J.-J. Park and Y. J. Chong, "Field-measurement-based received power analysis for directional beamforming millimeter-wave systems: Effects of beamwidth and beam misalignment," *ETRI Journal*, vol. 40, no. 1, Feb. 2018.

[31] J. Brady, J. Hogan and A. Sayeed, "Multi-beam MIMO prototype for real-time multiuser communication at 28 GHz," *IEEE Globecom Workshop on Emerging Technologies for 5G*, Dec. 2016.

[32] W. G. Newhall and T. S. Rappaport, "An antenna pattern measurement technique using wideband channel profiles to resolve multipath signal components," *Proceedings of the Antenna Measurement Techniques Association 19th Annual Meeting Symposium*, Nov. 1997, pp. 17–21.

[33] S. Salous, "Multi-band multi-antenna chirp channel sounder for frequencies above 6 GHz," *10th European Conference on Antennas and Propagation (EuCAP 2016)*, Apr. 2016, pp. 1–4.

[34] S. Salous, X. Raimundo and A. Cheema, "Path loss model in typical outdoor environments in the 50–73 GHz band," *11th European Conference on Antennas and Propagation (EuCAP 2011)*, 2017, pp. 1–5.

[35] S. Sun, G. R. MacCartney, Jr., M. K. Samimi and T. S. Rappaport, "Synthesizing omnidirectional antenna patterns, received power and path loss from directional antennas for 5G millimeter-wave communications," *2015 IEEE Global Communications Conference (GLOBECOM)*, Dec. 2015, pp. 3948–3953.

[36] G. R. MacCartney, Jr., M. K. Samimi and T. S. Rappaport, "Omnidirectional path loss models in New York City at 28 GHz and 73 GHz," *2014 IEEE 25th Annual International Symposium on Personal Indoor and Mobile Radio Communications (PIMRC)*, Sept. 2014, p. 227.

[37] K. Haneda, S. Katsuyuki, J. Nguyen, J. Järveläinen and J. Putkonen, "Estimating the omnidirectional pathloss from directional channel sounding," *10th European Conference on Antennas and Propagation (EuCAP 2016)*, 2016, pp. 1–5.

[38] NIST/SEMATECH, "Chi-square goodness-of-fit test," online: www.itl.nist.gov/div898/handbook/eda/section3/eda35f.htm.

[39] A. F. Molisch and M. Steinbauer, "Condensed parameters for characterizing wideband mobile radio channels," *International Journal of Wireless Information Networks*, vol. 6, no. 3, pp. 133–154, 1999.

[40] International Telecommunication Union, Recommendation (ITU-R), Study Group 3, "P. 1407-6, Multipath propagation and parameterization of its characteristics," online: www.itu.int/rec/R-REC-P.1407/en.

[41] P. Vouras, K. A. Remley, B. Jamroz, et al., "Over-the-air testing with synthetic-aperture techniques," *Metrology for 5G and Emerging Wireless Technologies*, IET: Stevenage, UK, 2021.

4 Verification Techniques

Alenka Zajić, Kate A. Remley, Theodore S. Rappaport,
George MacCartney, Jr., Yunchou Xing, Shu Sun, Camillo Gentile,
Jeanne T. Quimby, Jelena Senic, Ruoyu Sun, Peter Papazian,
Russell W. Krueger, Reiner Thomä, Robert Müller, Christian Schneider,
Diego Dupleich, Juyul Lee, Myung-Don Kim, Jae-Joon Park,
Hyun-Kyu Chung, Robert W. Heath, Jr., Vutha Va, David Michelson,
Yvo de Jong, Mustapha Bennai, Sana Salous, Akbar Sayeed,
Ismail Guvenc, Ozgur Ozdemir and Andreas F. Molisch

4.1 Introduction to Measurement Verification

Channel sounder verification ensures that participants measure and report channel characteristics that are due to the environment as opposed to measurement artifacts arising from the use of a suboptimal configuration, from nonidealities in the sounder hardware or from errors in analysis and/or post-processing. Examples of a suboptimal configuration might be the use of a signal whose duration is so long that the environment changes significantly before its transmission has completed, or the use of an incorrect filter on the transmitted signal. Examples of hardware-induced measurement artifacts include spurs or excessive noise floor introduced into the received signal by nonideal electronic components, as well as nonflat frequency characteristics over the typically wide bandwidth. Hardware-induced artifacts tend to occur more commonly in millimetre-wave (mmWave) channel sounders than for sounders operating in microwave frequency bands. This is, in part, because mmWave channel sounders are newer technology often designed with extremely challenging operating characteristics in terms of center frequency, bandwidth and speed. Also, the mmWave hardware from which the channel sounder is constructed is operating closer to the state of the art and, consequently, may exhibit less-ideal behavior. While "ideal performance" may be seen as a noble objective, sometimes it is enough to know and model the nonideal behavior as we sometimes can compensate for its effect. Characterizing and, where possible, correcting (or calibrating out) such non-idealities, is the goal of channel sounder verification.

Channel sounder verification allows Alliance participants to confidently utilize data from different but nominally similar environments to develop channel models. For example, participants in the Alliance's Measurement Subgroup have recently been conducting measurements in a class of propagation environments termed "large-office space," which is similar to the indoor hotspot (InH) environment proposed for use above 6 GHz in [1]. Such large office areas can take on many physical configurations with a variety of propagation characteristics. In order for the Alliance's Channel Modeling Subgroup to utilize data from these related but different environments, it is essential that the group can confidently extract the statistics of the environmental characteristics while understanding limitations due to hardware impairments.

For these reasons, the participants in the 5G mmWave Channel Model Alliance have established a channel sounder verification program. The program allows labs to compare their measured, processed data to theory or to an artifact having known characteristics. Three types of verification have been studied: "in-situ," "controlled condition" and "comparison-to-reference" verification.

In-situ verification may be conducted during field tests to provide confidence that the channel sounder is behaving as expected. Such verification is conducted in environments that are expected to provide known propagation conditions, such as a relatively open area that exhibits free-space or two-ray propagation path-loss behavior. A second example of in-situ verification includes the prediction of power delay (angle) profile characteristics such as individual multipath component (MPC) time delays or angles of arrival from map-based knowledge of an environment.

Controlled-condition verification involves channel sounder measurements in which channel conditions are determined by design. Controlled environments may simulate free-space conditions in, for example, an anechoic chamber or an open-area test site. The time and/or angular response of a channel sounder may be verified by placing reflectors at known locations or suppressing unintended multipath through the use of RF absorbers. Verification artifacts may also be used to assess the channel sounder's performance by providing an engineered channel such as an attenuator to simulate path loss or an artifact made from coaxial cables of different lengths to simulate multipath by splitting the transmitted signal, propagating each replica through coaxial cables providing different path delays, and then recombining. Channel emulators could fall under the verification artifact category.

Finally, the comparison-to-reference verification technique compares channel sounder measurements to those of a well-characterized "golden" reference instrument for the same channel. Often, the channel to be measured is static, such as an indoor room without movement, or very well controlled, such as a verification artifact. An example of a reference instrument is a vector network analyzer (VNA) having a complete uncertainty analysis. Agreement between the channel sounder and VNA reference measurements provides confidence in the channel sounder's hardware and postprocessing without requiring any assumptions about the ideality of the channel.

These three verification techniques assess the ability of each channel sounder to measure metrics such as path loss, multipath delay spread, angle-of-arrival and Doppler by comparison to known or assumed conditions. These techniques allow users to study channel sounder configuration, hardware nonidealities and postprocessing.

4.2   "In-Situ" Verification Method

In-situ verification is typically conducted during field tests, and the goal is to provide confidence that the channel sounder is behaving as expected. Measurements for in-situ verification are often conducted in a relatively open area that exhibits free-space or two-ray path-loss behavior. The goal of these measurements is to verify that the path loss agrees with "ground truth" computed from Friis transmission formula [2, 3] and that measured power delay profile (PDP) characteristics such as individual MPC timing delays agree with the map-based knowledge of the environment. An additional level of confidence is added if it is verified that measured angles of arrival (AoA) also agree with the map-based knowledge of a room. Note that map-based verification assumes that precise maps of the environment and sufficient knowledge of the material parameters are available. Also, the position of the sounding equipment relative to the environment has to be precisely recorded.

4.2.1 Verification of Path-Loss Measurements

In-situ verification methods compare measured path-loss with a theoretical path-loss model that assumes a line-of-sight (LoS) link between the transmitter and receiver and propagation in free space (e.g., a large open area). Examples of in-situ measurement scenarios used for path-loss exponent estimation at mmWave frequencies are shown in Figure 4.1.

Examples of In-Situ Path-Loss Verification Results

NIST performed "in-situ" verification measurements of their 83 GHz switched-array correlation-based channel sounder in the lobby area shown in Figure 4.1(a). The measurements were performed over several antenna-separation distances between 4 m and 13 m and the path-loss exponent was estimated from the Friis equation (eq. 2.4). Data were collected while the channel sounder receiver was in motion. The temporal variation of the path loss is shown in Figure 4.2(a) and the difference between the theoretical values and measurements are shown in Figure 4.2(b) [5]. The standard deviation of the difference between the measurements and the theoretical path-loss values corresponds to the uncertainty in these channel sounder measurements. This standard deviation was 1.18 dB, and the resulting path-loss exponent was 1.93. The nominal transmitter/receiver antenna patterns from the manufacturer were used in the SAGE algorithm [6] since the antennas have not yet been calibrated in an anechoic chamber. We attribute deviations of the data points from the free-space path-loss line to this lack of antenna pattern knowledge, nonidealities in the position measurements and antenna orientations (difficult at 83 GHz) and reflections from the surfaces in the environment.

NYU performed path-loss verification of their wideband sliding correlator channel sounder at a center frequency of 73.5 GHz [8–12], where the sliding correlator channel sounder architecture and applications are detailed in [7, 12]. Received power was measured in the open laboratory shown in Figure 4.1(b) for transmitter–receiver (TX–RX) separation distances of 1, 2, 3, 4 and 5 m, using boresight-aligned 20 dBi, 15° half-power beamwidth (HPBW) horn antennas at both the TX and RX. Additionally, narrowband continuous-wave (CW) measurements were conducted for

(a)

(b)

Figure 4.1 (a) NIST lobby area, © 2017 IEEE. Reprinted, with permission, from [4]; (b) NYU open lab space.

TX–RX separation distances of 2, 3 and 4 m in the same open laboratory, and also at 33 m during a rural outdoor measurement campaign [13]. The path-loss exponent was estimated using the 1 m close-in (CI) free-space reference distance for the CI path-loss model [14]. The estimated path-loss exponent was 1.99 and a standard deviation of 0.2 dB was observed [8].

Figure 4.3 shows both wideband and narrowband path-loss measurements as a function of distance. The results show that measured path-loss values are very similar to free-space path loss. The results also demonstrate the usefulness of a 1 m CI

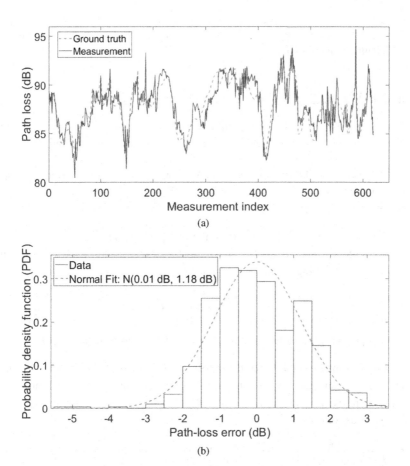

Figure 4.2 (a) Temporal variation of the path loss; (b) the error in path-loss measurements. © 2017 IEEE. Reprinted, with permission, from [4].

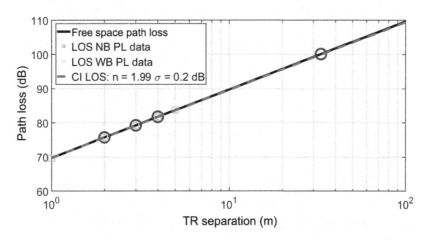

Figure 4.3 Wideband and narrowband path-loss measurements at 73.5 GHz as a function of distance. © 2017 IEEE. Reprinted, with permission, from [7].

free-space reference distance for the CI path-loss model when using either CW or wideband measurements.

4.2.2   In-Situ Verification of Power Delay Profile Measurements

Two methods to verify multipath propagation include (1) free-space map-based multipath verification and (2) two-ray map-based multipath verification.

Free-space map-based multipath verification. This method assumes an open LoS environment, such as the ones shown in Figure 4.1. The verification steps are as follows [2]:

- Compute the measured delay from the first peak in the PDP.
- Compute the theoretical delay from the TX–RX geometry (3D distance between the two) assuming free-space propagation, e.g.,

$$\tau_{TOA} = d/c. \tag{4.1}$$

- Compare empirical and theoretical delays and estimate a measurement error.

Two-ray map-based multipath verification. This method assumes an open LoS environment such as the ones shown in Figure 4.1, but where, in addition to the LoS path, there exists a specular ground reflection path, as illustrated in Figure 4.4. The steps are as follows [15–17]:

- Record the measured PDP and verify that any measurement artifacts are more than 20 dB below the maximum peak in a pure LoS environment without reflection.
- Compute delay times of the first arriving (LoS) and the second arriving (ground-reflected) path from measurements.
- Compute theoretical delay times of the LoS and reflected paths from the TX–RX geometry.
- Compare the empirical and theoretical delays and estimate a measurement error.

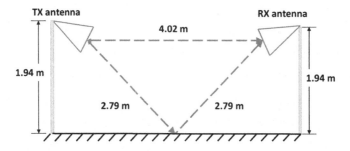

Figure 4.4 Illustration of the two-ray multipath verification method using the sliding correlator channel sounder. © 2017 IEEE. Reprinted, with permission, from [7].

(a)

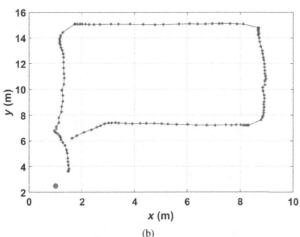

(b)

Figure 4.5 (a) Conference room environment at the NIST Boulder Labs used for PDP verification; (b) RX antenna trajectory with measured data points (stars), and TX antenna location (dot). © 2017 IEEE. Reprinted, with permission, from [4].

Examples of In-Situ PDP Verification Results

NIST performed in-situ PDP verification measurements of their 83 GHz switched-array correlation-based channel sounder in the conference room area shown in Figure 4.5(a) [4]. The RX antenna positioner was moving during the measurements along the path shown in Figure 4.5(b), while the TX antenna was stationary in the corner of the room at a height of 2.5 m. Each star corresponds to the location where data were acquired. These points were used to calculate the delay. The measured and computed delays are compared in Figure 4.6(a) and the difference between them is plotted in Figure 4.6(b). Data for a single run are shown.

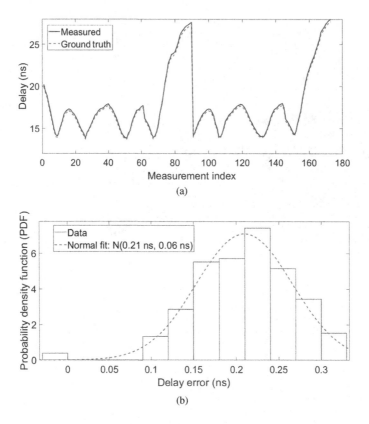

Figure 4.6 (a) Comparison of measured and computed delay; (b) the error in delay. © 2017 IEEE. Reprinted, with permission, from [4].

NYU has performed PDP verification measurements in the open lab shown in Figure 4.1(b) using a wideband sliding correlator channel sounder at a center frequency of 73.5 GHz [8–12]. The channel sounder was programmed to operate with a 500 Mcps (2 ns baseband chip width) signal using a PN code length of 2,047 chips, centered at 73.5 GHz for an RF null-to-null bandwidth of 1 GHz [7, 12]. The measurement setup is illustrated in Figure 4.4. The 73.5 GHz TX and RX both employed 20 dBi, 15° HPBW horn antennas at heights of 1.94 m each. The 2D TX–RX separation distance was 4.02 m, with each antenna adjusted for approximately 46° down-tilt to point toward the lab floor. Figure 4.4 also shows that the LoS path (first arriving signal) should travel a distance of 4.02 m whereas the reflected path (second arriving signal) from the ground should travel a distance of $2.794 + 2.794 = 5.588$ m. The measurement accuracy of the measured distances was within 1 mm, as measured with both a metal measuring tape and an electronic laser range finder.

Figure 4.7 shows the measured PDPs and absolute timing [7, 12]. The theoretical difference in distance and the travel time between the two paths is 1.567 m and 5.223 ns, respectively [11]. The measured distance and time differences were calculated using the maximum peak time index. As shown in Figure 4.7(b). The MPC in the

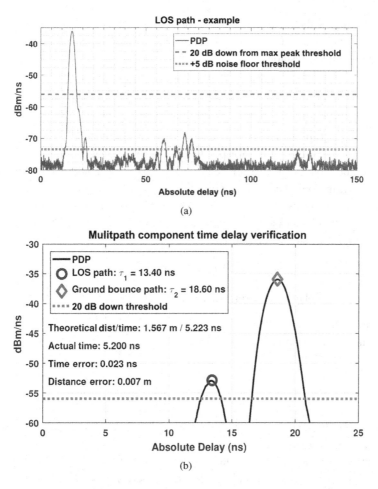

Figure 4.7 (a) LoS PDP shows that the sidelobes are below 20 dB down from the maximum peak; (b) PDP of two-ray verification model. © 2017 IEEE. Reprinted, with permission, from [7].

first path arrived at $\tau_1 = 13.40$ ns (LoS path), and the component in the second path arrived at $\tau_2 = 18.60$ ns (ground-reflected path). The time dilation property of the sliding correlator resulted in an effective sampling rate of 20 samples per nanosecond, which is a temporal resolution of 0.05 ns, or a distance resolution of 15 mm. The difference between theoretical and measured time/distance results in a small error of 0.023 ns and 0.007 m, well within the precision of the channel sounder sampling resolution of 0.05 ns/15 mm.

4.2.3 Standard Uncertainty in Path-Delay Measurements for a Two-Ray Environment

The uncertainty related to the measured estimate of the time difference in two path delays extracted from a single PDP will consist of two components, one related to the repeatability of the measurement and another related to the channel sounder's finite sampling increment [18]. These two components would be combined in a root-sum-of-squares (RSS) fashion as

$$u_{\Delta\tau_{\text{meas}}} = \sqrt{\left(u^2_{\text{repeatability}} + u^2_{\text{resolution}} \right)}. \tag{4.2}$$

For the case presented in Figure 4.7(b), where the channel consisted of a direct path and a single reflection corresponding to a ground bounce from the floor, the "true" value for both paths can be estimated from the measured geometry of the environment. Ultimately, it would be more complete to include uncertainties in the measurements of the path distances, but for now we assume that the time delays corresponding to the paths are known exactly. For the case presented in Figure 4.7(b), the difference in time delay between the two paths is

$$\Delta\tau_{\text{theory}} = 5.223 \text{ ns}.$$

The measured delay values from the channel sounder are:

- direct path: $\tau_1 = 13.40$ ns,
- ground-bounce path: $\tau_2 = 18.60$ ns,

leading to

$$\Delta\tau_{\text{meas}} = 5.20 \text{ ns}.$$

As noted in the text above, the effective sampling interval of the channel sounder is $d_{\text{sounder}} = 0.05$ ns.

The first component of uncertainty corresponds to noise introduced by the instrumentation. Thus, we assume only random errors here. A systematic offset (bias) would be treated differently. It would be estimated from repeat measurements of the channel. This component of uncertainty can be computed from the variance of the measured difference to the known difference for N repeat measurements; that is,

$$s^2 = \frac{\sum_{i=1}^{N}(\tau_{\text{meas},i} - 5.223)^2}{N}, \tag{4.3}$$

and

$$u^2_{\text{repeatibility}} = \frac{s^2}{N} \text{ ns}. \tag{4.4}$$

We divide by N in eq. (4.4) because we are estimating our quantity of interest, $\Delta\tau_{\text{meas}}$, from N repeat measurements. Note that a single measurement ($N = 1$) would correspond to the difference between the measured and true values squared, which yields the value reported in Figure 4.7(b): 0.023 ns.

The second component of uncertainty, capturing the limited resolution of the channel sounder, is based on the assumption that the true value can occur with uniform probability anywhere within the sampling interval. For this case

$$u^2_{\text{resolution}} = \frac{d^2_{\text{sounder}}}{12} \text{ ns,} \qquad (4.5)$$

where if

$$x \approx \text{uniform}((-d)/2, (+d)/2),$$

then

$$u^2(x) = d^2/12,$$

which corresponds to the variance of the uniform distribution. Combining the two components of uncertainty yields

$$u_{\Delta\tau_{\text{meas}}} = \sqrt{\left(u^2_{\text{repeatibility}} + u^2_{\text{resolution}} \right)} = \sqrt{\frac{s^2}{N} + \frac{d^2_{\text{sounder}}}{12}} \text{ ns.} \qquad (4.6)$$

For the example presented here, this corresponds to

$$u_{\Delta\tau_{\text{meas}}} = \sqrt{(5.20 - 5.223)^2 + 0.05^2/12} = \sqrt{5.290 + 2.083} \times 10^{-2} = 0.027 \text{ ns.} \qquad (4.7)$$

This value would be reported as the standard uncertainty in the path-delay measurements.

4.2.4 In-Situ Verification of Angular-Resolved Power Delay Measurements

In addition to verifying time delay of MPCs, it is useful to verify AoA for the observed multipath rays. The free-space map-based AoA multipath verification method assumes an open LoS environment such as the ones shown in Figure 4.1. The verification steps are as follows:

- Compute the measured azimuthal angle from the first peak in the PDP for each RX element.
- Compute the ground-truth AoA from the TX–RX geometry (3D angular distance between the two) assuming free-space propagation.
- Compare the empirical and theoretical AoAs and estimate a measurement error.

An Example of Free-Space Map-Based AoA Verification Result

NIST has performed in-situ AoA verification measurements of their 83 GHz switched-array, correlation-based channel sounder in the conference-room area shown in Figure 4.5(a). The RX antenna consisted of an array of 16 scalar-feed-horn antennas spaced in a ring, with approximately 22.5° azimuth separation. The SAGE algorithm was used to extract AoA data on a finer grid [5, 6]. The measurements were conducted along the path shown in Figure 4.5(b). Eight of the antennas were oriented toward the horizon, while the other eight were oriented 45° upward.

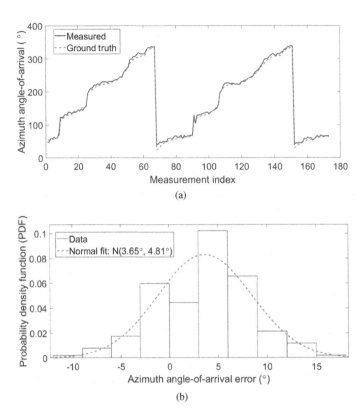

Figure 4.8 (a) Comparison of measured and computed azimuth angles of arrival; (b) the measurement error. © 2017 IEEE. Reprinted, with permission, from [4].

The measured and computed azimuth AoA values are compared in Figure 4.8(a) and the measurement error is plotted in Figure 4.8(b). The results show agreement with an error of approximately $3.65° \pm 4.81°$ between the measured and calculated angles. While these values may seem high, it is the final use of these metrics that dictates the significance of a particular error. For example, a $4°$ angular positioning error may be significant for a pencil-beam antenna, but this value corresponds to an error in distance on the order of 7 mm at 3 m, which may not be of significance for a path-loss calculation.

4.3   Controlled-Condition Verification

A "controlled-condition" level of verification involves channel sounder measurements in an environment or of an artifact having known characteristics. This can be accomplished by conducting measurements in an anechoic chamber, putting absorbers, reflectors or transmitters in strategic places in the room, or by using specially designed artifacts such as an artificial-multipath channel made from coaxial cables of different lengths that can test the ability of each channel sounder to resolve well-characterized MPCs.

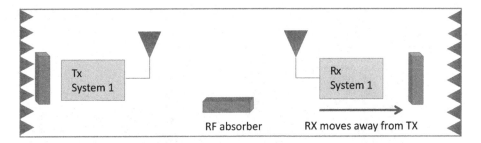

Figure 4.9 Illustration of the path-loss measurement setup for controlled-environment validation.

4.3.1   Verification of Path-Loss Measurements

In this section we describe verification of path-loss measurements using a controlled environment, as illustrated in Figure 4.9. The goal of this setup is to position absorbers around the room such that they prevent any multipath effect in the LoS environment. The verification steps are as follows:

- Orient the TX and RX antennas toward each other (if directional).
- Conduct LoS measurements starting at a known distance with an increasing separation, starting at $1/4$ of $1/BW$, with at least 10 distances.
- The environment should introduce as few reflections as possible (a big room or outdoors, or use RF absorbers).
- Process the path-loss measured data as described in Section 4.2.1.
- Use one of the statistical path-loss models to estimate the path-loss exponent.
- Compare the estimated path loss exponent to the theoretical value 2.

An Example of Controlled-Condition Path Loss Verification Results
Georgia Tech has performed path-loss verification of their VNA-based channel sounders at 26–43 GHz and 110–170 GHz (D-band) [19–24]. Received power was measured in the open laboratory shown in Figure 4.10. We can observe that both TX and RX are covered with absorbers and the floor and surrounding elements are also covered with absorbers. The goal was to eliminate all possible MPCs and create a true LoS environment. The TX–RX separation distances ranged from 20 to 180 cm for 30 GHz band with at least 10 measurement points. For the D-band channel sounder, the distances varied from 30 to 85 cm. This range was limited by the transmitted power. The path-loss exponent was estimated using both the floating intercept (FI) and CI free-space reference models, and the estimated path-loss exponent was 2.001 for 30 GHz band and 1.98 for the D-band measurements. Additionally, a standard deviation of 0.2 dB in the path-loss slope was observed for both sets of measurements [24].

Figure 4.11 shows path loss as a function of frequency for several distances across two frequency bands (a) 26–43 GHz and (b) 110–170 GHz. It also compares the theoretical (Friis formula (eq. 2.4)) with the measured data. Very good agreement was observed at 30 GHz, while the D-band data fluctuate around the theoretical value. The fluctuation with frequency becomes even more dominant at 300 GHz [20].

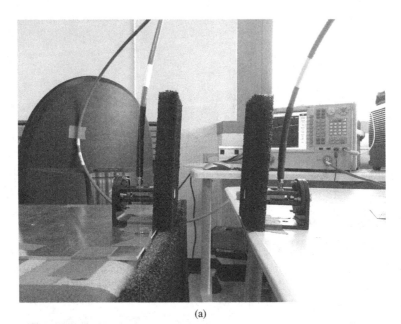

(a)

(b)

Figure 4.10 Georgia Tech channel sounder. (a) 30 GHz and (b) 110–170 GHz measurement setups. © 2017 IEEE. Reprinted, with permission, from [24].

4.3.2   Controlled-Condition Verification of PDP Measurements

In this section we describe PDP verification methods using controlled environments. There are several methods to achieve this goal. We will describe two possible approaches: (1) a lab setup multipath verification method; and (2) a conducted

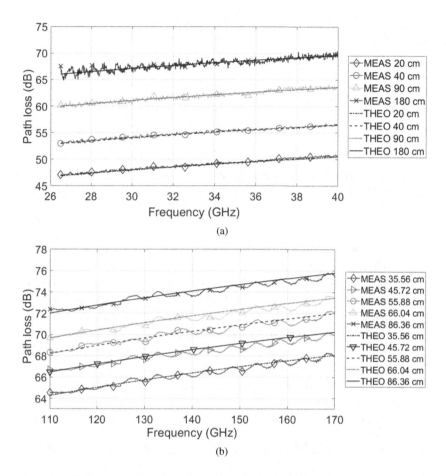

Figure 4.11 Path loss as a function of frequency for several separation distances for (a) 30 GHz, and (b) D-band measurements. © 2017 IEEE. Reprinted, with permission, from [24].

verification-artifact method. Method 1 utilizes over-the-air testing, while method 2 uses a conducted channel to replicate specific multipath conditions.

Lab-setup Multipath Verification. An example of a lab setup for multipath verification is shown in Figure 4.12. The goal of the setup is to position reflective surfaces in strategic places within an anechoic chamber and measure delay caused by specular reflections. The PDP derived from measured results is then compared to the theoretical PDP, which is calculated from the positions of the reflective materials placed in the chamber. As with in situ methods, positioning errors can be a significant source of uncertainty in this method. The verification steps are as follows:

- In an unloaded anechoic chamber, record the measured PDP and verify that the sidelobes are well below 20 dB from the maximum peak (see Figure 4.7(a)).
- Introduce two metal sheets between the TX and RX. One is placed vertically and one is placed flat on the ground surface, as shown in Figure 4.12. Size and

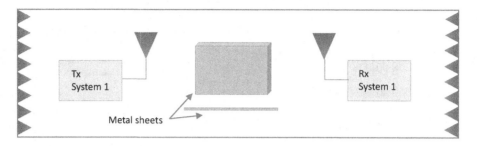

Figure 4.12 Illustration of a multipath verification setup for controlled-condition verification of PDP. The vertical plate is rotated until it blocks the LoS path. The horizontal plate produces a reflected path.

orientation of the sheets will depend on the antenna type, with smaller sheets used with directional antennas.

- Measure the PDP with the channel sounder. The vertical plate should be rotated until the LoS component in the PDP is reduced well below the level of the reflected signal. This blockage is difficult to achieve at lower frequencies, but is usually obtainable for mmWave measurements, especially if directional antennas are used.
- Record the measured PDP produced by only specular reflection from the horizontal sheet (only reflected path is significant).
- Compute the delay time from the reflected path.
- Compute the theoretical delay time of reflected path from the TX–RX geometry.
- Compare the empirical and theoretical delays and estimate a measurement error.

Note that if the time resolution of the channel sounder is sufficient, the two delays (direct and reflected) may be measured simultaneously and compared to theoretical ones. In this case, only the horizontal sheet is needed to produce the reflected path. This would be similar to the two-ray method described in Section 4.3.3, without the calculation of the phase center.

An Example of Controlled-Condition PDP Verification Results
Georgia Tech has performed PDP verification of their VNA-based channel sounder at 110–170 GHz (D-band) [19, 21]. The PDP was measured in an open laboratory space as shown in Figure 4.13(a) and an aluminum plate of size 30.5 × 30.5 × 0.3 cm was used as a reflector. The angular position of the RX was varied while the TX position was fixed at 45°. The angles ϕ_T and ϕ_R represent the offset angles of the TX and RX measured from the LoS position, as shown in Figure 4.13(a).

The peaks that appear in the PDP shown in Figure 4.13(b) correspond to delay times of $\tau = 2.7$ ns and $\tau = 3.6$ ns, respectively. After computing the theoretical delay time of the direct and reflected path from the TX–RX geometry, we can confirm that these two peaks correspond to the LoS and reflected paths from the aluminum plate, with errors on the order of 5%.

Multipath Verification Artifacts. For this verification approach, participants conduct measurements of an emulated multipath channel. Because the artifact is physically

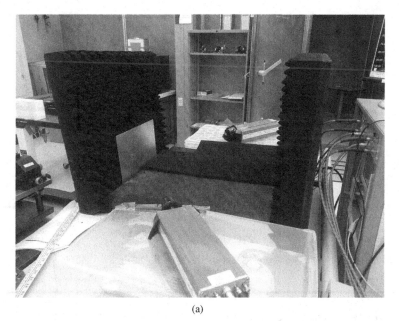

(a)

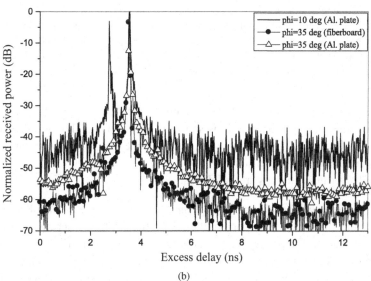

(b)

Figure 4.13 (a) Georgia Tech's 110–170 GHz measurement setup with a metal sheet as a reflective surface. (b) 110–170 GHz PDP measurements for different RX antenna angles with aluminum and cardboard plates as reflecting surfaces. © 2015 IEEE. Reprinted, with permission, from [19].

connected between the transmitter and receiver, this approach can only be used if the RF ports of the channel sounder are accessible. Several coaxial cables of different lengths may be used to create a "verification artifact." The cables are connected through power dividers at the TX side and power combiners at the RX side.

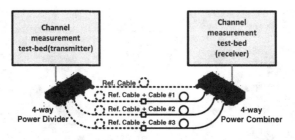

* a percentage of the wave velocity in free space

| Test cable | Cable | Mark | Vp (%)* | Vp (m/s) | Time delay | | Length (m) | Expected Time Delay (ns) | Measured Time Delay (ns) |
					n/m	n/ft			
Ref. Cable	UT-85-Form	UT	70	2.10E+08	4.76	1.45	1.00	-	-
Cable #1	SS405	SS	70	2.10E+08	4.76	1.45	0.52	2.49	2.0
Cable #2	IW2301-10	IW10	84	2.52E+08	3.97	1.21	10.09	40.04	40.0
Cable #3	IW2301-20	IW20	84	2.52E+08	3.97	1.21	20.17	80.04	80.0

Figure 4.14 A multipath verification artifact or "test-bed" created by the ETRI 5G Giga-Communication Research Lab consisting of three coaxial cables having known time delays. The measured time delay is compared to the expected time delays. Note that for wideband signals, pulse arrival times would depend on group velocity rather than phase velocity, although in high-quality cables these should be close to each other. Reproduced from [25] with permission by the Electronics and Telecommunications Research Institute.

An example of this type of verification artifact, created by the ETRI 5G Giga-Communication Research Lab, is shown in Figure 4.14. These verification artifacts are used to assess system hardware performance and to verify the multipath delay resolution of the channel sounder.

The verification steps are as follows:

- Remove the channel sounder's antennas and connect the TX and RX output ports to the input and output ports of the verification artifact, respectively. Add external attenuation if necessary to prevent overdriving the receiver.
- Conduct measurements of the channel and postprocess to obtain the PDP.
- Compare the measured PDP to theory, based on the estimated time delay through the various multipath cables.

An Example of PDP Verification Results with a Verification Artifact

The ETRI 5G Giga-Communication Research Lab verified the MPC delay resolution of their 28 GHz correlation-based channel sounder with the verification artifact displayed in Figure 4.14. The results are shown in Figure 4.15. The resolution of the channel sounder was estimated to be 2 ns, and the measured time delays were within this resolution to the estimated values, as shown in the table in Figure 4.14. Note that uncertainty in the manufacturer's reported value of v_p could impact the uncertainty in the reported time delays, especially at mmWave frequencies.

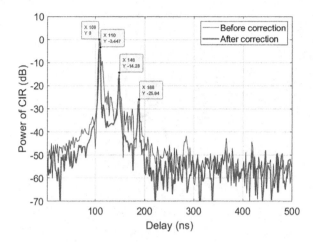

Figure 4.15 Measured, processed results of the PDP for the verification artifact. Multipath resolution is 2 ns. Reproduced from [25] with permission by the Electronics and Telecommunications Research Institute.

4.3.3   Precision Antenna Positioner Measurements to Estimate Path Loss and Antenna Phase Centers

By aligning TX and RX antennas at boresight and increasing the separation with a 1D antenna positioner, it is possible to measure the two-ray lobing pattern of the path loss caused by the change in the direct and reflected signal path lengths. This change in path length causes the signals to add constructively or destructively over several phase rotations or cycles, as shown in Figure 4.16(b). From the known geometry of the path of the 1D positioner (and accounting for the antenna patterns), a comparison of predicted and measured can be used to estimate the path loss and phase center of the antennas, as described in [26] and summarized below.

Steps:

- Place the transmitter and receiver over a metal ground plane with RF absorber placed vertically around the periphery of the measurement range to eliminate undesired MPCs (see Fig. 4.16(a)).
- Align the antenna boresights (only one TX–RX antenna pair) and measure the separation distance between antenna apertures.
- Using the precision positioner, move the receive antenna over a distance of 60 λ in 1 λ increments.
- Measure channel impulse responses (CIRs) and plot path loss versus separation distance.
- Compare the results using the theoretical two-ray formula and antenna pattern data, calibrating the measured path loss in postprocessing to match the theoretical results.
- The separation offset gives the phase center locations, and the first-order polynomial fit of measured path loss should agree with the Friis formula.

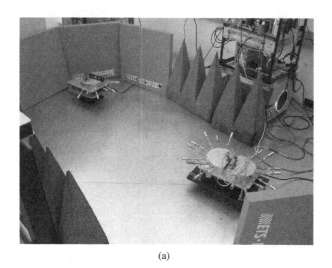

(a)

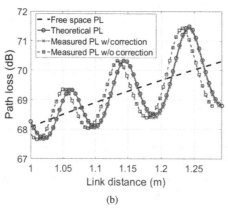

(b)

Figure 4.16 Two-ray verification test: (a) photo of the two-ray measurement with fixed TX (left) and RX (right) on a 1D positioner, with absorber surrounding the setup to eliminate undesired reflections; (b) path-loss results. © 2017 IEEE. Reprinted, with permission, from [26].

A two-ray verification measurement was conducted for the NIST 60 GHz switched array, correlation-based channel sounding system to calibrate path loss and to find the phase center of the horn antennas. The vertically oriented TX and RX antennas were placed on a conducting ground plane. The center of the horn antennas was approximately 19.84 cm above the ground plane. The RX antenna was mounted on a 1D precision positioner that could move with a manufacturer-specified accuracy of 0.006 mm. The RX antenna was moved with steps of one wavelength for 60 steps. The distance between the edge of the horn antennas ranged from 0.99 m to 1.283 m. Note that only one antenna in each of the TX or RX antenna arrays was used. As shown in Figure 4.16(b), the two-ray path-loss lobes were observed, but they are slightly different from the theoretical two-ray model. In the theoretical two-ray model, the direct path signal is attenuated according to the Friis formula and its phase is a function of frequency and the LoS path length. The ground is attenuated according to the Friis formula (with the reflected path length) multiplied by the square of the reflection coefficient; its phase is calculated from the reflected path length. The reflection coefficient for vertical polarization Γ_v is

$$\Gamma_v = \frac{[\epsilon_r - j\sigma/(2\pi f_c \epsilon_0)]\sin(\psi) - \sqrt{\epsilon_r - j\sigma/(2\pi f_c \epsilon_0) - \cos^2(\psi)}}{[\epsilon_r - j\sigma/(2\pi f_c \epsilon_0)]\sin(\psi) + \sqrt{\epsilon_r - j\sigma/(2\pi f_c \epsilon_0) - \cos^2(\psi)}}, \qquad (4.8)$$

where ϵ_0 denotes the permittivity of free space (8.854×10^{-12} F/m); ϵ_r is the relative dielectric constant of the metal ground, which is 1; σ is the metal ground conductivity (1.45×10^6 S/m); and ψ denotes the grazing angle. Note that the antenna gain for the reflected path is given by the antenna pattern, which is smaller than that of the direct path. The received signal is the vector combination of the two complex signals. Details about the two-ray model are given in [27].

The phase center is inside the horn antenna aperture, which extends the link distance. For the results shown in Figure 4.16(b), the link distance was increased by 1 cm to compensate for the phase-center offsets from the antenna apertures. With this increase, the estimated path loss, given by the curve with crosses in Figure 4.16(b), matched the theoretical one, given by the curve with the circles, well. We concluded that the phase center in each of the TX and RX horn antennas was offset from the aperture by 5 mm. Estimating the uncertainty in these measurements is the topic of future research.

4.3.4 Controlled Condition Verification of AoA Measurements

In this method, an antenna fixture provides known values of AoA and polarizations. The verification steps are as follows:

- Emulate a well-defined, coherent two-path scenario with approximately equal path lengths, where the "two propagation paths" are defined by a suitable arrangement of two transmit antennas. See, for example, the fixture in Figure 4.17. The angular separation should be adjustable.
- Ensure that the two paths show a clearly defined phase and polarization difference. The best case for resolution is typically a 90° phase difference, whereas the worst case would be equal and opposite phases. Applying orthogonal polarization would reveal if the radiation pattern calibration is complete [28].
- Record the received signal that emanates from the transmission fixture described above. Depending on the antenna array design, there should be some angular coverage considered (e.g., boresight $\pm x°$ for linear arrays, 360° for circular arrays; azimuth and elevation for 2D arrays). Note that circular arrays should show uniform resolution behavior, but linear/planar arrays should have highest resolution for boresight and reduced-dimension calibration.
- Note that because the receiver noise figure may not be easy to define for comparison, it might be appropriate to define the effective isotropic radiated power (EIRP) transmit power.

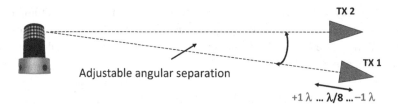

Figure 4.17 Emulation of a coherent two-path scenario.

- Two cases are typically considered: (1) the assumption of two paths of equal power, and (2) no a-priori assumption of the number of paths and amplitudes (perhaps clipping at −20 dB and for 10 estimated paths). The latter is used to assess the real environment in which the measurements are conducted.

Because this verification procedure utilizes multipath, one must consider the coherence of the received waves. Coherence arises if MPCs fall within the same resolution cell and if their phase difference is fixed. Therefore, to fully evaluate the performance of the setup for angular resolution verification, the received signal should contain at least two wideband impinging waves in the same delay cell, ideally with adjustable angular separation and adjustable phase difference [28, 29].

If antenna arrays are used to estimate AoD, coherent processing of the output signals acquired by the array is applied. The resulting angular resolution performance depends on the application of correct propagation and device data models. The latter includes calibration issues and phase stability of the device, especially related to antenna characteristics. See further discussion in [27, 30–34]. Moreover, some high-resolution parameter estimation (HRPE) procedures (ESPRIT, MUSIC, ML) may only allow restricted device data models and are especially sensitive to coherency of impinging waves [35].

Note that if rotated antennas are used for AoD estimation, the resulting angular resolution is basically limited by the directivity of the antennas that were used. The advantage is a simple and robust technique because no coherent-array signal processing is required. Still, one is faced with some accuracy issues such as the proper choice of the angular step size. The angular step size may have an influence on the results if the user intends to synthesize an equivalent omnidirectional PDP. Another potential influence on accuracy may be changing noise levels if the system's automatic gain control (AGC) changes when rotating the antenna.

An Example of Controlled-Condition AoA Verification Results
Refer to Figure 4.17 for the following discussion.

- Use horn antennas to emulate two paths. Apply the same transmit signal to each, such as the sounder signal distributed by a splitter.
- Emulate various phase differences by rotating the antenna cantilever (see Figure 4.18). For example, create orthogonal polarization by rotating the antenna by 90° or by switching ports if dual-polarized horns are available.
- Rotate the antenna under test (AUT) in azimuth and elevation.

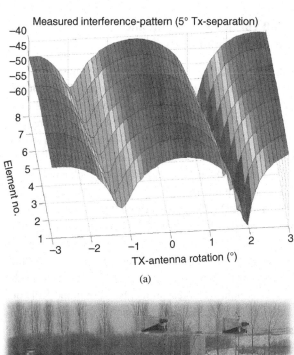

(b)

Figure 4.18 (a) Coherent multipath causing a spatially static interference pattern. This measurement shows the importance of evaluating AoA as a function of antenna element and the phase difference introduced by rotating the TX antenna. The rotation (designated along the lower axis) should introduce the same interference pattern for each antenna element (designated along the vertical axis). However, the interference pattern is not identical for each element. (b) Photograph of the measurement setup.

- Choose between synchronized and nonsynchronized TX–RX in order to keep phase-drift effects separate. The photograph in Figure 4.18(b) shows the rotated antenna cantilever. Note that the distance between the transmitter and receiver should not be too small, in order to avoid any near-field curvature effects. As shown in Figure 4.18, use of high-gain horns allows free-field test (no chamber).

4.3.5   Controlled Condition Verification of Antenna Cross-Polarization Discrimination

Recent wireless systems have employed dual-polarized antenna architectures to obtain channel diversity with orthogonally polarized propagating signals [36]. Dual polarization will be of significant advantage in wireless communications as frequencies move higher and the channel becomes more sparse and antennas with greater directionality are used [37, 38]. Thus, characterizing the cross-polarization discrimination (XPD) of antennas is a vital task for channel sounders that use orthogonally polarized or dual-polarized TX and RX antennas [17]. Several participants' channel sounders have dual polarization capability, as noted in Table 3.2.

Verification of antenna XPD may be performed by first measuring free-space path loss at various closely spaced distances with co-polarized antennas, followed by measurements at the same distances but using cross-polarized antennas. Each of the antennas is sequentially rotated by 90° from the other either electrically or mechanically so that the two antenna polarizations are orthogonal. The difference in path loss between the co-polarized and cross-polarized measurements result in the XPD of the antennas, which should be constant over various distances, and no matter which of the antennas is oriented orthogonally from the other. Because it is unlikely the test range is ideal, we recommend that the method described below be performed at three or more closely spaced distances in the far-field region to ensure measurement reliability, accuracy and reciprocity for determining the XPD between the TX and RX antennas.

For the sake of simplicity, the following discussion regarding characterization of a channel sounder's antenna XPD assumes vertically polarized (electric field perpendicular to the ground plane) and horizontally polarized (magnetic field perpendicular to the ground plane) pyramidal horn antenna, but the method may be generally applied to any type of dual-polarized antenna.

The measurements must be performed in a controlled, open and static environment that satisfies the following three constraints (refer to Figure 4.19). First, the measurement should be in LoS free space with the TX and RX antenna-separation distance beyond the far-field or Fraunhofer distance of the antennas while also ensuring that the TX and RX antennas are boresight aligned. A general rule-of-thumb to ensure plane-wave propagation is to set the RX antenna at least five or more Fraunhofer distances from the radiating TX antenna [16], where the Fraunhofer distance was defined in eq. (2.6). Thus, the TX and RX antenna-separation distances D_{TR} should obey $D_{TR} \geq 5 \times D_f$. Note that boresight alignment of the TX and RX antennas is necessary for both co-polarized and cross-polarized measurements.

Second, the propagation path should be free from nearby reflectors or obstructions that might cause multipath reflections or induce fading in the measurement. Specifically, the heights and distances should be selected in conjunction with the HPBW of the TX and RX antennas such that the projected ground bounce or other reflection sources from the TX antennas are far outside of the HPBW angular spread of the TX antenna and should not arrive anywhere near the HPBW viewing angle of the RX antenna.

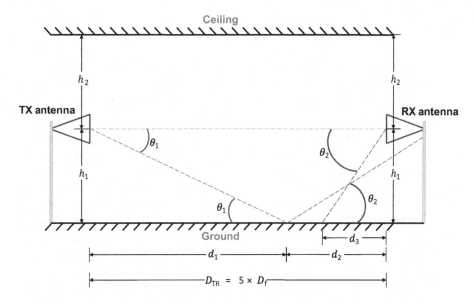

Figure 4.19 Sketch of geometry and test setup for accurately measuring antenna cross-polarization discrimination between two orthogonally polarized antennas for channel sounder verification. © 2018 IEEE. Reprinted, with permission, from [17].

Third, the heights of the antennas and the TX and RX antenna-separation distance should be such that ground bounces and ceiling bounces do not provide reflection or scattering that may confuse the determination of the received power level within or just outside the HPBW of the main lobe of the TX antenna pattern, or within or just outside of the HPBW viewing angle of the RX antenna pattern [9–11]. Figure 4.19 shows a sketch and geometry of a typical measurement setup. See [17] for more detail.

By solving a set of geometry equations pertaining to the sketch in Figure 4.19 to ensure far-field radiation (beyond the Fraunhofer distance) and to ensure that there are no MPCs induced by the surrounding environment, the heights of the antennas above the ground (h_1) and below the ceiling (h_2) must satisfy:

$$h_1, h_2 > \frac{D_{TR}}{\frac{1}{\tan \theta_1} + \frac{1}{\tan \theta_2}}, \tag{4.9}$$

where D_{TR} is the TX and RX antenna-separation distance based on five Fraunhofer distances of the TX antenna, h_1 is the height of the TX and RX antennas above the ground, and h_2 is the distance to the antennas from the ceiling and any obstructions or walls on either side of the straight line between the TX and RX antennas.

To calculate the XPD of the channel sounder antennas, one must follow the three rules outlined above and then measure the path loss between the TX and RX antennas at several different distances d_i that are greater than D_{TR} ($D_{TR} \geq 5 \times D_f$) when they are co-polarized (e.g., both vertically polarized) and then removing the antenna gains such that

$$PL_{VV}(d_i)(\text{dB}) = P_{(t-V)}(\text{dBm}) - P_{(r-V)}(d)(\text{dBm}) + G_{TX}(\text{dBi}) + G_{RX}(\text{dBi}),$$

$$(4.10)$$

where $P_{(t-V)}$ is the transmit power into the TX antenna in dBm, $P_{(r-V)}$ is the received power at the output of the RX antenna in dBm at a distance d in meters, G_{TX} is the gain of the TX antenna in dBi, G_{RX} is the gain of the RX antenna in dBi, and PL_{VV} is the measured path loss in dB at distance d in meters. Note that antenna gains are not necessary to calculate the XPD, but are required for accurately measuring and calibrating far-field free-space path loss with the channel sounder.

Next, path loss for cross-polarized antennas at the same distances should be measured when the TX antenna is vertically polarized and the RX antenna is horizontally polarized. Note that the heights and TX–RX separation distances d_i for both co- and cross-polarized measurements should be as similar as possible for this comparison. Rotating by 90° is commonly performed via a 90° waveguide twist for pyramidal horn antennas with waveguide flanges. The cross-polarized path loss is calculated as:

$$PL_{VH}(d_i)(\text{dB}) = P_{(t-V)}(\text{dBm}) - P_{(r-H)}(d_i)(\text{dBm}) + G_{TX}(\text{dBi}) + G_{RX}(\text{dBi}),$$

$$(4.11)$$

where the transmit power and antenna gains are identical to the values in eq. (4.9), but where $P_{(r-H)}(d_i)$ is the received power in dBm at distance d_i at the output of the horizontally polarized RX antenna, and $PL_{VH}(d_i)$ is the cross-polarized path loss in dB at distance d_i.

To ensure reciprocity and to further ensure accurate characterization, an additional set of measurements should be made for the same measurement distances and heights, etc., where the TX antenna is rotated by 90° with the RX antenna fixed, if the first set of measurements was made while rotating the RX antenna, or vice versa. Doing so ensures that $PL_{VH}(d_i)(\text{dB})$ is equivalent to $PL_{VH}(d_i)(\text{dB})$ in eq. (4.11). Collecting these measurements will ensure that the cross-polarization is similar at the same nominal distances no matter which antenna is cross-polarized.

The XPD between the antennas at all distances d_i (in dB) is then found by subtracting $PL_{VV}(d_i)$ (in dB) from $PL_{VH}(d_i)$ (in dB) at the different distances:

$$XPD(d_i)(\text{dB}) = PL_{VH}(d_i)(\text{dB}) - PL_{VV}(d_i)(\text{dB}),\qquad(4.12)$$

where $XPD(d_i)$ is typically a positive value in dB that should be a constant, regardless of distance. Since the TX–RX antenna separation distances are nominally identical for the co- and cross-polarized free-space path-loss measurements at a particular location, the difference between the two values may be considered the cross-polarization discrimination between the arriving signals, induced by the differences in antenna polarization. For measurement confidence, measurements made at all distances d_i should be compared to ensure the XPD value is consistent. The mean of all i measurements and their standard deviation may be reported.

Because the experimental procedure outlined above assumes multipath or environmental reflections that may confuse whether the determination of received power is significant, the measurements can be performed with a channel sounder in either narrowband or wideband operation. An additional step when using a wideband channel sounder can be used to ensure that no additional multipath or scattering is used to calculate the XPD value, by measuring received power as the power in the first arriving (free-space/boresight-to-boresight) MPC for both co- and cross-polarized antenna measurements. Ideally, there is only one significant path if the requirements and steps outlined above are followed correctly.

4.4   Comparison-to-Reference Verification

In addition to the in-situ and controlled-condition verification methods, a third method compares channel sounder measurements to those made by a reference instrument. Note that for this technique the channel is not necessarily "controlled" in the sense of providing specific characteristics, but rather it remains unchanged between the reference system and channel sounder measurements. For the examples reported here, a VNA serves as the reference instrument, providing traceable measurements with uncertainties characterized by the NIST Microwave Uncertainty Framework (MUF) [39, 40]. Uncertainties are propagated through the various steps in the measurements and postprocessing to the final channel parameters such as path loss, PDP and RMS delay spread.

In conjunction with the uncertainty analysis, the comparison-to-reference verification technique requires an appropriate RF environment. A static RF environment, either free-space or conducted, is a suitable choice when a VNA is the reference instrument. For this case, limiting the movement in the channel during the measurement is generally the main source of error for this technique. For a conducted channel, types of movement can be disconnecting and connecting of systems to the conducted channel during the course of the measurement campaign, cable movement and temperature or humidity changes. For a free-space RF propagation channel, types of movement can be personnel motion, wind, antenna and the cable movement as in the conducted channel.

Two examples of comparison-to-reference channel verification are demonstrated using NIST channel sounders. The first example illustrates how a portable channel verification artifact [40, 41] is used to verify the performance of the NIST 60 GHz switched array, correlation-based channel sounder, as shown in Figure 4.20. The portable channel verification artifact simulates a direct path and up to two additional MPCs using different lengths of coaxial cables and attenuators. The difference here is that a VNA is used to characterize the artifact, rather than using the manufacturer's specifications for the cables, providing a reference measurement. In addition, a temperature controller is used here to maintain phase stability and control drift for the frequency range of 10–60 GHz.

The portable channel verification artifact is measured using a calibrated VNA. The VNA measurement of the artifact uses "before" and "after" calibrations and physical models of the calibration standards, combined with repeat measurements to obtain an uncertainty analysis of the artifact. Next, the portable channel verification artifact is

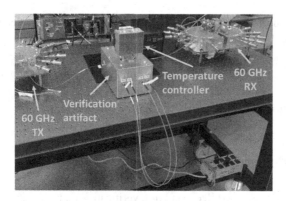

Figure 4.20 "Comparison-to-reference" channel sounder verification using a portable channel verification artifact with the NIST 60 GHz channel sounder. © 2019 IEEE. Reprinted, with permission, from [40].

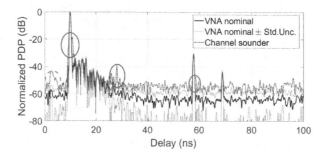

Figure 4.21 PDP of the verification artifact in double multipath configuration, measured by the NIST 60 GHz channel sounder (dash-dotted line) and with a NIST VNA (solid line). The standard uncertainty in the VNA measurement is shown in the dotted line. © 2019 IEEE. Reprinted, with permission, from [40].

measured using the NIST 60 GHz channel sounder or other channel sounders. The two results are then plotted together and compared. Figure 4.21 shows a PDP comparison with uncertainty between the NIST 60 GHz channel sounder and VNA measurement of the portable channel verification artifact in a double multipath configuration. The circles drawn with a solid line indicate locations in delay where the channel sounder results were significantly different from the VNA results. The dashed-line circle shows an artifact that was traced to an imperfect coax-to-waveguide adapter.

The comparison-to-reference verification technique can also be performed in any stable free-field RF propagation channel. This channel might be quite ideal or could include a great deal of multipath. As an example, Figure 4.22 shows the NIST 83 GHz switched array, correlation-based channel sounder and a LoS RF propagation channel [41]. For this RF propagation channel, the antennas were pointed directly at one another with approximately 1 m of separation. The surface of the optical table was covered by a rubber sheet, which was not absorbing. However, because the same channel is measured by both the VNA and the channel sounder, the presence of

Figure 4.22 Comparison-to-reference 83-GHz channel sounder verification for an LoS propagation channel. A rubber sheet covers the measurement surface. © 2021 IEEE. Reprinted, with permission, from [41].

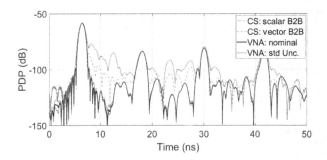

Figure 4.23 PDP comparison of reference VNA and 83 GHz channel sounder measurements using both scalar and complex back-to-back calibration techniques for an LoS RF propagation channel with metal sheets placed behind the antennas. © 2021 IEEE. Reprinted, with permission, from [41].

reflections should not substantially affect the verification comparison. To minimize hardware movement, mechanical switches having over 55 dB of isolation were used to connect the channel sounder and VNA to the antennas.

In the comparison-to-reference technique, the channel sounder takes a measurement of the RF channel and then the user switches to the VNA and allows the VNA to take a measurement of the test channel with as little movement of the hardware such as cables and antennas as possible. The VNA measurements are calibrated in postprocessing and the VNA's reference planes are translated to those of the channel sounder, so that the two instruments effectively measure the same channel. This requires not only calibrating the VNA but also de-embedding the switches and other linear components in the path [41].

Figure 4.23 shows a PDP comparison result from the measurement of a controlled multipath propagation channel in the 83 GHz band. The TX and RX antennas were aimed toward each other with approximately 1 m of separation. To study channel sounder performance in a controlled multipath environment, highly conducting metal

Table 4.1 Comparison of metrics derived from an 83 GHz correlation-based channel sounder and VNA measurements for several different multipath threshold (Mth) values for an LoS channel with metal sheets placed directly behind the antennas.

LOSPEC		RMS delay spread (ns)	Delay window (ns)	No. MPCs
Mth = 0 dB	VNA	0.00 ± 0.00	1.085 ± 0.001	1
	NIST CS: Vector B2B	0.00	1.045	1
	NIST CS: Scalar B2B	0.00	1.353	1
Mth = −10 dB	VNA	0.294 ± 0.00	1.085 ± 0.001	1
	NIST CS: Vector B2B	0.283	1.045	1
	NIST CS: Scalar B2B	0.314	1.353	1
Mth = −20 dB	VNA	0.315 ± 0.00	1.085 ± 0.001	1
	NIST CS: Vector B2B	0.301	1.045	1
	NIST CS: Scalar B2B	0.381	1.353	2
Mth = −30 dB	VNA	2.743 ± 0.054	1.085 ± 0.001	4
	NIST CS: Vector B2B	2.953	1.045	5
	NIST CS: Scalar B2B	3.123	1.353	8
Mth = −40 dB	VNA	2.864 ± 0.049	1.085 ± 0.001	5
	NIST CS: Vector B2B	3.071	1.045	8
	NIST CS: Scalar B2B	3.31	1.353	11
Mth = −50 dB	VNA	2.899 ± 0.05	1.085 ± 0.001	11
	NIST CS: Vector B2B	3.114	1.045	18
	NIST CS: Scalar B2B	3.38	1.353	17

sheets were placed directly behind the channel sounder's TX and RX antennas. The metal sheets were separated by approximately 1.6 m. The channel sounder PDP was normalized to that of the VNA and time aligned with the VNA measurement of the channel. Results are shown for both scalar (magnitude only) and vector back-to-back calibrations of the channel sounder.

Metrics derived from these PDPs are given in Table 4.1. The uncertainty bounds from the sensitivity analysis for RMS delay spread, delay window and number of MPCs were derived for several different multipath threshold values ranging from 0 dB to −50 dB. The results show that use of a scalar (power only) calibration can result in significant errors in the estimation of these metrics.

4.5   Illustration of Verification Methods

In this section we illustrate methods that are currently used by 5G mmWave Channel Alliance participants to verify the performance of their channel sounders. Note that

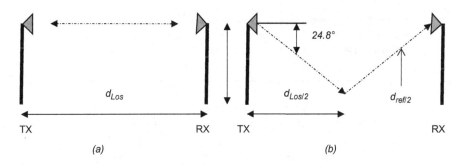

Figure 4.24 Antenna configuration for multipath verification of CRC's channel sounder. Controlled-condition measurements were made in an anechoic chamber at frequencies of 26 GHz and 38 GHz.

we often provide only the values of the metrics obtained and do not include their uncertainties. One of the goals of this book is to illustrate best practices for verifying sounder performance, rather than to judge the utility of the specific values obtained in these illustrative examples. For instance, in practice, one would typically use an in-situ verification approach during field tests to determine whether a sounder was generally operating as expected. An uncertainty analysis would not be applied to these data; however, the user would have increased confidence in their measured results. The final data analysis based on the measured results would likely contain uncertainties, but these might have been found from previous controlled-condition measurements. Thus, the lack of uncertainties reported in the following examples should not be construed to imply that uncertainties are not necessary.

Also, because results are presented from the current version of each group's channel sounders, this section is, essentially, a snapshot in time, reflecting the current activities of the groups listed below. It is anticipated that groups will extend their channel sounders and methods of verification in future work. The motivation for this section is to provide real-world illustrative examples of techniques and typical results achieved by state-of-the-art labs for today's mmWave channel sounding measurements.

4.5.1 Communications Research Centre

The CRC VNA-based channel sounder [42] was verified using a combination of controlled condition, free-space map-based multipath and two-ray map-based multipath methods. See [43] for additional detail on these measurements.

The verification method used an open LoS environment, as shown in Figure 4.24(a), and a specular ground reflection path, as shown in Figure 4.24(b). In the LoS case, both antennas were pointing at each other. For the ground-reflection case, the antennas were adjusted for 24.8° down-tilt to point toward the anechoic chamber floor.

The 2D TX–RX separation distance, which is also the LoS signal propagation distance, is $d_{\text{LoS}} = 6.5$ m, whereas the ground path signal should travel a distance

Table 4.2 The "empirical results" columns show the difference between direct and reflected signals in terms of physical distance and time of arrival for the 26 GHz and 38 GHz measurements. The "channel sounder" columns provide the sounder's expected resolution.

Frequency	Difference between direct and reflected rays (empirical results)		Channel sounder resolution	
(GHz)	d (m)	τ (ns)	d (m)	τ (ns)
26	0.059	0.19	0.24	0.8
38	0.089	0.29	0.30	1

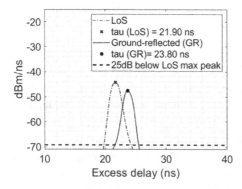

Figure 4.25 38 GHz PDP of free-space LoS and ground-reflection map-based MPCs.

of $d_{\text{Ref}} = 2\sqrt{h^2 + (d_{\text{LoS}}/2)^2} = 7.16$ m, where $h = 1.5$ m is the height of the antennas and d_{LoS} is the 2D separation distance between the TX and RX antennas. Note that these distances have been rounded for clarity of presentation. The measurement accuracy of the measured distances was actually within a few millimeters, as measured with a ribbon measuring tape supported by a lead wire. Both the LoS and ground measurements were taken separately in the order mentioned.

Figure 4.25 shows measured results for the 38 GHz measurement, where the symbols show the theoretical direct and ground-reflected ray delay times and the lines show the measured results. The difference between the theoretical and measured values of (1) distance (estimated from measurement) and (2) excess time delay for the direct and multipath components is shown in Table 4.2. The agreement is well within the channel sounder's expected resolution.

Results for the path-loss exponent for various polarizations, calculated from the 1 m CI free-space reference model,

$$L(d)[dB] = L_i(d_0) + 10\eta \log_{10}(d/d_0) + X_\sigma, d > d_0, \tag{4.13}$$

with $d_0 = 1$ m are shown in Table 4.3. Uncertainties in the path-loss exponent are not illustrated in this example.

Table 4.3 Path-loss exponents for VV and HH polarizations.

Frequency (GHz)	Polarization	Path-loss exponent
26	VV	1.96
	HH	1.92
38	VV	2.02
	HH	1.99

4.5.2 Durham University

Verification for the Durham frequency swept (chirp-based) channel sounder is illustrated with results in the 60 GHz band in both outdoor and indoor environments [44]. Similar verification techniques are also illustrated in the 25–28 GHz band and the 50–73 GHz band, as-yet unpublished. In [44], the outdoor environment was a low-rise urban area, and the indoor was a 38 m long office corridor containing tables and chairs.

An in-situ path loss verification was carried out using least squares regression, from which a path-loss exponent of 1.9 for the outdoor environment was estimated. This exponent approximates the free space value of 2. PDP verification is illustrated by in-situ comparison to a map-based ray-tracing model. Ray-tracing predicted a significant MPC corresponding to a 30 m reflection. The measured PDP produced a signal component at approximately 100 ns for the indoor environment, which, assuming free-space propagation, agrees with the 100 ns delay for a 30 m reflection.

4.5.3 ETRI

ETRI's correlation-based channel sounder verification technique for path loss and multipath delay at 28 GHz is illustrated in [45, 46]. Path-loss measurements were conducted along a straight stretch of road with fields on each side over distances ranging from 1.4 km to 1.6 km. The measurements were replicated using horn and omnidirectional antennas, comparing the path-loss exponent of the results to that obtained from a single frequency, CI model for a theoretical two-ray reflection (path-loss exponent = 1.92). The path-loss exponent for the horn antenna was 1.87 and the exponent was 1.90 for the omnidirectional antenna. Uncertainties are not reported in these illustrative examples.

To verify multipath delay, three cables (0.5 m, 10 m, 20 m in length) were excited from the output port of the channel sounder's transmitter through power splitters, as illustrated in the example on controlled-condition verification techniques and in [46]. Knowing the wave velocity in the cables allowed for comparison between the measured time delay for each cable and the expected time delay. The maximum difference between the measured and expected time delays was 0.49 ns in the 0.5 m cable (2.0 ns measured and 2.49 ns expected), with the measured time delay in the other two cables both 0.04 ns longer than expected. The differences may be due to the actual versus theoretical values of wave velocity in the cables; however, this is as yet undetermined.

Table 4.4 Path-loss exponents for the Ka-, D-, and 300 GHz bands, determined using the Georgia Tech channel sounder [24].

	CIF PLE	ABG PLE
Ka-band	2.001	1.982
D-band	1.983	1.997
300 GHz	1.997	2.005

ETRI's channel sounder verification techniques for the AoA values are also illustrated using an in-situ technique at 28 GHz. In an open environment, the transmit antenna remained stationary and the receive antenna was rotated. Two transmit antennas were used; one with a HPBW of 30° and another with 60° HPBW. The directions of the three strongest MPCs were compared, with the directions reported in the receive antenna's specifications. For the strongest MPC direction, the measured AoA underestimated the specified value by $1.3°-1.7°$. The measured AoA of the second strongest MPC exactly matched the specified angle. The measured angle for the third strongest MPC direction was overestimated by 0.2° with the 60° HPBW TX antenna, and by 1.9° using the 30° HPBW TX antenna.

4.5.4   Georgia Tech

Georgia Tech's VNA-based channel sounder in-situ verification technique was illustrated through comparisons with theoretical values of the free-space path-loss exponent and time delay determined from multipath peak positions in the PDP [24]. Three frequency ranges were considered in the free-space path-loss verification: Ka-band (26–43 GHz), D-band (110–170 GHz), and 300 GHz (300–316 GHz) [19].

Kim et al. measured path loss at various distances for the different frequency ranges in [24]. The multifrequency CI frequency (CIF) and alpha–beta–gamma (ABG) models were used to obtain path-loss exponent values for a comparison with the theoretical free-space path-loss exponent of 2. Table 4.4 displays the path-loss exponent values for each frequency range for the different multifrequency models.

Directional antennas and an aluminum reflector were used to illustrate controlled-condition verification of time-delay measurements in the D-band [19]. The receive antenna was rotated to separately obtain direct path and reflected signals. The peak positions in the computed PDP were compared to theoretical delay calculations based on direct measurement. The measured peak positions for the direct and reflected paths occurred at 2.7 ns and 3.6 ns, respectively. Both measured times were within 1 ns of the ideal theoretical delay times (2.54 ns for direct and 3.1 ns for reflected).

4.5.5   TU Ilmenau

The in-situ correlation-based channel sounder verification techniques used by the Technical University of Ilmenau have been illustrated in different frequency bands and measurement scenarios. The path-loss verifications were primarily conducted in an indoor entrance hall scenario and an outdoor rooftop-to-street environment. The measured distances were between approximately 5 m and 70 m for the indoor scenario and up to 100 m for the outdoor scenario [47, 48]. Using a threshold of 20 dB [49] for the entrance hall scenario, the path loss exponent from measurement for a 7 GHz center frequency was 1.76. For 30 GHz, the estimated path-loss exponent was 1.71 and for 60 GHz the estimated exponent was 1.77. Assuming that the directivity characteristics of the antennas are approximately the same at the three frequencies (HPBW of 30°), the results may be compared with the theoretical path-loss exponent of 2.0. The delay for these environments was verified in-situ by comparing calculated delay to laser distance measurements, which had a maximum error of 90 mm. This error was based on measurements from antenna front to antenna front. After correcting the laser measurement, the signal delay mean error was 0.14 ns, which corresponded to a distance of around 42 mm.

For the 70 GHz channel sounder, in-situ path-loss and delay verification was illustrated by comparing the measured results with ray-tracing results in a small office environment [50]. A 3D ray-tracing tool designed for indoor environments was used. The ray-tracing tool included scattering effects from walls, openings and visible elements, including utility installations and furniture. Two transmitter locations and seven receiver positions were considered. The measured and predicted path loss from both transmitter locations were compared. The root-mean-square error (RMSE) values were 2.8 dB and 0.7 dB, averaged over all receiver positions. The predicted RMS delay spread from the transmitter locations had average relative errors of 6.9% and 2.3%, with worst-case errors of -15% to 37% over all of the receiver positions.

4.5.6 Keysight

Verification of Keysight's correlation-based channel sounder was performed at 28 GHz using a 4×4 MIMO array with a stimulus signal of 1,024 samples. Path loss, path delay, AoA and AoD were considered [13].

With the antennas placed 2 m apart, path loss and path delay were measured and compared to theoretical free space path loss and LoS path delay values of -67.4 dB and 6.67 ns, respectively. The measured path loss was -66.15 dB, and the measured path delay was 6 ns; these corresponded to differences of 1.25 dB and 0.67 ns (2% and 10% different from the theoretical values).

To test the AoA and AoD, the receive antenna array was rotated in the azimuthal plane from $-45°$ to 45° using a 5° step size. The average measured AoA measurement error was 1.87°, while the average AoD measurement error was 1.58°.

4.5.7 NIST

In-Situ Verification [4, 51]

In-situ verification of the 83 GHz NIST switched array, correlation-based channel sounder was conducted for path loss and signal delay in an indoor open lobby area. Measurements were taken at separation distances ranging from 4 m to 13 m, and the path-loss exponent estimation was 1.93, compared to the ideal free-space path-loss exponent of 2. Additionally, the measured path-loss values for the separation distances were compared to predicted losses (again calculated with the Friis equation). The errors between the measured and theoretical path loss values had a mean of 0.01 dB and a standard deviation of 1.18 dB found from a Normal distribution fit.

The signal delay was measured at each separation distance and compared to values estimated from a map of the lobby area produced by the channel sounder's mobile positioning robot. The signal delay mean error was 0.21 ns, with a standard deviation of 0.06 ns based on a Normal distribution fit.

In-situ AoA verification was also carried out for the NIST channel sounder. Measurements were collected in an office environment using a receive antenna consisting of 16 scalar-feed-horn antennas. AoA results were extracted using the SAGE algorithm for each location along a measurement path that circled the room. These values were compared with calculated AoA values of the direct path for each measurement location. The mean error between the measured and theoretical values was 3.65°, with a standard deviation of 4.81°.

The theoretical Doppler frequency shift was compared to that estimated from in-situ measurements made by the NIST mobile 83 GHz channel sounder. The receive antenna, attached to a mobile robot, was moved approximately 6 m along a linear trajectory toward and away from the fixed transmit antenna. The path was aligned perpendicular to a wall so that the Doppler shift of the reflected path was also measured. The velocity was tracked internally by the robot. Measured frequency shifts of the direct and reflected paths were 210 Hz and −210 Hz, respectively. The theoretical Doppler shift for the recorded velocity (750 mm/s) was 209 Hz.

Comparison-to-Reference Verification [40]

The NIST 60 GHz correlation-based channel sounder was verified by comparing the RMS delay spread of a verification artifact as measured by the channel sounder and a VNA. The artifact contains a direct path and up to two MPCs. Acceptable delay spread values are those that do not fall outside the VNA error bounds.

Several multipath thresholds ranging from −10 dB to −45 dB were considered in the verification (Table 4.5). The channel-sounder-measured RMS delay spread was generally higher than that of the VNA measurements, and exceeded the uncertainty in the VNA for threshold values of −35 dB and below. The broadening of peaks due to internal reflections in the channel sounder and the presence of an additional peak at 28 ns occurring 40 dB below the highest peak contributed to the higher RMS delay spread values for the channel sounder measurements.

4.5.8   North Carolina State University

North Carolina State University's controlled-condition correlation-based channel sounder verification technique is illustrated by means of PDP measurements in a large room, as shown in Figure 4.26. The transmitter and receiver were placed on tripods at a height of 1.5 m each. The delays and signal strengths for the LoS and NLoS paths were measured. The NLoS path consisted of a ground reflection, where two settings were considered: a ground reflection without a metallic reflector and a ground reflection when a metallic reflector was placed on the ground. The dimensions of the measurement setup are illustrated in Figure 4.27. The LoS path is approximately

Table 4.5 Wireless channel metrics for different multipath thresholds calculated from NIST VNA and channel sounder (CS) measurements at 60.5 GHz.

Multipath threshold (dB)	τ_{RMS} VNA (ns)	τ_{RMS} CS (ns)	Number MPCs VNA	Number MPCs CS
−10	0.15 ± 0.00	0.15	1	1
−20	0.15 ± 0.00	0.18	1	2
−30	0.16 ± 0.00	0.19	1	1
−35	1.20 ± 0.07	1.37	4	3
−40	1.30 ± 0.06	1.46	9	7
−45	1.35 ± 0.07	1.49	13	11

Figure 4.26 NC State's controlled-condition verification of PDP measurements is illustrated in a large meeting room on the campus.

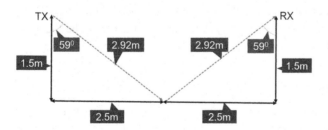

Figure 4.27 Measurement setup with a ground-reflected component. Transmitter and receiver are each placed at 1.5 m height.

5.0 m (measured with a ruler), whereas the NLoS path was approximately 5.8 m (calculated from the geometry of the setup). The measurements were repeated with and without the metallic reflector on the ground, as shown in Figure 4.26.

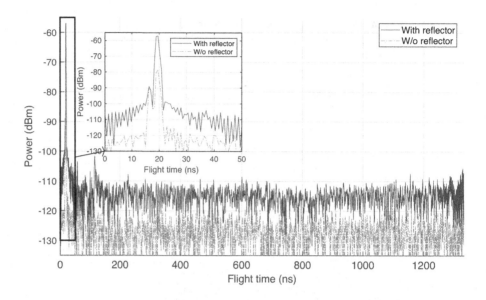

Figure 4.28 Power delay profile with and without the metallic reflector on the ground.

The results are shown in Figure 4.28. For both cases, the LoS component arrived at 16.3 ns and the NLoS component arrived at 19.5 ns. The corresponding LoS distance calculated from the measurements was 4.88 m and the NLoS calculated distance was 5.85 m, which are close to the distances illustrated in Figure 4.27. The LoS received power was −86.5 dBm with the reflector and −87.6 dBm without the reflector, a difference of 1.1 dB. The NLoS component received power was −54 dBm with the reflector and −75.9 dBm without the reflector, a difference of 21.5 dB. Additional results with metallic reflectors using the NCSU channel sounder can be found in [52, 53].

Figure 4.29 illustrates NCSU's free-space path-loss verification technique performed in an engineering building corridor. The measurements were repeated twice from approximately 1.0 to 5.0 m distance every 30 cm, again measured with a ruler. For this example, the path-loss exponent was measured to be 1.9 and 1.89 for measurement 1 and 2, respectively. The results are close to the theoretical path-loss exponent of 2.

4.5.9   NYU WIRELESS

Free-space path-loss and time delay measurements for the NYU WIRELESS correlation-based channel sounder were verified at a 73.5 GHz center frequency (see Figures 4.3 and 4.7). Narrowband (CW) and wideband (1,000 MHz null-to-null bandwidth) LoS free-space path-loss measurements were collected in an open, indoor environment at ranges up to 5 m and up to 33 m in an open outdoor environment. The path-loss exponent for these measurements was 1.99, and the standard deviation from the free-space theoretical values was 0.2 dB. This path-loss exponent may be compared to the ideal value of 2.0 based on the Friis equation [2, 7].

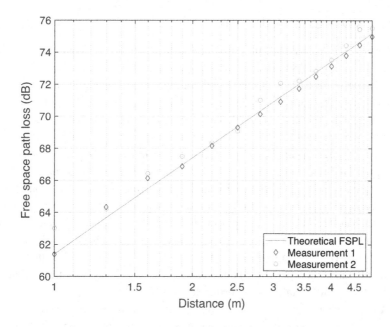

Figure 4.29 Example of NC State's free-space path loss verification measurements at 28 GHz.

To verify the accuracy of multipath time delays, the antennas were angled downward to produce a ground bounce. The expected excess time delay was 5.223 ns, based on the path difference between the LoS and ground-bounce paths. The time delay determined from the measured PDP was 5.200 ns; the difference between the values is within the uncertainty of the antenna positions and sampling interval of 0.05 ns [7].

NYU's methods for verifying AoD and AoA were also illustrated at 28 and 73 GHz by synthesizing an omnidirectional antenna pattern from measurement data using directional horn antennas [54, 55] and considering the resulting omnidirectional received power and path loss. Using directional antennas with different beamwidths (narrowbeam and widebeam horn antennas) yielded nearly identical received power and path loss synthesized over a systematic incremental scan of the entire azimuth plane. Similar verification was also performed at 38 GHz [56–58]. This method for synthesizing omnidirectional power from directional measurements can be used in place of sophisticated algorithms like SAGE/RiMAX. Also, at the time of development, high-gain adaptive arrays at mmWave frequencies were not available. The use of narrowbeam high-gain rectangular horn antennas with highly accurate 3D mechanical rotation at both the transmitter and receiver is a simple way of determining AoD and AoA information with a resolution equal to the HPBW of the antennas as described in [54]. This method may be used as opposed to more complicated algorithms that may induce artifacts and that attempt to create super-resolution by means of an antenna array.

4.5.10   University of British Columbia

The UBC VNA-based channel sounder has been verified through measurements conducted using the NIST channel verification artifact [59], LoS path-loss measurements conducted at street level along linear paths in street canyon environments [60] and fully double-directional channel measurements conducted on the roof of the MacLeod building in the presence of a wall and several discrete scatterers on the roof [61].

Comparison-to-reference verification measurements conducted using the NIST channel verification artifact were used to assess dynamic range and temporal (delay) accuracy, and to compare and optimize postprocessing strategies and procedures [59]. Illustrative results showed that dynamic range was reduced by about 5 dB when the frequency upconverter used in the transmit section of the channel sounder was operated at maximum output power. Backing off by about 6 dB to 30 dBm was required to achieve an optimal dynamic range of approximately 45 dB with Kaiser windowing ($\beta = 7$) applied. Averaging was not performed as it resulted in a reduced dynamic range (approx. 10 dB) due to phase offsets in successive traces. These results did not account for antenna performance, however.

In-situ LoS path-loss measurements conducted at street level were used to demonstrate general system performance verification including antenna performance [60]. The results confirmed that the measurement system returned an approximately free-space path loss characteristic with errors of less than 1 dB over the range of distances (up to 6 m) in which the directional receiving antenna effectively rejected multipath reflections from the ground and surrounding structures, as shown in Figure 4.30. Beyond that range, the path loss diverged from the free-space characteristic, as shown by the black crosses in the graph. The results also demonstrated that antenna alignment poses significant challenges when conducting measurements when the transmitter and receiver are deployed on uneven ground.

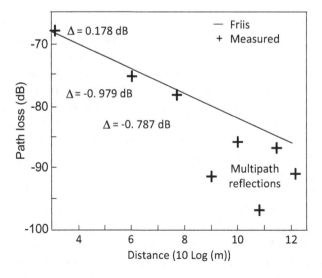

Figure 4.30 Measured and theoretical (Friis) path loss for the 28 GHz UBC channel sounder [60].

Table 4.6 UBC rooftop measurement results [61].

Beam type	Path length (m)	Meas. delay* (ns)	Calc. delay (ns)
Direct beam	5.95	20	19.8
Side wall reflection	11.90	40	39.7
Utility pole reflection	39.4	129	131

*Meas. delay is extracted from the impulse response corresponding to the LoS scenario where the antennas are pointed directly at each other.

To address such antenna alignment issues, UBC utilizes an in-house-developed automated levelling platform upon which the azimuth-elevation positioner and receiver front end are mounted. The platform is supported by a ball joint and two vertically mounted linear actuators. A three-axis inclinometer is used to determine the orientation of the platform. An Arduino-based controller samples the inclinometer a few times per second and adjusts the extension of the linear actuators as required to level the platform. The automatic levelling platform significantly speeds up the measurement process by eliminating time-consuming manual levelling as the receiver is moved from location to location.

The fully double-directional channel measurements conducted on the roof of the MacLeod building were used to further illustrate UBC in-situ verification techniques for AoA, AoD and delay associated with both discrete and extended reflectors when compared to map-based predictions [61]. Discrepancies in measured and calculated delays were around 0.3 ns for distances up to 12 m and 2 ns at 40 m, as shown in Table 4.6. A minor issue with the antenna rotator was identified and easily corrected.

4.5.11 University of Southern California

USC's correlation-based channel sounder in-situ verification techniques were illustrated using LoS path-loss measurements [62]. Path-loss measurements were conducted in an open area with the transmitter and the receiver placed on scissor lifts at the height of 5 m. The measurements were performed for distances ranging from 30 m to 122 m. The path-loss exponents were estimated at 1.997 and 2.051 for CI and ABG models, respectively. For both models, the observed path-loss exponents were almost equal to the free-space path-loss exponent of 2.

In-situ delay verification of the USC channel sounder was illustrated by comparing the observed multipath delays with the distances calculated from Google Maps [63]. By utilizing the AoA and AoD estimated from beam directionality (described below), and the delay information, it was possible to map the significant MPCs to the likely scatterers in the environment. The delay offsets between estimations from measurements and from the map of the environment vary from −1.2 ns to 1.1 ns. This is mainly caused by the 2.5 ns delay resolution of the sounder, due to its 400 MHz bandwidth.

The USC channel sounder uses phased-antenna arrays in both the transmitter and the receiver to form beams aiming toward different directions. During the initial

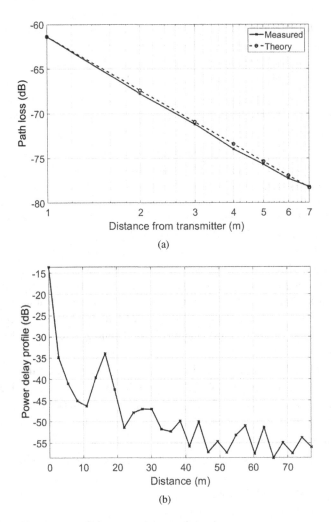

Figure 4.31 (a) Measured and theoretical path loss for the Univ. of Wisconsin CAP-MIMO system, according to the Friis formula; (b): measured PDP with the dominant multipath reflection located six samples from the LoS component. © 2016 IEEE. Reprinted, with permission, from [65].

investigations, this directional information was utilized to estimate the AoA and the AoD, as illustrated in [62]. The accuracy of the AoA estimations was validated via controlled-condition, in-chamber measurements when the receiver was rotated to different azimuth angles while the transmitter was fixed, and vice versa for the AoD. The validation was limited to 12° which was the 3 dB beamwidth.

4.5.12   University of Wisconsin

Illustration of in-situ path loss and multipath delay time verification techniques using the University of Wisconsin–Madison CAP-MIMO channel sounding system, which uses a lens array for multi-beamforming [64], was provided at 28 GHz [65, 66]. Indoor measurements were collected at 1 m intervals spanning a range of 1–7 m in a straight line. The measured path-loss values for each distance were compared to theoretical values obtained from the Friis formula and are shown in Figure 4.31(a).

During these measurements, the most significant MPC was observed in the PDP at a 16 m separation distance, as shown in Figure 4.31(b). This multipath was attributed to reflections from a metal cabinet and metal door frame located behind the receiver. The metal objects and the receiver were spaced approximately 8 m apart.

4.5.13   Summary of Verification Methods

Table 4.7 summarizes the verification methods used by Alliance participants.

Table 4.7 Summary of verification methods used by 5G mmWave Channel Model Alliance participants.

Group, location, contact	Sounder architecture, transmit signal	Path-loss verification	Multipath delay verification	Angle of arrival Verification
Communications Research Centre, Canada, Yvo de Jong	Mulitband, VNA-based, sinusoid	In-situ, Friis	Map-based	N/A
Durham University, Durham, UK, Prof. Sana Salous	Multiband, chirp, FMCW	In-situ, Friis	Map-based	N/A
ETRI, Daejeon, S. Korea, Juyul Lee, Myung-Don Kim	Correlation-based, PRBS	In-situ, two-ray	Controlled-condition, multipath artifact	In-situ, map-based
Georgia Tech, Atlanta, GA, Prof. Alenka Zajic	VNA-based, sinusoid	Controlled-condition, Friis	Controlled-condition	N/A
TU Ilmenau, Ilmenau, Germany, Prof. Reiner Thomä, Robert Müller	Correlation-based, PRBS	N/A	In-situ, map-based	N/A
Keysight, Santa Rosa, CA, Robin Wang, Sheri Detomasi	Correlation-based, Keysight proprietary sequence	In-situ, Friis	In-situ, map-based	In-situ, map-based
NIST, Boulder, CO, Peter Papazian, Camillo Gentile, Jeanne Quimby, Kate Remley	Correlation-based, PRBS	In-situ, Friis	In-situ, map-based, and comparison to reference	In-situ, map-based
North Carolina State University, Raleigh, NC, Prof. Ismail Guvenc, Ozgur Ozdemir	Correlation-based, National Instruments proprietary sequence	In-situ, Friis	Controlled-condition	N/A
NYU WIRELESS New York, NY, Prof. Ted Rappaport, Hangsong Yan, George MacCartney, Yunchou Xing	Correlation based, dual architecture (real time and PRBS) Direct-correlation/ real time	In-situ, Friis	In-situ, map-based	N/A

Table 4.7 (*Continued*)

University of British Columbia, Vancouver, Canada, Prof. Dave Michelson	VNA-based, sinusoid	In-situ, Friis	NIST artifact	In-situ, map-based
University of Southern California and Samsung, Prof. Andy Molisch	Correlation-based, multitone signal	Chamber, Friis	In-situ, map-based	Controlled-environment, anechoic chamber
University of Wisconsin–Madison, Akbar Sayeed	FPGA-based baseband processor sinusoid; PRBS, chirp FMCW, OFDM	In-situ, Friis	In-situ, map-based	N/A

References

[1] 3GPP, "Technical specification group radio access network: Study on channel model for frequencies from 0.5 to 100 GHz (Release 14)," 3rd Generation Partnership Project (3GPP), TR 38.901 V14.2.0, Sept. 2017, online: www.3gpp.org/DynaReport/38901.htm, accessed Dec. 1, 2018.

[2] H. T. Friis, "A note on a simple transmission formula," *Proceedings of the IRE*, vol. 34, no. 5, pp. 254–256, May 1946.

[3] A. Zajic, *Mobile-to-Mobile Wireless Channels*, Artech House: Boston, MA, 2013.

[4] K. A. Remley, C. Gentile, A. Zajic and J. T. Quimby, "Methods for channel sounder measurement verification," *IEEE 86th Vehicular Technology Conference (VTC2017-Fall), Workshop W5 5G Millimeter-Wave Channel Measurement, Models, and Systems*, Sept. 2017.

[5] C. Gentile, J. Senic, P. B. Papazian, J.-K. Choi and K. A. Remley, "Pathloss models for indoor hotspot deployment at 83.5 GHz," *IEEE Global Communications Conference: The First International Workshop on 5G Millimeter-Wave Channel Models (GLOBECOM 2016)*, pp. 1–6, Dec. 2016.

[6] K. Hausmair, K. Witrisal, P. Meissner, C. Steiner and G. Kail, "SAGE algorithm for UWB channel parameter estimation," *Proceedings of COST 2100 Management Committee Meeting*, Feb. 2010.

[7] G. R. MacCartney, Jr. and T. S. Rappaport, "A flexible millimeter-wave channel sounder with absolute timing," *IEEE Journal on Selected Areas in Communications*, vol. 35, no. 6, pp. 1402–1418, June 2017.

[8] T. S. Rappaport, G. R. MacCartney, Jr., M. K. Samimi and S. Sun, "Wideband millimeter-wave propagation measurements and channel models for future wireless communication system design (invited paper)," *IEEE Transactions on Communications*, vol. 63, no. 9, pp. 3029–3056, Sept. 2015.

[9] W. G. Newhall, T. S. Rappaport and D. G. Sweeney, "A spread spectrum sliding correlator system for propagation measurements," *RF Design*, Apr. 1996, pp. 40–54.

[10] W. G. Newhall, K. Saldanha and T. S. Rappaport, "Using RF channel sounding measurements to determine delay spread and path loss," *RF Design*, pp. 82–88, Jan. 1996.

[11] W. G. Newhall and T. S. Rappaport, "An antenna pattern measurement technique using wideband channel profiles to resolve multipath signal components," *Antenna Measurement Techniques Association 19th Annual Meeting and Symposium*, pp. 17–21, Nov. 1997.

[12] G. R. MacCartney, Jr. and T. S. Rappaport, "A flexible wideband millimeter-wave channel sounder with local area and NLOS to LOS transition measurements," *2017 IEEE International Conference on Communications (ICC)*, May 2017, pp. 1–7.

[13] G. R. MacCartney, Jr., S. Sun, T. S. Rappaport, Y. Xing, H. Yan, J. Koka, R. Wang and D. Yu, "Millimeter wave wireless communications: New results for rural connectivity," *All Things Cellular' 16, in conjunction with ACM MobiCom*, Oct. 7, 2016, pp. 31–36.

[14] S. Sun, T. S. Rappaport, T. Thomas, A. Ghosh, H. Nguyen, I. Kovacs, I. Rodriguez, O. Koymen and A. Partyka, "Investigation of prediction accuracy, sensitivity, and parameter stability of large-scale propagation path loss models for 5G wireless communications," *IEEE Transactions on Vehicular Technology*, vol. 65, no. 5, pp. 2843–2860, May 2016.

[15] T. S. Rappaport, R. W. Heath, Jr., R. C. Daniels and J. N. Murdock, *Millimeter Wave Wireless Communications*, Pearson/Prentice Hall: Upper Saddle River, NJ, 2015.

[16] T. S. Rappaport, *Wireless Communications: Principles and Practice*, 2nd ed., Prentice Hall: Upper Saddle River, NJ, 2001.

[17] Y. Xing, O. Kanhere, S. Ju, T. S. Rappaport and G. R. MacCartney Jr., "Verification and calibration of antenna cross-polarization discrimination and penetration loss for millimeter wave communications," *IEEE 88th Vehicular Technology Conference (VTC2018-Fall)*, Aug. 2018, pp. 1–6.

[18] J. Hannig, H. K. Iyer and C. M. Wang, "Fiducial approach to uncertainty assessment accounting for error due to instrument resolution," *Metrologia*, vol. 44, pp. 476–483, 2007.

[19] S. Kim, W. T. Khan, A. Zajic and J. Papapolymerou, "D-band channel measurements and characterization for indoor applications," *IEEE Transactions on Antennas and Propagation*, vol. 63, no. 7, pp. 3198–3207, July 2015.

[20] S. Kim and A. Zajic, "Statistical characterization of 300-GHz propagation on a desktop," *IEEE Transactions on Vehicular Technology*, vol. 64, no. 8, pp. 3330–3338, Aug. 2015.

[21] W. T. Khan, S. Kim, A. Zajic and J. Papapolymerou, "D-band indoor path loss measurements," *Proceedings of IEEE International Symposium on Antennas and Propagation*, July 2014, pp. 1–2.

[22] S. Kim and A. Zajic, "Path loss model for 300-GHz wireless channels," *Proceedings of IEEE International Symposium on Antennas and Propagation*, July 2014, pp. 1–2.

[23] S. Kim and A. Zajic, "300 GHz path loss measurements on a computer motherboard," *Proceedings of the 10th European Conference on Antennas and Propagation (EuCAP)*, Apr. 11–15, 2016, pp. 1–5.

[24] C.-L. Cheng, S. Kim and A. Zajic, "Comparison of path loss models for indoor 30 GHz, 140 GHz, and 300 GHz channels," *Proceedings of the 11th European Conference on Antennas and Propagation (EuCAP)*, Mar. 19–24, pp. 1–5.

[25] H.-K. Kwon, M.-D. Kim and J. Liang, "Evaluation of multi-path resolution for millimeter-wave channel sounding system," *Proceedings of International Conference on ICT (ICTC)*, Oct. 2015, pp. 1037–1039.

[26] R. Sun, P. B. Papazian, J. Senic, Y. Lo, J. Choi, K. A. Remley and C. Gentile, "Design and calibration of a double-directional 60 GHz channel sounder for multipath component tracking," *11th European Conference on Antennas and Propagation (EuCAP 2017)*, Mar. 19–24, 2017, pp. 1–5.

[27] J. D. Parsons, *The Mobile Radio Propagation Channel*, 2nd ed., Wiley: New York, 2000.

[28] M. Landmann, A. Richter and R. S. Thomä, "Performance evaluation of antenna arrays for high-resolution DOA estimation in channel sounding," *2004 International Symposium on Antennas and Propagation (ISAP04)*, Aug. 17–21, 2004.

[29] R. S. Thomä, M. Landmann, A. Richter and U. Trautwein, *Multidimensional High-Resolution Channel Sounding*, Hindawi Publishing Corporation: London, 2006.

[30] M. Landmann, W. Kotterman and R. S. Thomä, "On the influence of incomplete data models on estimated angular distributions in channel characterisation," *The Second European Conference on Antennas and Propagation (EuCAP 2007)*, 2007, pp. 1–8.

[31] M. Landmann, M. Käske and R. S. Thomä, "Impact of incomplete and inaccurate data models on high resolution parameter estimation in multidimensional channel sounding," *IEEE Transactions on Antennas and Propagation*, vol. 60, no. 2, pp. 557–573, Feb. 2012.

[32] M. Landmann, A. Richter and R. Thomä,"DoA resolution limits in MIMO channel sounding," *2004 IEEE International Symposium on Antennas and Propagation and USNC/URSI National Radio Science Meeting*, June 20–26, 2004.

[33] G. Sommerkorn, D. Hampicke, R. Klukas, A. Richter, A. Schneider and R. Thomä, "Uniform rectangular antenna array calibration issues for 2-D ESPRIT application," *4th European Personal Mobile Communications Conference EPMCC 2001*, Feb. 20–22, 2001.

[34] J. C. Liberti and T. S. Rappaport, *Smart Antennas for Wireless Communications: IS-95 and Third Generation CDMA Applications*, Prentice Hall: Upper Saddle River, NJ, 1999.

[35] R. Haardt, R. Thomä and A. Richter, "Multidimensional high-resolution parameter estimation with applications to channel sounding," *High-Resolution and Robust Signal Processing*, Yingbo Hua (Ed.), Marcel Dekker: New York, 2003, pp. 253–337.

[36] J. J. A. Lempianen, J. K. Laiho-Steffens and A. F. Wacker, "Experimental results of cross polarization discrimination and signal correlation values for a polarization diversity scheme," *1997 IEEE 47th Vehicular Technology Conference*, 1997, pp. 1498–1502.

[37] T. S. Rappaport, Y. Xing, O. Kanhere, S. Ju, A. Madanayake, S. Mandal, A. Alkhateeb and G. C. Trichopoulos "Wireless communications and applications above 100 GHz: Opportunities and challenges for 6G and beyond," *IEEE ACCESS*, pp. 78729–78757, June, 2019.

[38] S. Ju, Y. Xing, O. Kanhere and T. S. Rappaport, "Millimeter wave and sub-terahertz spatial statistical channel model for an indoor office building," *IEEE Journal on Selected Areas in Communications*, vol. 39, no. 6, pp. 1561–1575, June 2021.

[39] D. F. Williams and A. Lewandowski, "NIST microwave uncertainty framework," online: www.nist.gov/ctl/rf-technology/related-software.cfm, accessed 2011.

[40] J. N. H. Dortmans, J. T. Quimby, K. A. Remley, D. F. Williams, J. Senic, and R. Sun, "Design of portable verification artifact for millimeter-wave-frequency channel sounders," *IEEE Transactions on Antennas and Propagation*, vol. 67, pp. 6149–6158, Sept. 2019.

[41] J. T. Quimby, D. F. Williams, K. A. Remley, P. B. Papazian, D. Ribeiro and J. Senic, "Channel sounder performance verification using a vector network analyzer (VNA) in a controlled RF propagation environment," *Transactions on Antennas and Propagation*, forthcoming.

[42] Y. L. C. De Jong, J. A. Pugh, M. Bennai and P. Bouchard, "2.4 to 61 GHz Multiband Double-Directional Propagation Measurements in Indoor Office Environments," *IEEE Transactions on Antennas and Propagation*, vol. 66, no. 9, pp. 4806–4820, Sept. 2018, doi: 10.1109/TAP.2018.2851279.

[43] M. Bennai, "CRC channel sounder verification and validation," white paper, online: https://sites.google.com/a/corneralliance.com/5g-mmwave-channel-model-alliance-wiki/home, accessed Apr. 9, 2018.

[44] S. Salous, S. Feeney, X. Raimundo and A. Cheema, "Wideband MIMO channel sounder for radio measurements in the 60 GHz band," *IEEE Transactions on Wireless Communications*, vol. 15, no. 4, pp. 2825–2832, 2016.

[45] J. Lee, J. Liang, M.-D. Kim, J.-J. Park, B. Park and H. K. Chung, "Measurement-based propagation channel characteristics for millimeter-wave 5G Giga communication systems," *ETRI Journal*, vol. 38, no. 6, pp. 1031–1041, Dec. 2016.

[46] H.-K. Kwon, M.-D. Kim, and J. Liang, "Evaluation of multi-path resolution for millimeter-wave channel sounding system," *Proceedings of the International Conference on ICT (ICTC)*, Oct. 2015, pp. 1037–1039.

[47] R. Müller, J. Luo, Y. Xin, G. Calcev and K. Zeng, "Channel measurement summary of the rooftop to street and entrance hall scenario for 11ay," *Contribution IEEE 802.11-15/0956r*, 2016.

[48] R. Müller, S. Häfner, D. Dupleich, R. S. Thomä, G. Steinböck, J. Luo, E. Schulz, X. Lu and G. Wang, "Simultaneous multi-band channel sounding at mm-Wave frequencies," *Proceedings of the EuCAP 2016*, 2016, pp. 1–5.

[49] D. Dupleich, N. Iqbal, C. Schneider, S. Haefner, R. Müller, S. Skoblikov, J. Luo and R. S. Thöma, "Investigations on fading scaling with bandwidth and directivity at 60 GHz," *Proceedings of the EuCAP 2017*, Mar. 2017.

[50] F. Fuschini, S. Hafner, M. Zoli, R. Muller, E. M. Vitucci, D. Dupleich, M. Barbiroli, J. Luo, E. Schulz, V. Degli-Esposti and R. S. Thomä. "Analysis of in-room mm-wave propagation: Directional channel measurements and ray tracing simulations." *Journal of Infrared Millimeter and Terahertz Waves*, vol. 38, pp. 727–744, 2017.

[51] P. B. Papazian, C. Gentile, K. A. Remley, J. Senic and N. D. Golmie, "A radio channel sounder for mobile millimeter-wave communications: System implementation and measurement assessment," *IEEE Transactions on Microwave Theory and Technology*, vol. 64, no. 9, pp. 2924–2932, Sept. 2016.

[52] W. Khawaja, O. Ozdemir, F. Erden, Y. Kakishima, I. Guvenc and M. Ezuma, "Effect of passive reflectors for enhancing coverage of 28 GHz mmWave systems in an outdoor setting," *Proceedings of the IEEE Radio Wireless Symposium (RWS)*, Jan. 2019.

[53] W. Khawaja, O. Ozdemir, Y. Yapici, I. Guvenc and Y. Kakishima, "Coverage enhancement for mmWave communications using passive reflectors," *Proceedings of the IEEE Global Symposium on Millimeter Waves (GSMM)*, May 2018.

[54] S. Sun, G. R. MacCartney, M. K. Samimi and T. S. Rappaport, "Synthesizing omnidirectional antenna patterns, received power and path loss from directional antennas for 5G millimeter-wave communications," *2015 IEEE Global Communications Conference (GLOBECOM)*, 2015, pp. 1–7.

[55] G. R. MacCartney and T. S. Rappaport, "Rural macrocell path loss models for millimeter wave wireless communications," *IEEE Journal on Selected Areas in Communications*, vol. 35, no. 7, pp. 1663–1677, July 2017.

[56] T. S. Rappaport, E. Ben-Dor, J. N. Murdock and Y. Qiao, "38 GHz and 60 GHz angle-dependent propagation for cellular and peer-to-peer wireless communications," *2012 IEEE International Conference on Communications (ICC)*, 2012, pp. 4568–4573.

[57] A. I. Sulyman, A. T. Nassar, M. K. Samimi, G. R. MacCartney, T. S. Rappaport and A. Alsanie, "Radio propagation path loss models for 5G cellular networks in the 28 GHZ and 38 GHz millimeter-wave bands," *IEEE Communications Magazine*, vol. 52, no. 9, pp. 78–86, Sept. 2014.

[58] G. R. MacCartney, Jr., T. S. Rappaport, M. K. Samimi, and S. Sun, "Millimeter-wave omnidirectional path loss data for small cell 5G channel modeling," *IEEE Access*, vol. 3, pp. 1573–1580, Sept. 2015.

[59] A. Bhardwaj, S. Bonyadi and D. G. Michelson, "NIST verification of the UBC channel sounder," technical memorandum, June 10, 2017.

[60] D. G. Michelson, "Extremely high frequency channel characterization and modeling project update," technical report, University of British Columbia, Jan. 10, 2017.

[61] A. Bhardwaj, P. Zarei, C. Loo, Y. Liu and D. G. Michelson, "Measurement plan: D2D beam quality," technical report, University of British Columbia, May 23, 2017.

[62] C. U. Bas, R. Wang, D. Psychoudakis, T. Henige, R. Monroe, J. Park, J. Zhang and A. F. Molisch, "A real-time millimeter-wave phased array MIMO channel sounder," *IEEE Vehicular Technology Conference*, Sept. 24–27, 2017, pp. 1–6.

[63] C. U. Bas, R. Wang, S. Sangodoyin, S. Hur, K. Whang, J. Park, J. Zhang and A. F. Molisch, "28 GHz microcell measurement campaign for residential environment," *IEEE Globecom* 2017.

[64] J. Brady, J. Hogan and A. Sayeed, "Multi-beam MIMO prototype for real-time multiuser communication at 28 GHz," *IEEE Globecom Workshop on Emerging Technologies for 5G*, Dec. 2016.

[65] A. Sayeed, J. Brady, P. Cheng and U. Tayyab, "Indoor channel measurements using a 28 GHz multi-beam MIMO prototype," *Workshop on Millimeter-Wave Channel Models, IEEE VTC Fall 2016*, Sept. 2016.

[66] A. Sayeed and J. Brady, "Beamspace MIMO channel modeling and measurement: Methodology and results and 28 GHz," *IEEE Globecom Workshop on Millimeter-Wave Channel Models*, Dec. 2016.

5 Introduction to Millimeter-Wave Channel Modeling

Andreas F. Molisch, Camillo Gentile, Theodore S. Rappaport, Alenka Zajić, Ozge Hizir Koymen and Kate A. Remley

The purpose of any propagation channel model is to represent the essential physical propagation effects *that influence system design and performance*, without getting swamped by irrelevant details. Thus, while the propagation channel itself is independent of any system that operates in it, channel *models* do depend on the system. To be more specific, both the parameterizations of a channel model, and even the fundamental effects described in the model, depend on the system parameters. For example, a propagation model for an AM-radio operating at <1 MHz carrier frequency will need to consider over-the-horizon propagation while not concerning itself with the surface roughness of buildings, while the reverse is true for millimeter-wave (mmWave) cellular propagation models. Thus, while many channel models exist for the cellular and Wi-Fi communications channels, much less is available for mmWave channels.

Another important relationship exists between the channel models and channel measurements. Of course, the channel models have to be parameterized by measurements. On the other hand, the model structures used need to inform the measurement campaigns about what parameters are particularly relevant, thus indirectly determining *how* the measurements should be done, to what accuracy they should be done, etc. For this reason, it is essential that measurement and modeling groups retain constant contact. Often the interface between the two groups is a list of multipath components (MPCs) or similar description of the channel.

The main point of this part of the book is to give a survey of the state-of-the-art in channel modeling, covering the main modeling methods that have been presented in the literature. Over the years, a variety of fundamental models have been developed, which can be grouped approximately into (1) tapped delay line stochastic models, (2) geometry-based stochastic channel models and (3) quasi-deterministic models. Each of these groups has been considered for mmWave channel modeling by various academic and industrial groups, in particular for the description for the dispersion in the delay and spatial domains. Additionally, the clustering of MPCs into groups with similar behavior is important for all of these types of models. Finally, various path-loss and shadowing models have also been proposed, and it is interesting that a seemingly old topic like generic path-loss modeling has drawn new interest in the past years.

We stress here that the following chapters are a compendium of existing methods. This book refrains from any listing of pros and cons of different techniques.

The subsequent chapters are organized as follows. The remainder of Chapter 5 covers general channel description methods and establishes notation. Chapter 6 describes path-loss and shadowing modeling methods. Chapter 7 covers various methods for clustering of MPCs, either based on similar long-term behavior or on grouping in the delay/angle domain. Chapter 8 discusses the three main approaches to describing dispersion – that is tapped delay lines, GSCM and quasi-deterministic models. The unique aspects of modeling of peer-to-peer networks are covered in Chapter 9. Chapter 10 investigates description methods for temporal variations of the channel, in particular over time intervals in which the wide-sense stationarity (WSS) assumption is not fulfilled. Chapter 11 discusses modeling efforts at sub-THz frequencies. Finally, Chapter 12 covers the connection between the measurements and modeling.

5.1   Deterministic Description Methods

In this subsection we will list various methods for deterministic description of the channels – that is, quantifying the impact of wireless propagation for a given time/location of the transmitter (TX), receiver (RX), and interaction objects (IOs). Such descriptions are obviously key descriptions, as they are essentially the outputs from the measurements and form the input for all of the various modeling activities. The connections of these quantities to the measurements is discussed in Chapter 12.

We start out with simple systems where the TX and RX have a single antenna each, so that the system is completely described by the input $x(t)$ and the output $y(t)$ of the channel. We then progress to more complicated, multi-antenna channels, which are highly relevant for mmWave systems. The description follows closely the one in [1].

5.1.1   Time-Variant Impulse Response

In single-antenna systems, input and output of the channel are related by the "time-variant impulse response" $h(t, \tau)$ by [2]

$$y(t) = \int_{-\infty}^{\infty} x(t - \tau) h(t, \tau) d\tau. \qquad (5.1)$$

This relationship is analogous to the well-explored input–output relationship in linear time-invariant (LTI) systems; the difference lies in the fact that now the impulse response is time-variant.

An intuitive interpretation is possible if the impulse response changes only slowly with time. Then we can consider the behavior of the system at a particular time t, like that of an LTI system. The variable t can thus be viewed as "absolute" time that tells us which impulse response $h(\tau)$ is currently valid. Such a system is also called *quasi-static*. Such an interpretation is meaningful if the timescale on which the channel impulse response (CIR) changes is much larger than the duration of the impulse response, and furthermore much larger than the transmitted symbol duration. These conditions are fulfilled in many practical wireless systems: Times over which the channel stay constant are on the order of milliseconds or larger, while the duration

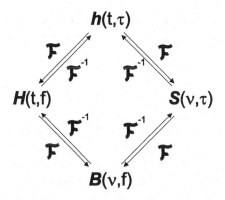

Figure 5.1 Interrelation between the deterministic system functions. © 2011 IEEE Press/John Wiley. Reprinted, with permission, from [1].

of impulse responses (and typical symbol durations) are microseconds or smaller. More rapidly time-varying channels are discussed in detail in [3].

As the impulse response of a time-variant system, $h(t, \tau)$, depends on two variables, τ and t, we can perform Fourier transformations with respect to either (or both) of them. This results in four different, but equivalent, representations (Figure 5.1). Fourier-transforming the impulse response with respect to the variable τ results in the *time-variant transfer function* $H(t, f)$:

$$H(t, f) = \int_{-\infty}^{\infty} h(t, \tau) \exp(-j2\pi f \tau) d\tau. \tag{5.2}$$

The input–output relationship is given by

$$y(t) = \int_{-\infty}^{\infty} X(f)H(t, f) \exp(j2\pi f t) df. \tag{5.3}$$

The interpretation is straightforward for the case of the quasi-static system – the spectrum of the input signal is multiplied with the spectrum of the "currently valid" transfer function, to give the spectrum of the output signal. For quasi-static channels, the transfer function calculus $Y(f) = H(f)X(f)$ is valid.

If we do a Fourier transformation with respect to t we obtain the *delay–Doppler function*, also known as *spreading function* $S(v, \tau)$:

$$S(v, \tau) = \int_{-\infty}^{\infty} h(t, \tau) \exp(-j2\pi v t) dt. \tag{5.4}$$

This function describes the spreading of the input signal in the delay/Doppler domains.

The *Doppler-variant transfer function* $B(v, f)$ is obtained by transforming $S(v, \tau)$ with respect to τ:

$$B(v, f) = \int_{-\infty}^{\infty} S(v, \tau) \exp(-j2\pi f \tau) d\tau. \tag{5.5}$$

The CIR can be related to the representation of the MPCs. Assuming that the propagation channel can be completely represented by summing up the contributions of the different MPCs:

$$h(t, \tau) = \sum_{n=1}^{N_p} \tilde{a}_n(t)\delta(\tau - \tau_n), \qquad (5.6)$$

where $\tilde{a}_n$ is the complex amplitude of the nth MPC; note that this amplitude of the MPC is the amplitude response of the "radio channel" that includes the antennas; in other words, the channel from the TX antenna connector to the RX antenna connector (we will discuss below the definition of the pure propagation channel, which is defined as not including the antennas). τ_n is the delay associated with it.[1] Note that this impulse response is time-variant – at different absolute times t, a different impulse response $h(\tau)$ characterizes the channel. The actual value of this impulse response is determined by the value of the complex attenuation of the MPCs at time t. Equation (5.6) is valid for systems with large bandwidth, so that we can resolve all the MPCs. Thus, within a "region of stationarity" (see below for a definition) there is no small-scale fading, and the only temporal change of the MPCs is a phase shift due to the change of runtime between the TX and RX as the user equipment (or the interacting objects) move around. A system with finite bandwidth produces an impulse response that is the filtering of the CIR with the system response.

5.1.2   Directional Description and MIMO Matrix

For multi-antenna systems the directional characteristics of the MPCs play a major role. Thus, the impulse response eq. (5.6) should be replaced by the *double-directional impulse response* (DDIR) [5], which consists of a sum of contributions from the MPCs:

$$h(t, \mathbf{x}^T, \mathbf{x}^R, \tau, \theta^T, \theta^R) = \sum_{n=1}^{N_p} h_n(t, \mathbf{x}^T, \mathbf{x}^R, \tau, \theta^T, \theta^R) \qquad (5.7)$$

$$= \sum_{n=1}^{N_p} a_n(t)\delta(\tau - \tau_n)\delta(\theta^T - \theta_n^T)\delta(\theta^R - \theta_n^R),$$

where the locations of transmitter is $\mathbf{x}^T$ and receiver is $\mathbf{x}^R$, the *direction-of-departure (DoD)* is θ^T and the *direction-of-arrival (DoA)* is θ^R. The a_n are the complex amplitudes of the physical MPCs (i.e., without any antenna effects). Just like in the case of eq. (5.6), the phases of the a_n change quickly, while all other parameters (i.e., absolute amplitude $|a|$, delay, DoA and DoD) vary slowly with the transmit and receive

---

[1] The use of delta functions in the impulse response is a common didactic tool in the derivation of impulse responses for "infinite bandwidth" channels, and also employed here. However, it gives rise to both physical problems (since MPCs in reality are frequency-dependent), and mathematical issues in the definition of the power delay profile (squares of delta functions are not defined). For a more exact derivation, see [4].

locations (over many wavelengths). For this reason, the time dependence is written explicitly only for $a_n(t)$. The AoD (angle of departure) and AoA (angle of arrival) are spatial angles, which can be described by the azimuth/elevation pair. When only propagation in the horizontal plane occurs, then representation by the azimuth alone is sufficient. Note, however, that neglecting a nonzero elevation in measurements not only eliminates information about the elevation, but also leads to errors in the estimated azimuth. Similarly, neglecting polarization information leads to errors in the overall estimated parameters [6].

The above description is not the most general one, though we will use it in the following frequently to explain points without an excess of notation. For a completely general formulation, the following items need to be taken into account:

- *Diffuse multipath:* While the "finite sum of discrete multipath" model described above is very popular, it does not reflect the complete physical reality. Diffuse scattering, as well as wavefront curvature from nearby scatterers, can give rise to other components that (irrespective of their physical origin) are commonly called diffuse multipath components (DMCs). The total impulse response is then

$$h(t,\mathbf{x}^{\mathrm{T}},\mathbf{x}^{\mathrm{R}},\tau,\theta^{\mathrm{T}},\theta^{\mathrm{R}}) = \sum_{n=1}^{N_p} h_n(t,\mathbf{x}^{\mathrm{T}},\mathbf{x}^{\mathrm{R}},\tau,\theta^{\mathrm{T}},\theta^{\mathrm{R}}) + h_{\mathrm{DMC}}(t,\mathbf{x}^{\mathrm{T}},\mathbf{x}^{\mathrm{R}},\tau,\theta^{\mathrm{T}},\theta^{\mathrm{R}}).$$

$$(5.8)$$

The DMC, due to its nature, is most efficiently not described by a sum of delta functions in the delay/angle domain, but rather by a continuous version of the delay–angle–Doppler dispersion profile. While in principle a deterministic representation (using such continuous functions) is possible, the DMC is commonly interpreted as a random process, and its description is limited to the parameters of this random process; these are also the parameters extracted by high-resolution algorithms that take the DMC into account explicitly (see Chapter 12). These random processes can be described either by their delay–angle–Doppler characteristics, or their spatiotemporal autocorrelation function. It is common to assume a Kronecker model for the angle–delay–Doppler (i.e., the delay, DoA, and DoD are independent of each other).[2] An alternative approach models the parameters of the DMC as a function of the discrete components (i.e., each discrete MPC has a diffuse "tail" associated with it). Measurements in the centimeter-wave region have shown that 10–50% of all energy is in the DMC; similar measurements currently have not yet been made for mmWave frequencies. It must be noted that DMC can be expanded into a sum of plane waves when a very large number of discrete MPCs is allowed. For example, [7] measured more than 3,000 components with extreme-sized arrays in the mmWave regime.

---

[2] Note that this is a Kronecker model only for the diffuse component, while the discrete components might have a non-Kronecker structure.

- *Polarization:* The above description is for a single polarization direction. For a general description, two orthogonal polarizations (assuming we are in the far-field) should be described. In that case, the amplitude a_n becomes a matrix:

$$\begin{pmatrix} a_n^{V,V} & a_n^{V,H} \\ a_n^{H,V} & a_n^{H,H} \end{pmatrix}, \tag{5.9}$$

where superscripts V and H refer to vertical and horizontal polarization for concreteness, but any other set of orthogonal polarizations could be used as well. Obviously, measurement of the polarization-resolved amplitudes requires a dual-polarized array in the underlying channel sounder. More details are provided in the next subsection.

- *Doppler shift:* In the case of movement, an additional phase-shift term occurs for each of the MPCs, representing the Doppler shift of the component:

$$a_n \delta(\tau - \tau_n) \delta(\theta^T - \theta_n^T) \delta(\theta^R - \theta_n^R) \exp(j 2\pi \nu_n t), \tag{5.10}$$

where ν_n describes the Doppler shift of the nth MPC. It can be seen that when describing the spreading function instead of the time-variant impulse response, a completely symmetric formulation is achieved:

$$s_n(\nu, \mathbf{x}^T, \mathbf{x}^R, \tau, \theta^T, \theta^R) = a_n \delta(\tau - \tau_n) \delta(\theta^T - \theta_n^T) \delta(\theta^R - \theta_n^R) \delta(\nu - \nu_n). \tag{5.11}$$

This is the most general description that makes no statements about possible interrelationship between the Doppler and other parameters. For the (practically important) case that the Doppler shifts are created only by the movement of the user equipment, the Doppler shift in this case has a strict mapping to the DoA: $\nu_n = (v/c_0)\cos(\gamma_n)$, where γ_n is the angle between the velocity vector $\mathbf{v}$ (with magnitude v) and the DoA of the nth MPC (implicitly assuming here that the user equipment is the receiver). Equivalently, the Doppler shift experienced by the nth MPCs during a time interval t is $\mathbf{k_n} \cdot \mathbf{v}t$, where $\mathbf{k_n}$ is the wavevector of the nth MPC (note that this includes a direction, and is thus different from the scalar *wave number*).[3]

The discrete multipath model makes several implicit assumptions:

- *Narrowband assumption:* This typically means that the bandwidth is within 10% of the carrier frequency.[4] The assumption shows itself in the statement that the impulse response of each separate MPC is a delta impulse, or equivalently, that the transfer function of each MPC is frequency-flat. Common propagation effects such as reflection at dielectric layers and diffraction have a frequency dependence and thus lead to a distortion of the transfer function for each separate MPC (as opposed to the frequency selectivity of the total transfer function created by interference

---

[3] The wavevector is defined as a vector of magnitude $2\pi/\lambda$ in the direction of propagation of the wave. It can be defined in an arbitrary coordinate system; it is just required that the velocity vector is in the same coordinate system.

[4] Note that the narrowband assumption is different from the *flat-fading* assumption; see [1].

of the MPCs in a wideband system [1]). Furthermore, the narrowband assumption is required for the Doppler effect to lead to a Doppler *shift*, not a scaling. The narrowband assumption cannot be considered as a channel property alone, but depends on the observation system, namely the bandwidth of transmission.

- *Far-field assumption:* This means that the distance between the TX and RX and the scattering objects is sufficiently large that the propagation over each path can be represented as a plane wave *over the area of interest*, which typically will be the size of the antenna array aperture, and thus related to the observation system.

- Furthermore, we assume that the runtime of the signal over the size of the antenna array is smaller than the inverse system bandwidth, which is reflected in the assumption that τ is not a function of α or the angles of incidence.

All the parameters in the physical model depend on the location of the transmitter and receiver measurements $\mathbf{r}_{TX}, \mathbf{r}_{RX}$. Furthermore, the AoAs and AoDs assume a reference frame at the TX and RX locations. Most of the parameters vary slowly with the location: The absolute amplitude of a single MPC changes only over the range at which noticeable distance-dependent pathloss occurs, which is typically around 10% of the distance between TX and RX. Change in delay and angles are tied to the conditions of narrowband and far-field assumption, as discussed above, and their impact on system design and performance is related to the system bandwidth and spatial resolution. The parameter that changes most rapidly with location is the phase of the MPC, which (when there is relative motion) changes in a completely deterministic way in direct proportion to the change in the length of a propagation path, relative to wavelength, in the direction of motion. Note also that the model above furthermore assumes that the strengths of the MPCs stays constant over the area of interest. From this follows that the "area of interest" must be within the stationarity region.

It is noteworthy that in this representation, the MPC amplitudes reflect the complex gain of the propagation channel only, without any consideration of the antennas. The conventional impulse response of the radio channel (including the antennas) can be recovered by weighting the DDIR with the antenna pattern, and then integrating over all angles:

$$h(t,\tau) = \int \int h(t,\tau,\theta^T,\theta^R) g^T(\theta^T) g^R(\theta^R) d\theta^T d\theta^R, \qquad (5.12)$$

where g^T and g^R are the complex amplitude antenna patterns (in linear units) for the TX and RX antenna (note that here we assume the antenna patterns as scalar functions; the dual-polarization antenna pattern will be discussed below).

For multiple-antenna systems we are also often interested in the impulse response or channel transfer function of the radio channel (i.e., including the antenna characteristics) from each TX antenna element to each RX antenna element. This is given by the impulse response matrix. We denote the transmit and receive element coordinates, relative to an arbitrary reference position on each array, as $\mathbf{x}_1^T, \mathbf{x}_2^T, \ldots, \mathbf{x}_{N^T}^T$, and

$\mathbf{x}_1^R, \mathbf{x}_2^R, \ldots, \mathbf{x}_{N^R}^R$, respectively, so that the impulse response from the ith transmit to the kth receive element becomes

$$h_{k,i}(t, \tau) = h\left(t, \tau, \mathbf{x}_i^T, \mathbf{x}_k^R\right)$$

$$= \int d\tau' \int d\theta^T \int d\theta^R h(t, \tau, \theta^T, \theta^R) g_i^T(\theta^T) g_k^R(\theta^R) f_{i,k}(\tau - \tau'), \quad (5.13)$$

where $f_{i,k}(\tau)$ is the convolution of the transmit filter at the ith transmit antenna with the receive filter at the kth receive antenna; in some cases this is actually independent of i and k.

Note also that the complex antenna patterns $g_i^T(\theta^T)$ contain the effect of a phase shift if their location is offset from the reference location at the transmitter (and similarly at the receiver). In other words, $g_i^T(\theta^T)$ can be written as $g_i^{T,0}(\theta^T) \exp\left(j \langle \mathbf{k}(\theta_n^T), (\mathbf{x}_i^T - \mathbf{x}_1^T) \rangle\right)$, that is, the antenna pattern one would measure with the antenna phase center in the reference location, plus a phase offset given by the offset between the reference location and the actual element location. The location vector for a specific antenna element is, in principle, arbitrary, but commonly uses either the phase center of the array or simply the first element in the array. If all antenna elements are identical in type and orientation, $\mathbf{g_i}^{T,0}(\theta^T)$ becomes independent of i. When the array itself is moving, the reference point moves with it (i.e., the term above describes the *phase difference* between two elements on an array at any given time), since the phase shift due to the antenna movement is taken into account by a separate term $\exp(2j\pi\nu)$; alternatively the two terms can be combined.

For the case that the TX and RX arrays are uniform linear arrays with element spacing d_a and *steering vectors* $\mathbf{a}^T(\theta^T) = \frac{1}{\sqrt{N_t}}[1, \exp(-j2\pi\frac{d_a}{\lambda}\sin(\theta^T)), \ldots$ $\exp(-j2\pi(N_t - 1)\frac{d_a}{\lambda}\sin(\theta^T))]^T$ (and analogously defined $\mathbf{a}^R(\theta^R)$; θ^T and θ^R are measured from antenna broadside), the impulse response matrix becomes

$$\mathbf{H}(t) = \int\int h(t, \tau, \theta^T, \theta^R) g^{T,0}(\theta^T) g^{R,0}(\theta^R) \mathbf{a}^R(\theta^R) \mathbf{a}^{T\dagger}(\theta^T) d\theta^R d\theta^T, \quad (5.14)$$

where we have omitted the $f_{i,k}$ for brevity. Note that the steering vectors take on the form of a discrete Fourier transform (DFT) column. A more detailed discussion of this special case can be found in Chapter 12.

5.1.3   Polarization

A further refinement of the propagation model takes the polarization characteristics of channel and antennas into account [8]. Consider a situation in which both TX and RX have dual-polarized antennas, that is, antennas that are capable of independently transmitting and receiving orthogonally polarized waves (for the sake of simplicity, we henceforth use vertically and horizontally polarized waves, V and H, though alternative characterizations are possible). We note that the antennas are characterized by two (complex) antenna patterns, $g^{T,V}(\theta^T)$ and $g^{T,H}(\theta^T)$ AoA for the transmitter, and analogously for the receiver. The propagation is characterized by four polarization channels to be considered, HH, HV, VH, and VV, respectively. This gives rise to four impulse responses to be modeled. A generalization of eq. (5.7) reads

$$\mathbf{h}(t, \mathbf{x}^{\mathrm{T}}, \mathbf{x}^{\mathrm{R}}, \tau, \theta^{\mathrm{T}}, \theta^{\mathrm{R}}) = \sum_{n=1}^{N_p} \begin{pmatrix} a_n^{\mathrm{V},\mathrm{V}} & a_n^{\mathrm{V},\mathrm{H}} \\ a_n^{\mathrm{H},\mathrm{V}} & a_n^{\mathrm{H},\mathrm{H}} \end{pmatrix}$$

$$\delta(\tau - \tau_n)\delta(\theta^{\mathrm{T}} - \theta_n^{\mathrm{T}})\delta(\theta^{\mathrm{R}} - \theta_n^{\mathrm{R}})\exp(j2\pi\nu_n t), \quad (5.15)$$

and the generalization of eq. (5.13) thus reads (suppressing time dependence for ease of notation):

$$h_{k,i} = h\left(\mathbf{r}_{\mathrm{T}}^{(m)}, \mathbf{r}_{\mathrm{R}}^{(l)}\right)$$

$$= \sum_n \begin{bmatrix} g^{\mathrm{T},\mathrm{V}}(\theta_n^{\mathrm{T}}) \\ g^{\mathrm{T},\mathrm{H}}(\theta_n^{\mathrm{T}}) \end{bmatrix}^T \begin{bmatrix} a_n^{\mathrm{V},\mathrm{V}} & a_n^{\mathrm{V},\mathrm{H}} \\ a_n^{\mathrm{H},\mathrm{V}} & a_n^{\mathrm{H},\mathrm{H}} \end{bmatrix} \begin{bmatrix} g^{\mathrm{R},\mathrm{V}}(\theta_n^{\mathrm{R}}) \\ g^{\mathrm{R},\mathrm{H}}(\theta_n^{\mathrm{R}}) \end{bmatrix} \quad (5.16)$$

$$\exp\left(j\langle\mathbf{k}(\theta_n^{\mathrm{T}}), (\mathbf{x}_i^{\mathrm{T}} - \mathbf{x}_1^{\mathrm{T}})\rangle\right)\exp\left(-j\langle\mathbf{k}(\theta_n^{\mathrm{R}}), (\mathbf{x}_k^{\mathrm{R}} - \mathbf{x}_1^{\mathrm{R}})\rangle\right).$$

References

[1] A. F. Molisch, *Wireless Communications*, 3rd ed., IEEE Press and Wiley: Piscataway, NJ, 2022.

[2] P. Bello, "Characterization of randomly time-variant linear channels," *IEEE Transactions on Communications Systems*, vol. 11, pp. 360–393, Dec. 1963.

[3] F. Hlawatsch and G. Matz, *Wireless Communications Over Rapidly Time-Varying Channels*, Academic Press: New York, 2011.

[4] A. F. Molisch, L. J. Greenstein and M. Shafi, "Propagation issues for cognitive radio," *Proceedings of the IEEE*, vol. 97, no. 5, pp. 787–804, 2009.

[5] M. Steinbauer, A. F. Molisch and E. Bonek, "The double-directional radio channel," *IEEE Antennas and Propagation Magazine*, vol. 43, no. 4, pp. 51–63, 2001.

[6] M. Landmann, W. Kotterman and R. Thomä, "On the influence of incomplete data models on estimated angular distributions in channel characterisation," *The Second European Conference on Antennas and Propagation (EuCAP 2007)*, pp. 1–8, 2007.

[7] J. Medbo, H. Asplund and J.-E. Berg, "60 GHz channel directional characterization using extreme size virtual antenna array," *2015 IEEE 26th Annual International Symposium on Personal, Indoor, and Mobile Radio Communications (PIMRC)*, pp. 176–180, 2015.

[8] M. Shafi, M. Zhang, A. L. Moustakas, P. J. Smith, A. F. Molisch, F. Tufvesson and S. H. Simon, "Polarized MIMO channels in 3-D: Models, measurements and mutual information," *IEEE Journal on Selected Areas in Communications*, vol. 24, no. 3, pp. 514–527, 2006.

6   Path Loss/Shadowing

Camillo Gentile, Jelena Senic, Ruoyu Sun, Peter Papazian,
Theodore S. Rappaport, George MacCartney, Jr., Yunchou Xing, Shu Sun,
Andreas F. Molisch, Alenka Zajić and Kate A. Remley

6.1   Instantaneous Path Loss versus Average Path Loss

Path-loss models are the most widely used channel propagation models. This stems both from their simplicity and their direct application to link-layer analysis. This section provides an overview of path-loss models with concentration on models specific to millimeter-wave (mmWave) systems.

Instantaneous path loss is defined in eq. (6.1) as the ratio of transmit power at position $\mathbf{x}^T$ to receive power at position $\mathbf{x}^R$ at some time t [1]. It represents the amount of power loss that is incurred on the signal by the channel. The symbol $\rightarrow$ explicitly denotes that the quantity as linear:

$$\vec{PL}(t,\mathbf{x}^T,\mathbf{x}^R) = \frac{P_{TX}(t,\mathbf{x}^T)}{P_{RX}(t,\mathbf{x}^R)}. \tag{6.1}$$

Given the double-directional channel impulse response (CIR) in eq. (5.7), the RX power can be written in terms of the TX power as

$$P_{RX}(t,\mathbf{x}^R) = P_{TX}(t,\mathbf{x}^T) \cdot \int \int \int |h(t,\mathbf{x}^T,\mathbf{x}^R,\tau,\theta^T,\theta^R)|^2 d\tau d\theta^T d\theta^R \tag{6.2}$$

$$= P_{TX}(t,\mathbf{x}^T) \cdot \sum_{n=1}^{N_p} |a_n(t)|^2. \tag{6.3}$$

The simplification in eq. (6.3) also follows from eq. (5.7). Finally, by substituting eq. (6.3) into (6.1), we obtain

$$\vec{PL}(t,\mathbf{x}^T,\mathbf{x}^R) = \frac{1}{\sum_{n=1}^{N_p} |a_n(t)|^2}. \tag{6.4}$$

Hence, the omnidirectional path loss – omnidirectional because the N_p paths from all angles of departure (AoD) and arrival (AoA), θ^T and θ^R respectively, are included – can be written as the inverse of the sum over the individual MPC powers [1].

Note that the absolute or average amplitude, $|a_n|$, will vary "slowly," but the complex amplitude, $a_n(t)$, will vary "quickly" due to phase fluctuation on the order of wavelengths. Real channel sounders have finite resolution, and so at least some paths will inevitably overlap in field measurements. The constructive or destructive interference of their phases will cause the signal to fade in a *small-scale* sense. To factor out small-scale fading, we compute the average path loss:

$$\vec{PL}(\mathbf{x}^{\mathrm{T}}, \mathbf{x}^{\mathrm{R}}) = \frac{1}{T} \int_{t=0}^{T} \vec{PL}(t, \mathbf{x}^{\mathrm{T}}, \mathbf{x}^{\mathrm{R}}) dt. \tag{6.5}$$

The period, T, over which the average path loss is computed typically corresponds to movement up to 30–40 wavelengths [2]. Local time averaging may also be performed over a few hundred milliseconds to a few seconds in order to determine the average path loss in a dynamic channel environment [3]. Note that for the case of infinite delay/angle resolution, there is no difference between eqs. (6.4) and (6.5).

6.2   The α–β Model

In this section we consider variation of the average path loss over many orders of wavelengths, which is referred to as *large-scale* fading. Large-scale fading is most commonly modeled as a function of the distance $d = ||\mathbf{x}^{\mathrm{T}} - \mathbf{x}^{\mathrm{R}}||$ between the TX and RX according to some norm $|| \cdot ||$. Because the distance may be very large, the path loss may vary by several orders of magnitude and as such is expressed in decibels as $PL(d)$ (without $\rightarrow$).

A popular parameterized model for path loss, recently referred to as the floating-intercept (FI) or α–β model within the context of 3GPP is [4–17]:

$$PL(d) = \alpha \cdot 10 \log_{10}\left(\frac{d}{d_0}\right) + \beta + S. \tag{6.6}$$

As the average path loss is expressed in decibels, it is a relative value – relative to a reference value, β, at some reference distance, d_0. The reference distance is typically set as $d_0 = 1$ m [16] or as the Rayleigh (Fraunhofer) distance of the antenna $d_0 = \frac{2D^2}{\lambda}$, where D is the physical antenna length (largest dimension of radiator) and λ is the signal wavelength. The Rayleigh distance demarcates the boundary between the near- and far-fields. Alternatively, d_0 may be left as a floating parameter for fitting.

The path-loss exponent, α, gauges the rate of change of the path loss over distance. The expression $\alpha \cdot 10 \log_{10}\left(\frac{d_s}{d_0}\right) + \beta$ can be interpreted as the path loss at the distance d_s, which is the "starting distance" for the validity of the model, often the smallest distance at which the measurement values are available. When non-line-of-sight (NLoS) measurements alone are used for fitting, the α–β model can extrapolate back to a negative value of β, in particular in harsh propagation environments for which α is high. If quasi-free-space conditions – the RX is in line-of-sight (LoS) and in the far-field of the TX and the direct path is much more dominant than ambient paths – at the reference distance hold (which they usually do), the interpretation of negative β is that more power is sensed at d_0 than in free space. And so, if a lower limit is not set for the validity of the model, the model will be inconsistent with physical reality for $d < d_s$ (i.e., outside the range for which underlying measurements exist). Analogously, an upper limit corresponding to the largest distance over the measurements should also be set.

The stochastic nature of the α–β model (i.e., how for a fixed distance it varies in direction due to shadowing from diverse obstacles) is captured in the additive component, S, which is a zero-mean normally distributed random variable with standard deviation σ. Implicit in the model is *wide-sense stationarity* – that is, by definition it is only a function of the distance between the TX and RX and so it is independent of specific TX–RX locations. Hence the model is used to represent a wide range of different locations as well as a wide range of distances (vis-á-vis the distance-independent σ) without being in a specific city or tied to a specific base station. The fact that statistical models are statistical requires an amply diverse measurement set over an ensemble of many TX–RX locations from which to form a model.

6.2.1   Maximum-Power Path Loss versus Free-Space Path Loss

Millimeter-wave receivers (transmitters) will feature pencil beam antennas that can be electronically steered toward the direction-of-arrival (direction-of-departure) of the propagation path with maximum power, exploiting their high gain to compensate for the greater path loss witnessed at mmWave frequencies [18, 19]. This begs the question of why path-loss models assuming omnidirectional antennas are still prevalent in the mmWave literature. One answer is that omnidirectional path-loss models can be applied when coupled with the directional power spectrum at each measurement point [20]; that way the angular distribution of the power is also known. Another answer is that in order to compute the steering weights, channel-state information needs first to be estimated. This may be accomplished by generating an quasi-omnidirectional radiation pattern in order to see the full "picture" of the environment. Since the omnidirectional pattern will have minimal gain, the lowest bearer (modulation and coding scheme) will be in effect such that the maximum connectivity range can be established. Once the weights are computed, however, models for the maximum-power (BEST) path loss combined with the gain of the directional beam may be more relevant to determine the highest bearer attainable [21–25]. Of course, channel-state information can also be acquired by beamscanning techniques, making the omnidirectional radiation pattern obsolete.

This discussion of omnidirectional versus directional path loss brings up an important point about the use of omnidirectional path-loss models. One may not be able to simply apply an omnidirectional path-loss model with a directional antenna pattern gain since the multipath energy from all directions of departure and arrival will not be captured with the same amplification in the real world when using a directional antenna [26]. Therefore, directional path loss models are important in order to avoid underestimating path loss at mmWave bands and with directional antennas or beamforming antenna arrays [27]. The BEST path loss [5, 11, 22] is defined as

$$PL_{\text{BEST}}(d) = 10 \log_{10} \frac{1}{|a_1(d)|^2}, \tag{6.7}$$

where $|a_l(d)|$ denotes the distance-dependent average amplitude of path n and $n = 1$ indicates the BEST path, that is, $|a_1(d)|^2 > |a_n(d)|^2 \, \forall n, 2 \leq n \leq N_p$. Some methods to extract the strongest path – within the finite resolution of the systems – are provided in [5, 14]. The difference between the BEST and OMNI path loss is

$$PL_{\text{BEST}}(d) - PL(d) = 10 \log_{10} \left(1 + \frac{\sum\limits_{n=2}^{N_p} |a_n(d)|^2}{|a_1(d)|^2} \right), \qquad (6.8)$$

and so fitting a model to eq. (6.7) will translate to a different parameter set (α_{BEST}, β_{BEST}, σ_{BEST}).

The parameters of the free-space model are given through the Friis transmission equation [28]. In deriving the theoretical equation, the received power is computed as the TX power density – the power dissipated over the spherical surface area whose radius is d – captured by the effective antenna area (aperture size) of the RX [20, 29, 30]:

$$P_{\text{RX}} = \underbrace{\left(\frac{P_{\text{TX}}}{4\pi d^2} \right)}_{\text{TX power density}} \cdot \overbrace{\left(\frac{\lambda^2}{4\pi} \right)}^{\text{RX effective antenna area}}. \qquad (6.9)$$

This equation assumes omnidirectional antennas with unity gain. Note that in free space the power density is invariant to the wavelength and in turn invariant to the center frequency $f = \frac{c}{\lambda}$, where c is the speed of light. Rather, it is the effective antenna area that is frequency-dependent, and so the greater path loss at higher mmWave frequencies is not due to the channel itself, but due to the reduced physical antenna area that accompanies the effective antenna area.[1] In practice, the frequency dependency of path loss is not based on the effective antenna area alone, but also on the properties of the channel, such as penetration loss and reflection loss [11, 20, 31–37], oxygen-absorption loss at 60 GHz [38], etc. From eq. (6.9) it is easy to determine the α–β parameters of the free-space model, $PL_{\text{FS}}(d)$, as $\alpha_{\text{FS}} = 2.0$ and $\beta_{\text{FS}} = 20 \log_{10} \left(\frac{4\pi f}{c} \right)$.

In LoS conditions, the maximum-power path will be the direct path between the TX and RX and so $PL_{\text{BEST}}(d)|_{\text{LoS}} = PL_{\text{FS}}(d)$. By substituting into eq. (6.8), it is obvious that $PL_{\text{FS}}(d)$ will provide a good approximation for $PL(d)$ in quasi free-space conditions, that is, in LoS when the ambient MPCs, $n = 2 \ldots N_p$, are weak compared to the direct path. Figure 6.1 shows the OMNI path-loss model fit to measurements taken in a lobby/hallway environment at 83.5 GHz alongside the BEST and FS models. Note that indeed the BEST model agrees well with the free-space model in the LoS segment

---

[1] At higher frequencies, because the antennas are smaller, more units can be packed into a given physical space, increasing the effective area size of the RX antenna from a single unit alone. It follows that for a constant physical antenna size, not a constant effective antenna size, path loss in free space is indeed frequency-independent.

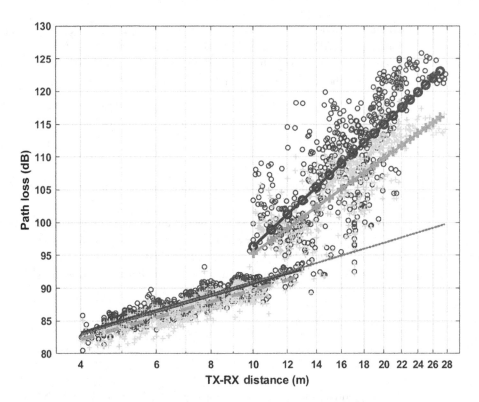

Figure 6.1 Comparison between the OMNI PL (crosses), maximum-power PL_{BEST} (circles), and free-space PL_{FS} (dashed) models at 83.5 GHz. In the LoS segment, ($\alpha = 1.96$, $\beta = 69.86$ dB, $\sigma = 0.86$ dB) and ($\alpha_{BEST} = 1.97$, $\beta_{BEST} = 71.18$ dB, $\sigma_{BEST} = 1.09$ dB) $\approx$ ($\alpha_{FS} = 2.00$, $\beta_{FS} = 70.86$ dB, $\sigma_{FS} = 0.00$ dB). In NLoS, from [5]. © 2017 IEEE. Reprinted, with permission, from [5].

and that – albeit only 2.4 dB – a finite difference between the FS (BEST) and OMNI models is observable in this environment. The 2.4 dB difference means that the direct path represents 58% of the total received power, while the ambient paths represent the remaining 42%. The difference would be larger in an environment with more reflective wall materials, making the ambient MPCs relatively stronger compared to the direct path. It is thus important to note that a received power larger than the power received in pure free-space conditions is physically reasonable and observed in experiments.

In NLoS, the maximum-power path will likely be a reflected path for two reasons: (1) the direct path will often go undetected due to high penetration losses at mmWave frequencies; and (2) diffracted paths are known to play a lesser role in mmWave propagation [20, 32, 38]. However, reflected paths can be significantly weaker than the direct path in free space due to the longer propagation path lengths and losses incurred by reflection. Hence, in NLoS the strongest reflected path will be more comparable in strength to that of the ambient paths; consequently the difference in eq. (6.8) will be much larger in NLoS compared to LoS. In fact, in the NLoS segment in Figure 6.1, we see that the gap between the BEST and OMNI models peaks at

7 dB, meaning that at maximum distance the strongest reflected path accounts for only 20% of the total received power, while the ambient paths combined account for the remaining 80%. In such cases when the power is limited, it may be beneficial for the antennas to generate multiple beams to exploit what power is available in the environment; this sort of information is useful for site planning [39–41].

6.2.2   Close-In Model

A variant of the α–β model is the close-in free-space reference distance (CI) path-loss model, which can be formulated as [1, 4, 8, 9, 11, 12, 15, 27, 42–46]:

$$PL(d) = \alpha_{CI} \cdot 10 \log_{10} \left(\frac{d}{d_0} \right) + \beta_{FS} + S. \tag{6.10}$$

The physical basis of the model is that path loss at any particular distance can be traced to the transmitted power through a CI free-space reference distance, where free-space propagation close to an unobstructed radiating antenna would hold true [1, 28, 47–49]. The model is premised on the use of a free-space reference distance $d_0 = 1$ m, where the RX will be in LoS and at that reference distance any ambient reflections will be insignificant compared to the direct path. Accordingly, the reference path loss is pinned to the free-space value, β_{FS}, and the path loss exponent, α_{CI}, is the sole, tunable model parameter. Figure 6.2 shows a comparison between the α–β and CI models for omnidirectional path loss in both LoS and NLoS conditions.

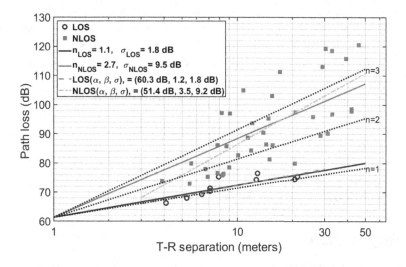

Figure 6.2 Comparison between the α–β and CI models for the indoor office environment at 28 GHz. © 2015 IEEE. Reprinted, with permission, from [11]. In LoS, ($\alpha = 1.2, \beta = 60.4$ dB, $\sigma = 1.8$ dB) and ($\alpha_{CI} = 1.1$, $\beta_{FS} = 61.38$ dB, $\sigma_{CI} = 1.8$ dB); in NLoS, ($\alpha = 3.5$, $\beta = 51.3$ dB, $\sigma = 9.3$ dB) and ($\alpha_{CI} = 2.7$, $\beta_{FS} = 61.38$ dB, $\sigma_{CI} = 9.6$ dB). (Note that in the legend α and β are reversed from the notation in this paper and $n = \alpha_{CI}$.)

The CI model has been widely utilized historically to model path loss in various propagation channels [1, 48], dating back to Friis and Bullington [28, 49], where the path-loss exponent parameter offers insight into path loss based on the environment, having a path loss exponent value of 2.0 in free space (as shown by Friis) [28]. It is noteworthy that $10 \cdot \alpha_{CI}$ describes path loss in dB in terms of decades of distances beginning at d_0 (making it very easy to compute power over distance in one's mind when d_0 is set to 1 m [11, 12, 15, 27, 38, 46]. The CI model inherently has an intrinsic frequency dependency of path loss already embedded within the free-space path loss (FSPL) term β_{FS}. The choice of $d_0 = 1$ m as the CI free-space reference distance has been shown to offer parameter stability and model accuracy for outdoor and indoor channels across a vast range of microwave and mmWave frequencies, and creates a standardized modeling approach [11, 15, 27, 38, 46]. Furthermore, the CI free-space anchor point in the CI model ensures that the path-loss model (regardless of transmit power) always has a physical tie and continuous relationship to the transmitted power over distance [11, 15]. More recent measurements and models above 100 GHz show this same trend [18, 50–52].

6.2.3   Distance-Dependent σ

It has been observed in [7, 10] through indoor measurements and outdoor ray-tracing that σ increases with distance, which is also validated through measured data for UMa LoS environments as shown in [53]. This makes intuitive sense because greater distances admit more diversity, both in the number and type of obstacles along the link, and hence more uncertainty. And so it also makes senses, at the expense of a more complex model, to cast σ as a function of distance. In [10], a different standard deviation is computed for each data point collected using a sliding window, with the window length equal to 10 points. Measurement results for NLoS indoors are shown in Figure 6.3(a). At 29 GHz, the standard deviation increases from 4 dB at 5.5 m to 14 dB at 70 m.

In [7], the distance dependence of σ is modeled explicitly as:

$$\sigma(d) = a \cdot 10 \log_{10}\left(\frac{d}{d_0}\right) + b, \tag{6.11}$$

and ray-tracing results for NLoS outdoors at 28 GHz are shown in Figure 6.3(b). The results show that the more complex model can lead to significantly different values not only in σ, but also in the other α and β parameters.

6.2.4   Weighted Fitting

The conventional method to determine path-loss model parameters is to perform a least-squares fit by minimizing the weighted objective function:

$$\sum_{i=1}^{N_s} w(d_i) \cdot \left(PL^{\text{data}}(d_i) - PL(d_i)\right)^2, \tag{6.12}$$

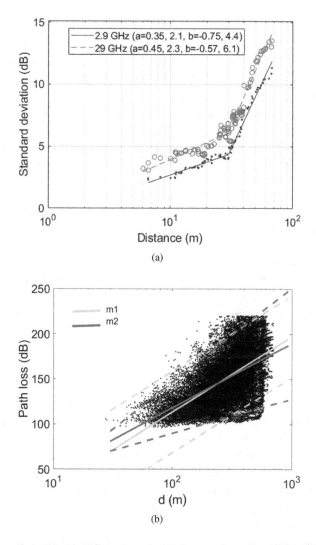

Figure 6.3 (a) $\sigma(d)$ computed with distance-dependent sliding window. Each point represents a window value. © 2015 IEEE. Reprinted, with permission, from [10]. (b) Distance-dependent model in eq. (6.11). © 2016 IEEE. Reprinted, with permission, from [7]. For the distance-independent model (light gray), ($\alpha = 8.37$, $\beta = -53.94$ dB, $\sigma = 22.58$ dB), while for the distance-dependent model (dark gray), ($\alpha = 7.77$, $\beta = -39.87$ dB, $a = 1.48$ dB, $b = -14.44$ dB) where ($\sigma_{\text{MIN}} = 6.52$ dB, $\sigma_{\text{MAX}} = 27.48$ dB).

where N_s denotes the number of collected data points, $PL^{\text{data}}(d_i)$, and $w(d_i)$ denotes the weights. When $w(d_i) = \frac{1}{N_s}$, all points are treated equally. While it may be ideal to collect path-loss data at uniform distances between the TX and RX, this may not be achievable due to practical limitations. For instance, elevator shafts or restricted areas may present impedances indoors; outdoors, collection in a street is impossible and points must circumvent buildings if only the outdoor environment is being modeled. These limitations force nonuniform sampling that will ultimately give rise to different parameter values. One means to mitigate for nonuniform sampling is to assign nonuniform weights to the points – $w(d_i) = \frac{1}{N_b}\frac{N_s}{N_i}$ – by binning the data, where N_b is the number of bins and N_i is the number of data points in point i's bin. This scheme assigns a heavier weight to distance bins with fewer points and vice-versa in the attempt to normalize the data. Details of the method are provided in [6, 7], as well as other weighting schemes.

The least-squares method above directly yields the values of the model parameters α and β. The value of σ, however, is only a byproduct of the minimization step given from the residuals of the fit. Because σ is not a fit parameter, it can lead to suboptimal results. Although more computationally taxing, minimizing the log-likelihood function,

$$ LLF = -\sum_{i=1}^{N_s} w(d_i)\left(\log\frac{1}{\sigma(d_i)^2} + \log\left(\phi\left(\frac{PL^{\text{data}}(d_i) - PL(d_i)}{\sigma(d_i)}\right)\right)\right), \quad (6.13) $$

through maximum likelihood estimation (MLE) enables obtaining a better fit because the search is over the σ space as well. Here, $\phi(\cdot)$ is the standard normal probability density function (PDF). Furthermore, MLE provides a means to handle more complex models such as the distance-dependent $\sigma(d)$ in eq. (6.11), extending the search space to a and b.

6.2.5   Censored Data

An important effect in fitting path-loss models to measured data arises from those locations at which the path-loss values are not available due to measurement noise, that is, no path loss can be recorded if the received signal power is below the noise (sensitivity) threshold. When the locations for such occurrences are known (but obviously the path-loss values are not), the data are called *censored* [54, 55]. The censored data are typically ignored, which introduces a selection bias. Consequently, the fitted model does not correctly represent the actual propagation conditions; the stronger is this effect, the smaller is the dynamic range of the data.

If censored data samples are considered, the log-likelihood for these samples have to be considered as well and are given as

$$ LLF^* = -\sum_{i=1}^{N_s^*} w(d_i)\left(\log\left(1 - \Phi\left(\frac{PL^* - PL(d_i)}{\sigma(d_i)}\right)\right)\right), \quad (6.14) $$

where $*$ refers to censored data, that is, N_s^* is the number of censored data points, the path-loss level for the censoring is PL^* and $\Phi(\cdot)$ is the cumulative distribution function (CDF) of the standard normal distribution. The path-loss parameters are then estimated as

$$ \underset{\alpha,\beta,\sigma}{\arg\min}\{LLF + LLF^*\}. \quad (6.15) $$

6.3 Breakpoint Model

The α–β model is based on the assumption that spatial variations are stationary, that is, that the mean and the variance of the path-loss fit equation depend only on the distance between TX and RX, but not the absolute position. This is enforced by creating a large ensemble of measurement points that are parameterized only by their Euclidean distance to the TX, and then providing a fit. Although the model can be used to represent a wide range of TX locations by aggregating data measured from each of them individually, it will inevitably lead to a larger standard deviation – oftentimes much larger – than what would otherwise be witnessed at any single base station alone. Thus, the stationarity is enforced by the procedure of the fitting. Several papers indicate that such stationarity does not hold in the microwave regime [56–58]. Recently, [59, 60] showed that similarly the assumption does not hold in mmWave and proposed an alternative model. Finally, since link distances will be much shorter than for sub 6 GHz, path loss will be more localized to the TX.

One way to provide a more accurate model when path loss is TX specific, the *dual-slope* or *breakpoint* (BP) model has been applied in some environments [5, 6, 42, 60, 61]:

$$PL^{\mathrm{BP}}(d) = \begin{cases} PL(d), & d \le d_1 \\ PL_1(d), & d > d_1 \end{cases} \tag{6.16}$$

for

$$PL_1(d) = \alpha_1 \cdot 10 \log_{10}\left(\frac{d}{d_1}\right) + \beta_1 + S_1, \tag{6.17}$$

where d_1 is known as the breakpoint distance. For example, it can be used when the path-loss exponent changes notably from obstructed line-of-sight conditions (OLoS) to harsh NLoS [10]. Alternatively, the breakpoint model can characterize the transition from LoS to NLoS [5] where d_1 marks the transition point. In general, the breakpoint can be found analytically by minimizing the sum of the least-squares errors associated with the two separate segments [10]; although in some environments the LoS–NLoS transition point is clear from the environment geometry. For example, in [5] the breakpoint falls at a corner indoors while outdoors in [60, 62, 63] the breakpoints fall at street intersections.

Often, continuity in path loss between the two segments is imposed [10, 60, 63], leading to a more constrained model for which $\beta_1 = PL(d_1; S = 0) = \alpha \cdot 10 \log_{10}\left(\frac{d_1}{d_0}\right) + \beta$, where $PL(d, S = 0)$ is the *expected* path loss, that is, in the absence of shadowing. Attention should be paid, however, when imposing continuity, especially in mmWave systems. This is because penetration losses for these systems can be particularly high, depending on the obstructing material(s), especially at frequencies in the upper band of the spectrum. As such, in the transition to NLoS the direct path may abruptly either go undetected or be severely attenuated compared to LoS, and in turn its contribution to the total received power severely diminished, yielding a transition loss $\beta_1 - PL(d_1; S = 0) > 0$ or equivalently $\beta_1 > \alpha \cdot 10 \log_{10}\left(\frac{d_1}{d_0}\right) + \beta$. In such cases, discontinuity in the path loss can be anticipated and so should be allowed [5].

The dual-slope model may apply in strictly LoS conditions as well, for example if beyond some breakpoint waveguiding conditions come into effect for which the exponent shrinks below that of free space. More than a single breakpoint is possible, as seen in [60] with ray-tracing simulations with unlimited dynamic range. To our knowledge, however, this is yet to be witnessed with real mmWave measurements because the higher path loss will permit only much shorter links compared to sub-6 GHz.

Generally speaking, a breakpoint needs to be based on some physical relationship that occurs repeatedly across many locations, cities, base stations, etc. For example, work in [64] showed that for LoS topographies in a microcell scenario, a single-slope path loss model was applicable prior to the first Fresnel zone clearance breakpoint, while a dual-slope model is necessary beyond that breakpoint, where the breakpoint was dependent on the TX and RX antenna heights and frequency [64]. A path-loss exponent value of 4.0 for the asymptotic two-ray ground-bounce propagation model beyond the distance for first Fresnel zone clearance in LoS was shown by in [49]. Note that the classical work by Bullington actually developed a breakpoint model and it was based on the physical signal propagation property that occurred in all large cells and which was consistently observable. Thus, a breakpoint will typically be visible in the data when a basic physical phenomenon relating to distance and height is repeatedly observed to impact the measurement data set (such as a ground-bounce phenomenon) over a wide range of distances and antenna heights, and is repeatable and visible in large data sets in many different specific locations across a vast range of locations [53]. It is recommended to make a vast number of measurements throughout many buildings and locations to reliably declare a breakpoint model to exist in general, and more importantly, should be able to point to a definitive physical rationale for the particular value of the breakpoint.

6.3.1   Floating-Breakpoint Model

Although the dual-slope model may provide a smaller fit error, it is often tied to the breakpoint associated with a particular TX position. In fact, if multiple TX positions are included in a campaign, separate parameters for each position are usually reported. Although it is informative to expose the range of parameter values than can be encountered, when it comes to application it may not be clear which model to use. Essentially, this problem is a reflection of the fact that a nonstationary channel (model with parameters that depend on the absolute position) is fitted by a stationary model (constant channel parameters).

A remedy for this is to expand eq. (6.16) into a *floating-breakpoint* model. Specifically, all data points before the respective breakpoints, d_1^j – where j indexes the TX positions – are combined and a single $PL(d)$ is fit to them, yielding a single set of parameters (α, β, σ) for the first segment. In the next step, an incremental path loss is computed in eq. (6.18) for each TX position. It is computed by subtracting the expected path loss at the breakpoint – its parameters (α, β) are now known – from the path loss of the second segment:

$$PL_1(d) - PL(d_1^j; S = 0) = \alpha_1 \cdot 10 \log_{10}\left(\frac{d}{d_1^j}\right) - \alpha \cdot 10 \log_{10}\left(\frac{d_1^j}{d_0}\right) + \beta_1 - \beta + S_1.$$

(6.18)

The incremental path loss represents only the additional path loss after the breakpoint and so its value is not tied to any specific breakpoint. As such, the data points after the respective breakpoints can be combined in eq. (6.18) across all TX positions, to which a single set of parameters $(\alpha_1, \beta_1, \sigma_1)$ for the second segment is fit.

As the floating-breakpoint model is breakpoint-independent, any value d_1 can be inserted when applying the model. This, together with the fact that data points from all TX's are utilized in the fitting, renders the model more generally applicable than separate models per TX position. Finally, it is worth pointing out that the model allows overlap between the first and second segments, as illustrated in Figure 6.1. As in other cases, this is practical when the first segment represents LoS and the second NLoS. For example, consider when the TX in some environment is located at the end of a hallway: The RX can continue in LoS down the hallway at a longer distance in one direction than rounding the corner in the opposite direction (the corner creates NLoS conditions).

6.3.2   Distance Metric

The default distance metric for path-loss models is the Euclidean distance. Given that at mmWave frequencies diffraction will be negligible, propagation will mostly take place through free-space transmission and specular reflection. This, combined with the geometry of some environments, such as hallways indoors or streets below the clutter outdoors, leads to the waveguiding effect. When waveguiding occurs, most of the energy will propagate along the direction of the virtual waveguide. Unless the TX is also in the waveguide, this direction will not correspond to the Euclidean direction. When applicable, more precise models can be constructed if the distance metric is oriented along the waveguide, akin to the Manhattan distance. An example is the dual-slope model in [5], in which the breakpoint demarcates the LoS–NLoS transition at d_1 from a lobby to a hallway. In the LoS segment, the Euclidean distance is used; in the NLOS segment, a piecewise distance $d = d_1 + \Delta d$ is used instead, where Δd is the *incremental* distance along the hallway from the breakpoint to the RX. Similar metrics are used in [63].

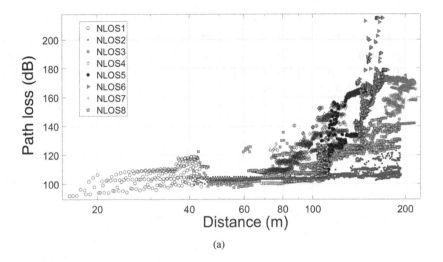

(a)

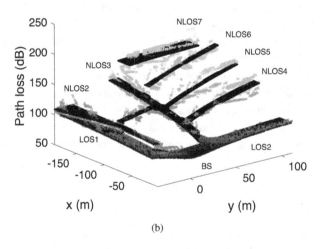

(b)

Figure 6.4 (a) Path loss as a function of Euclidean distance when seven streets in an urban-canyon environment in NLoS are grouped together in a single model. (b) By applying a different breakpoint model to each of the streets – each with its own path-loss exponent – in conjunction with the incremental distance along each of the streets, the standard deviation of each model is much smaller than that of the single model alone. © 2016 IEEE. Reprinted, with permission, from [60].

Another example is illustrated in Figure 6.4, in which waveguiding takes place in an urban-canyon environment. Different streets are signified by different colors. Thus, a mobile station moving along a trajectory within one street would only experience path loss of that particular color. We can clearly observe that different streets have greatly differing path-loss coefficients; for example, points on the NLoS2 street show a negligible slope, while the NLoS8 street corresponds to an almost vertical line in Figure 6.4; these effects can be explained physically in terms of waveguiding and diffraction [60]. Thus, the difference in path loss between two points clearly depends not only on the difference between the points, but also on the absolute position – most notably, which street the two points are in.

These insights can be described by a breakpoint model that is based on a combination of the typical street-by-street FI model and a corner coupling loss. Critically, the path loss does not depend on the Euclidean distance, but rather on the distance of the link along the street. The model bears some resemblance with the model of [65], which was also used in the COST 259 microcell channel model.

An assumption of this model is that the path loss along the different sections of the propagation path, down the street canyons, add up, and that there is an additional loss when the waves couple into a new street canyon. Specifically, let us first write the α–β model for street $n + 1$ using corner loss Δ_{n+1} defined as:

$$PL_{n+1}(d_{n+1}) = 10 \cdot \alpha_{n+1} \cdot \log_{10}(d_{n+1}) + \beta_{n+1} \tag{6.19}$$

$$\Delta_{n+1} = PL_{n+1}(d_n) - PL_n(d_n) \tag{6.20}$$

$$\Delta_{n+1} = 10 \cdot \alpha_{n+1} \cdot \log_{10}(d_n) + \beta_{n+1} - PL_n(d_n). \tag{6.21}$$

Substituting β_{n+1} to eq. (6.19) we get

$$PL_{n+1}(d_{n+1}) = 10 \cdot \alpha_{n+1} \cdot (\log_{10}(d_{n+1}) - \log_{10}(d_n)) + \Delta_{n+1} + PL_n(d_n), \tag{6.22}$$

where Δ_{n+1} is the corner coupling loss at the corner of street n and $n+1$ and $PL_n(d_n)$ is the PL model value at d_n along street n.

In order to maintain meaningful physical interpretation, the corner coupling loss is constrained:

$$\Delta_{n+1} \geq 0. \tag{6.23}$$

A detailed parameterization of this street-by-street model, with parameter values derived from extensive ray-tracing simulations in two different cities, is provided in [59].

6.4   Frequency-Dependent versus Height-Dependent Models

Free-space path loss is not only distance-dependent, but also frequency-dependent. Here, the frequency dependency is written out explicitly as

$$PL(d, f)_{FS} = \alpha_{FS} \cdot 10 \log_{10}\left(\frac{d}{d_0}\right) + \underbrace{\beta'_{FS} + \gamma_{FS} \cdot 10 \log_{10}\left(\frac{f}{f_0}\right)}_{\beta_{FS}}, \tag{6.24}$$

where β_{FS} is expanded into two parts and $\beta'_{FS} = 20\log_{10}\left(\frac{4\pi f_0}{c}\right)$, $\gamma_{FS} = 2.0$, and $f_0 = 1$ GHz is the reference frequency. Omnidirectional and maximum-path path loss will have the same form, however, with generalized parameter-set values $(\alpha, \beta', \gamma, \sigma)$ and $(\alpha_{BEST}, \beta'_{BEST}, \gamma_{BEST}, \sigma_{BEST})$, respectively, including shadowing. This model is referred to as the α–β–γ model [4, 15, 31, 42, 45].

An alternative version to the α–β–γ model is the close-in frequency (CIF) model [4, 31, 35, 42]:

$$PL_{CIF}(d, f) = \alpha_{CI} \cdot \left(1 + b_{CI} \cdot \left(\frac{f - f_0}{f_0}\right)\right) \cdot 10\log_{10}\left(\frac{d}{d_0}\right)$$

$$+ \beta'_{FS} + 20\log_{10}\left(\frac{f}{f_0}\right) + S_{CI}. \tag{6.25}$$

In this model, the β and γ parameters are pinned to the FS values ($\gamma_{FS} = 2.0$ is implicit) while α_{CI} remains the same tunable path-loss exponent as in the CI model. Furthermore, the reference frequency in general will not be 1 GHz, but rather the linear average of all frequencies considered. The additional parameter, b_{CI}, which is typically small, accounts for the joint $d - f$ dependency around f_0. Note that the CIF model reverts to the CI model when $f = f_0$. Figure 6.5 shows a comparison between the α–β–γ and CIF models. The CIF model effectively scales the path-loss exponent as a function of the carrier frequency, since certain scenarios such as indoor hotspot have shown that path loss increases with frequency aside from the initial difference in the first meter of free-space propagation [11, 16].

It is worth noting that the path loss may not be strictly distance- and frequency-dependent [11], but may also depend on the TX height [66]. For example, in the current 3GPP/ITU-R RMa path loss models [67, 68], the TX height is considered to be from 10 m to 150 m, which would introduce an average difference of 29 dB over all of TR distances from 10 m to 5,000 m, as shown in simulation results in figure 8 of [66]. The CI model with a height-weighted path-loss exponent (CIH model) is suitable in RMa scenarios for various TX heights [66, 69]. The CIH model uses the same mathematical form as the CIF model eq. (6.25) except that the path-loss exponent is a function of the base station height instead of frequency, as given by:

$$PL_{CIH}(d, f, h_{BS}) = \alpha_{CI}\left(1 + b_{TX}\left(\frac{h_{BS} - h_{B0}}{h_{B0}}\right)\right) 10\log_{10}\left(\frac{d}{d_0}\right)$$

$$+ \beta'_{FS} + 20\log_{10}\left(\frac{f}{f_0}\right) + S_{CI}, \tag{6.26}$$

where h_{BS} is the base station antenna height in meters, h_{B0} is a default RMa base station height, and b_{TX} is a model parameter that is optimized and which quantifies the linear base-station-height-dependent path-loss exponent about the default/average base station height h_{B0} [66, 69].

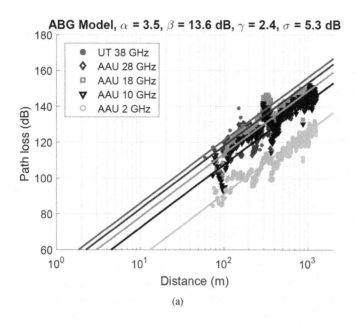

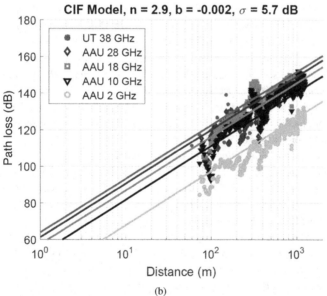

Figure 6.5 Comparison between the α–β–γ and CIF models for the Urban Micro Street Canyon (UMi SC) environment in NLoS. © 2016 IEEE. Reprinted, with permission, from [15]. In (a), $\alpha = 3.5$, $\beta' = 24.4$ dB, $\gamma = 1.9$, $\sigma = 8.0$ dB; in (b), $\alpha_{CI} = 3.1$, $\beta'_{FS} = 32.4$ dB, $\gamma_{FS} = 2.0$, $\sigma_{CI} = 8.1$ dB and $b_{CI} = 0.0$.

References

[1] T. S. Rappaport, *Wireless Communications: Principles and Practice*, 2nd ed., Prentice-Hall: Upper Saddle River, NJ, 2002.

[2] M. K. Samimi, G. R. MacCartney, S. Sun and T. S. Rappaport, "28 GHz millimeter-wave ultrawideband small-scale fading models in wireless channels," *2016 IEEE 83rd Vehicular Technology Conference (VTC Spring)*, May 2016, pp. 1–6.

[3] G. R. MacCartney and T. S. Rappaport, "Study on 3GPP rural macrocell path loss models for millimeter wave wireless communications," *2017 IEEE International Conference on Communications (ICC)*, May 2017, pp. 1–7.

[4] C.-L. Cheng, S. Kim and A. Zajic, "Comparison of path loss models for indoor 30 GHz, 140 GHz, and 300 GHz channels," *2017 11th European Conference on Antennas and Propagation (EuCAP)*, March 2017.

[5] J. Senic, C. Gentile, P. B. Papazian, J. K. Choi, K. A. Remley and J. K. Choi, "Analysis of E-band path loss and propagation mechanisms in the indoor environment," *IEEE Transactions on Antennas and Propagation*, vol. 65, no. 12, pp. 6562–6573, Dec. 2017.

[6] A. Karttunen, C. Gustafson, A. F. Molisch, R. Wang, S. Hur, J. Zhang and J. Park, "Path loss models with distance-dependent weighted fitting and estimation of censored path loss data," *IET Microwaves, Antennas Propagation*, vol. 10, no. 14, pp. 1467–1474, 2016.

[7] A. Karttunen, A. F. Molisch, R. Wang, S. Hur, J. Zhang and J. Park, "Distance dependence of path loss models with weighted fitting," *2016 IEEE International Conference on Communications (ICC)*, pp. 1–6, May 2016.

[8] M. D. Kim, J. Liang, J. Lee, J. Park and B. Park, "Path loss measurements and modeling for indoor office scenario at 28 and 38 GHz," *2016 International Symposium on Antennas and Propagation (ISAP)*, Oct. 2016, pp. 64–65.

[9] J. Ko, S. Hur, S. Lee, Y. Kim, Y.-S. Noh, Y.-J. Cho, S. Kim, S. Bong, S. Kim, J. Park, D.-J. Park and D.-H. Cho, "28 GHz channel measurements and modeling in a ski resort town in Pyeongchang for 5G cellular network systems," *2016 10th European Conference on Antennas and Propagation (EuCAP)*, Apr. 2016, pp. 1–5.

[10] O. H. Koymen, A. Partyka, S. Subramanian and J. Li, "Indoor mm-wave channel measurements: Comparative study of 2.9 GHz and 29 GHz," *2015 IEEE Global Communications Conference (GLOBECOM)*, Dec. 2015, pp. 1–6.

[11] G. R. MacCartney, T. S. Rappaport, S. Sun and S. Deng, "Indoor office wideband millimeter-wave propagation measurements and channel models at 28 and 73 GHz for ultra-dense 5G wireless networks," *IEEE Access*, vol. 3, pp. 2388–2424, Oct. 2015.

[12] G. R. MacCartney, J. Zhang, S. Nie and T. S. Rappaport, "Path loss models for 5G millimeter wave propagation channels in urban microcells," *2013 IEEE Global Communications Conference (GLOBECOM)*, Dec. 2013, pp. 3948–3953.

[13] M. Peter, W. Keusgen and R. J. Weiler, "On path loss measurement and modeling for millimeter-wave 5G," *2015 9th European Conference on Antennas and Propagation (EuCAP)*, May 2015, pp. 1–5.

[14] S. Sun, G. R. MacCartney, M. K. Samimi and T. S. Rappaport, "Synthesizing omnidirectional antenna patterns, received power and path loss from directional antennas for 5G millimeter-wave communications," *2015 IEEE Global Communications Conference (GLOBECOM)*, Dec. 2015, pp. 1–7.

[15] S. Sun, T. S. Rappaport, S. Rangan, T. A. Thomas, A. Ghosh, I. Z. Kovacs, I. Rodriguez, O. Koymen, A. Partyka and J. Jarvelainen, "Propagation path loss models for 5G urban micro- and macro-cellular scenarios," *2016 IEEE 83rd Vehicular Technology Conference (VTC Spring)*, May 2016, pp. 1–6.

[16] S. Sun, T. S. Rappaport, T. A. Thomas, A. Ghosh, H. C. Nguyen, I. Z. Kovacs, I. Rodriguez, O. Koymen and A. Partyka, "Investigation of prediction accuracy, sensitivity, and parameter stability of large-scale propagation path loss models for 5G wireless communications," *IEEE Transactions on Vehicular Technology*, vol. 65, pp. 2843–2860, May 2016.

[17] R. J. Weiler, M. Peter, W. Keusgen, H. Shimodaira, K. T. Gia and K. Sakaguchi, "Outdoor millimeter-wave access for heterogeneous networks: Path loss and system performance," *2014 IEEE 25th Annual International Symposium on Personal, Indoor, and Mobile Radio Communication (PIMRC)*, Sept. 2014, pp. 2189–2193.

[18] S. Ju, Y. Xing, O. Kanhere and T. S. Rappaport, "Millimeter wave and sub-terahertz spatial statistical channel model for an indoor office building," *IEEE Journal on Selected Areas in Communications*, vol. 39, no. 6, June 2021, pp. 1561–1575.

[19] S. Sun, T. S. Rappaport, R. W. Heath, A. Nix and S. Rangan, "MIMO for millimeter-wave wireless communications: Beamforming, spatial multiplexing, or both?" *IEEE Communications Magazine*, vol. 52, no. 12, pp. 110–121, Dec. 2014.

[20] I. Rodriguez, H. C. Nguyen, T. B. Sorensen, J. Elling, J. A. Holm, P. Mogensen and B. Vejlgaard, "Analysis of 38 GHz mmwave propagation characteristics of urban scenarios," *Proceedings of European Wireless 2015: 21th European Wireless Conference*, May 2015, pp. 1–8.

[21] G. R. MacCartney, M. K. Samimi and T. S. Rappaport, "Exploiting directionality for millimeter-wave wireless system improvement," *2015 IEEE International Conference on Communications (ICC)*, June 2015, pp. 2416–2422.

[22] T. S. Rappaport, G. R. MacCartney, M. K. Samimi and S. Sun, "Wideband millimeter-wave propagation measurements and channel models for future wireless communication system design (invited paper)," *IEEE Transactions on Communications*, vol. 63, pp. 3029–3056, Sept. 2015.

[23] A. I. Sulyman, A. Alwarafy, G. R. MacCartney, T. S. Rappaport and A. Alsanie, "Directional radio propagation path loss models for millimeter-wave wireless networks in the 28-, 60-, and 73-GHz bands," *IEEE Transactions on Wireless Communications*, vol. 15, pp. 6939–6947, Oct. 2016.

[24] A. I. Sulyman, A. T. Nassar, M. K. Samimi, G. R. MacCartney, T. S. Rappaport and A. Alsanie, "Radio propagation path loss models for 5G cellular networks in the 28 GHz and 38 GHz millimeter-wave bands," *IEEE Communications Magazine*, vol. 52, pp. 78–86, Sept. 2014.

[25] T. A. Thomas, F. W. Vook and S. Sun, "Investigation into the effects of polarization in the indoor mmwave environment," *2015 IEEE International Conference on Communications (ICC)*, June 2015, pp. 1386–1391.

[26] L. J. Greenstein and V. Erceg, "Gain reductions due to scatter on wireless paths with directional antennas," in *IEEE VTS 50th Vehicular Technology Conference, 1999. VTC 1999-Fall*, vol. 1, IEEE, 1999, pp. 87–91.

[27] G. R. MacCartney, T. S. Rappaport, M. K. Samimi and S. Sun, "Millimeter-wave omnidirectional path loss data for small cell 5G channel modeling," *IEEE Access*, vol. 3, pp. 1573–1580, 2015.

[28] H. T. Friis, "A note on a simple transmission formula," *Proceedings of the IRE*, vol. 34, pp. 254–256, May 1946.

[29] T. S. Rappaport, R. W. Heath, R. C. Daniels and J. N. Murdock, *Millimeter Wave Wireless Communications*, Prentice-Hall: Upper Saddle River, NJ, 2015.

[30] Y. Xing and T. S. Rappaport, "Propagation measurement system and approach at 140 GHz: Moving to 6G and above 100 GHz," *IEEE Global Communications Conference (GLOBECOM)*, Dec. 2018.

[31] Aalto University, AT&T, BUPT, CMCC, Ericsson, Huawei, Intel, KT Corporation, Nokia, NTT DOCOMO, New York University, Qualcomm, Samsung, University of Bristol and University of Southern California, "5G channel model for bands up to 100 GHz," White paper.

[32] S. Deng, G. R. MacCartney and T. S. Rappaport, "Indoor and outdoor 5G diffraction measurements and models at 10, 20, and 26 GHz," *2016 IEEE Global Communications Conference (GLOBECOM)*, Dec. 2016, pp. 1–7.

[33] F. Fuschini, S. Hafner, M. Zoli, R. Muller, E. M. Vitucci, D. Dupleich, M. Barbiroli, J. Luo, E. Schulz, V. Degli-Esposti and R. S. Thoma, "Item level characterization of mm-wave indoor propagation," *EURASIP Journal on Wireless Communications and Networking*, vol. 4, Dec. 2016.

[34] A. K. M. Isa, A. Nix and G. Hilton, "Material characterisation for short range indoor environment in the millimetre wave bands," *2015 IEEE 81st Vehicular Technology Conference (VTC Spring)*, May 2015, pp. 1–5.

[35] G. R. MacCartney, S. Deng and T. S. Rappaport, "Indoor office plan environment and layout-based mmwave path loss models for 28 GHz and 73 GHz," *2016 IEEE 83rd Vehicular Technology Conference (VTC Spring)*, May 2016, pp. 1–6.

[36] N. Moraitis and P. Constantinou, "Indoor channel measurements and characterization at 60 GHz for wireless local area network applications," *IEEE Transactions on Antennas and Propagation*, vol. 52, pp. 3180–3189, Dec. 2004.

[37] H. Zhao, R. Mayzus, S. Sun, M. Samimi, J. K. Schulz, Y. Azar, K. Wang, G. N. Wong, F. Gutierrez and T. S. Rappaport, "28 GHz millimeter wave cellular communication measurements for reflection and penetration loss in and around buildings in New York City," *2013 IEEE International Conference on Communications (ICC)*, June 2013, pp. 5163–5167.

[38] T. S. Rappaport, R. W. Heath, R. C. Daniels and J. N. Murdock, *Millimeter Wave Wireless Communications*, Prentice Hall: Upper Saddle River, NJ, 2015.

[39] T. S. Rappaport, S. Y. Seidel and K. R. Schaubach, *Site-Specific Propagation Prediction for PCS System Design*, Springer: New York, 1993.

[40] S. Y. Seidel and T. S. Rappaport, "Site-specific propagation prediction for wireless in-building personal communication system design," *IEEE Transactions on Vehicular Technology*, vol. 43, pp. 879–891, Nov. 1994.

[41] R. R. Skidmore, A. Verstak, N. Ramakrishnan, T. S. Rappaport, L. T. Watson, J. He, S. Varadarajan, C. A. Shaffer, J. Chen, K. Kyoon Bae, J. Jiang and W. H. Tranter,

"Towards integrated PSEs for wireless communications: Experiences with the S4W and SitePlanner®; projects," *SIGMOBILE Mobile Computing and Communications Reviews*, vol. 8, pp. 20–34, Apr. 2004.

[42] K. Haneda, L. Tian, H. Asplund, J. Li, Y. Wang, D. Steer, C. Li, T. Balercia, S. Lee, Y. Kim, A. Ghosh, T. Thomas, T. Nakamurai, Y. Kakishima, T. Imai, H. Papadopoulas, T. S. Rappaport, G. R. MacCartney, M. K. Samimi, S. Sun, O. Koymen, S. Hur, J. Park, J. Zhang, E. Mellios, A. F. Molisch, S. S. Ghassamzadeh and A. Ghosh, "Indoor 5G 3GPP-like channel models for office and shopping mall environments," *2016 IEEE International Conference on Communications Workshops (ICC)*, May 2016, pp. 694–699.

[43] S. Hur, Y. J. Cho, T. Kim, J. Park, A. F. Molisch, K. Haneda and M. Peter, "Wideband spatial channel model in an urban cellular environments at 28 GHz," *2015 9th European Conference on Antennas and Propagation (EuCAP)*, May 2015, pp. 1–5.

[44] J. Lee, J. Liang, J. J. Park and M. D. Kim, "Directional path loss characteristics of large indoor environments with 28 GHz measurements," *2015 IEEE 26th Annual International Symposium on Personal, Indoor, and Mobile Radio Communications (PIMRC)*, Aug. 2015, pp. 2204–2208.

[45] J. J. Park, J. Liang, J. Lee, H. K. Kwon, M. D. Kim and B. Park, "Millimeter-wave channel model parameters for urban microcellular environment based on 28 and 38 GHz measurements," *2016 IEEE 27th Annual International Symposium on Personal, Indoor, and Mobile Radio Communications (PIMRC)*, Sept. 2016, pp. 1–5.

[46] T. A. Thomas, M. Rybakowski, S. Sun, T. S. Rappaport, H. Nguyen, I. Z. Kovacs and I. Rodriguez, "A prediction study of path loss models from 2–73.5 GHz in an urban-macro environment," *2016 IEEE 83rd Vehicular Technology Conference (VTC Spring)*, May 2016, pp. 1–5.

[47] J. B. Andersen, "History of communications/radio wave propagation from Marconi to MIMO," *IEEE Communications Magazine*, vol. 55, pp. 6–10, Feb. 2017.

[48] J. B. Andersen, T. S. Rappaport and S. Yoshida, "Propagation measurements and models for wireless communications channels," *IEEE Communications Magazine*, vol. 33, pp. 42–49, Jan. 1995.

[49] K. Bullington, "Radio propagation fundamentals," *Bell System Technical Journal*, vol. 36, pp. 593–626, May 1957.

[50] Y. Xing and T. S. Rappaport, "Urban microcell radio propagation measurements and channel models for millimeter wave and terahertz bands (invited paper)," *IEEE Communications Letters*, July 2021.

[51] Y. Xing and T. S. Rappaport, "Propagation measurements and path loss models for sub-THz in urban microcells," *IEEE International Conference on Communications*, June 2021.

[52] Y. Xing, T. S. Rappaport and A. Ghosh, "Millimeter wave and sub-THz indoor radio propagation channel measurements, models, and comparisons in an office environment (invited paper)," *IEEE Communications Letters*, 2021.

[53] S. Sun, T. A. Thomas, T. S. Rappaport, H. Nguyen, I. Z. Kovacs and I. Rodriguez, "Path loss, shadow fading, and line-of-sight probability models for 5G urban macro-cellular scenarios," *2015 IEEE Globecom Workshops (GC Wkshps)*, Dec. 2015, pp. 1–7.

[54] C. Gustafson, T. Abbas, D. Bolin and F. Tufvesson, "Statistical modeling and estimation of censored pathloss data," *IEEE Wireless Communications Letters*, vol. 4, no. 5, pp. 569–572, 2015.

[55] A. Karttunen, C. Gustafson, A. F. Molisch, R. Wang, S. Hur, J. Zhang and J. Park, "Path loss models with distance-dependent weighted fitting and estimation of censored path loss data," *IET Microwaves, Antennas & Propagation*, vol. 10, no. 14, pp. 1467–1474, 2016.

[56] V. Erceg, L. J. Greenstein, S. Y. Tjandra, S. R. Parkoff, A. Gupta, B. Kulic, A. A. Julius and R. Bianchi, "An empirically based path loss model for wireless channels in suburban environments," *IEEE Journal on Selected Areas in Communications*, vol. 17, no. 7, pp. 1205–1211, Jul. 1999.

[57] S. Ghassemzadeh, R. Jana, C. Rice, W. Turin and V. Tarokh, "Measurement and modeling of an ultra-wide bandwidth indoor channel," *IEEE Transactions on Communications*, vol. 52, pp. 1786–1796, Oct. 2004.

[58] Z. Li, R. Wang and A. Molisch, "Shadowing in urban environments with microcellular or peer-to-peer links," *2012 6th European Conference on Antennas and Propagation (EUCAP)*, Mar. 2012, pp. 44–48.

[59] A. Karttunen, A. F. Molisch, S. Hur, J. Park and C. J. Zhang, "Spatially consistent street-by-street path loss model for 28-GHz channels in micro cell urban environments," *IEEE Transactions on Wireless Communications*, vol. 16, no. 11, pp. 7538–7550, 2017.

[60] A. F. Molisch, A. Karttunen, S. Hur, J. Park and J. Zhang, "Spatially consistent pathloss modeling for millimeter-wave channels in urban environments," *2016 10th European Conference on Antennas and Propagation (EuCAP)*, Apr. 2016, pp. 1–5.

[61] J. Medbo, K. Borner, K. Haneda, V. Hovinen, T. Imai, J. Jarvelainen, T. Jamsa, A. Karttunen, K. Kusume, J. Kyrolainen, P. Kyosti, J. Meinila, V. Nurmela, L. Raschkowski, A. Roivainen and J. Ylitalo, "Channel modelling for the fifth generation mobile communications," *The 8th European Conference on Antennas and Propagation (EuCAP 2014)*, Apr. 2014, pp. 3948–3953.

[62] J. Lee, M.-D. Lee, J. Liang, J.-J. Park and B. Park, "Frequency range extension of the ITU-R NLOS path loss models applicable for urban street environments with 28 GHz measurements," *2016 10th European Conference on Antennas and Propagation (EuCAP)*, April 2016, pp. 1–5.

[63] M. Sasaki, W. Yamada, T. Sugiyama, M. Mizoguchi and T. Imai, "Path loss characteristics at 800 MHz to 37 GHz in urban street microcell environment," *2015 9th European Conference on Antennas and Propagation (EuCAP)*, May 2015, pp. 1–4.

[64] M. J. Feuerstein, K. L. Blackard, T. S. Rappaport, S. Y. Seidel and H. H. Xia, "Path loss, delay spread, and outage models as functions of antenna height for microcellular system design," *IEEE Transactions on Vehicular Technology*, vol. 43, pp. 487–498, Aug. 1994.

[65] J.-E. Berg, "A recursive method for street microcell path loss calculations," *Sixth IEEE International Symposium on Personal, Indoor and Mobile Radio Communications, 1995. PIMRC'95. Wireless: Merging onto the Information Superhighway*, vol. 1, pp. 140–143, IEEE, 1995, pp. 140–143.

[66] G. R. MacCartney and T. S. Rappaport, "Rural macrocell path loss models for millimeter wave wireless communications," *IEEE Journal on Selected Areas in Communications*, vol. 35, pp. 1663–1677, July 2017.

[67] IUT-R, "Guidelines for evaluation of radio interface technologies for IMT-advanced," technical report, 2008.

[68] 3GPP, "Channel model for frequencies from 0.5 to 100 GHz," technical report, Rel. 14, V14.0.0, Mar. 2017.

[69] T. S. Rappaport, Y. Xing, G. R. MacCartney, A. F. Molisch, E. Mellios and J. Zhang, "Overview of millimeter wave communications for fifth-generation (5G) wireless networks – with a focus on propagation models," *IEEE Transactions on Antennas and Propagation*, vol. 65, pp. 6213–6230, Dec. 2017.

7 Multipath Component Clustering

Ruisi He, Reiner Thomä, Robert Müller, Christian Schneider,
Diego Dupleich, George MacCartney, Jr., Yunchou Xing, Shu Sun,
Theodore S. Rappaport, Camillo Gentile,
Andreas F. Molisch, Alenka Zajić and Kate A. Remley

In this chapter we present some recent progress of clustering and tracking algorithm designs for radio channels, which have been widely used in the cluster-based channel modeling for 4G and 5G communications. Most of the chapter is based on the work in [1–8].

7.1   Introduction

Accurate channel models are a prerequisite for the design and performance analysis of any wireless communication system. The main goal of channel modeling is to characterize the multipath components (MPCs) in different environments, with a consideration of the trade-off between model accuracy and complexity.

Since 3G, 4G and the next-generation systems have larger bandwidth as well as increasing number of multiple-input–multiple-output (MIMO) arrays, we have higher resolutions of MPCs in both delay and angle domains, and it is thus possible to characterize the behavior of MPCs in more detail. A large body of MIMO measurements has shown that the MPCs are generally distributed in groups (i.e., clustered) in real-world environments. Cluster-based channel modeling has been an important trend in the development of channel models [9–14].

To parameterize cluster-based MIMO channel models, the first step is to identify clusters from MPCs, which has been done manually by visual inspection, as the human brain is good at the detection of patterns and structures even in noisy data [5]. However, the procedure of visual inspection is cumbersome and tiring for a large amount of measurement data, and is thus not feasible for many practical clustering implementations. In addition, this approach is subjective, and different people may provide different clustering results.

Automatic clustering of MPCs overcomes some of the drawbacks of visual inspection and has been an active area of research in the past decade. The main challenges in automatic clustering of MPCs are as follows [5]: (1) the notion of clusters tends to be intuitive rather than well defined; (2) the number of clusters is usually unknown; (3) the similarity of MPCs is difficult to quantify; and (4) the cluster shapes assumed by a specific model are difficult to incorporate into the clustering algorithm.

In this chapter we briefly introduce some classical MPC clustering algorithms and also present some recent proposed algorithms with better performance. The results in this chapter can be used to cluster real-world measurement data, and can be further used for cluster-based channel modeling for 4G/5G communications.

7.2   Clustering Algorithms

In this section we present some clustering algorithms of MPCs, which are useful for cluster-based channel modeling.

7.2.1   K-Power Means-Based Clustering

The K-power-means (KPM) algorithm [2] is a popular algorithm that was used in the clustering of radio channels in the past. It is based on the KMeans algorithm, which is a hard partitional approach and directly divides data objects into some prespecified number of clusters. KMeans is typically used with a Euclidean metric for computing the distance between points and cluster centers; therefore, it can easily find spherical or ball-shaped clusters in data [5]. The KPM algorithm introduces the power of the MPCs to augment the standard KMeans concept. In the KPM algorithm, upper and lower bounds on the number of clusters have to be known a priori. The appropriate clustering result is finally determined based on some indices that emphasize the compactness of each cluster and isolation between the clusters. In [15], the MPC distance (MCD) is proposed to quantify the similarity between MPCs. A small value of MCD means that two MPCs are close to each other and can be grouped into the same cluster. It is found that using MCD as a distance measure can improve the performance of the KPM algorithm [1].

The main idea of KPM can be summarized as follows [2, 8]:

Clustering

1. Initialize M cluster centroids $\mu_1, \mu_2, \ldots, \mu_M$ randomly.
2. Assigning each MPC x (here, all MPCs include four parameters: angles of departure (AoD) $[\phi_1^T, \ldots, \phi_{N_p}^T]$, angles of arrivals (AoA) $[\phi_1^R, \ldots, \phi_{N_p}^R]$, delay $[\tau_1, \ldots, \tau_{N_p}]$ and power $[\alpha_1, \ldots, \alpha_{N_p}]$) to the reasonable cluster centroid μ_j: for each x, set

$$C^{(e)} := \arg\min_j \left\{ \alpha_x \cdot d_{\text{MPC}}(x, \mu_j^{(e)}) \right\}, \tag{7.1}$$

where superscript e represents the eth iteration. C represents the store indices of MPC clustering in the eth iteration. d_{MPC} is the MCD.
3. Update the cluster centroids: for each j, set

$$\mu_j^{(e+1)} := \frac{\sum\limits_{x \in \Phi} 1\left\{C^{(e)} = j\right\} \alpha_x \cdot x}{\sum\limits_{x \in \Phi} 1\left\{C^{(e)} = j\right\} \alpha_x}, \tag{7.2}$$

where Φ is the set of all the MPCs for one snapshot.
4. Repeat steps 2 and 3 until convergence.

Cluster Validation: CombinedValidate

For cluster validation, [2] uses a combination of two methods: the Calinski–Harabasz (CH) index and the Davies–Bouldin criterion (DB). A combination of the two introduced validation indices yields significant improvements of clustering performance. The basic idea of the CombinedValidate (CV) index is to restrict valid choices of the optimum number of clusters by a threshold set in the DB index. Subsequently, the CH index is used to decide on the optimum number out of the restricted set of possibilities.

Note that determination of the number of clusters is actually an important part of KPM. In [16], the performances of several cluster validity indices are evaluated and compared to select the best estimation of the number of clusters. It is found that the Xie–Beni index generally has the best performance, though none of the indices is able to always predict correctly the desired number of clusters. However, mostly people still need to use visual inspection to ascertain the optimum number of clusters when using KPM [17]. Furthermore, manual adjustments of algorithm parameters according to different data are usually required to improve the performance [8].

Cluster Pruning: ShapePrune

After successfully finding the optimum number of clusters, [2] uses the ShapePrune cluster-pruning algorithm for discarding outliers. This is achieved by removing data points that have the largest distance from their own cluster centroid. As a constraint, cluster power and cluster spreads must not change significantly. This last condition allows preserving the clusters' original power and shape, which is fundamental to achieving consistent results.

Figure 7.1 shows the measured MPCs of the MIMO channel in [2], where MPCs are color-coded with their power. Visual inspection gives the impression of nicely separated clusters in space. Applying the KPM clustering framework without user interaction to this data, we obtain the result depicted in Figure 7.2. The resulting

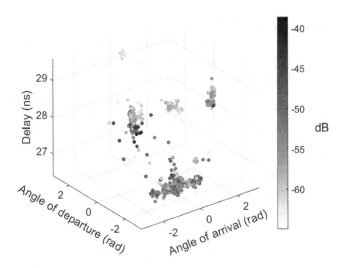

Figure 7.1 Unclustered MIMO measurement data in NLoS (no line of sight) scenario. © 2006 IEEE. Reprinted, with permission, from [2].

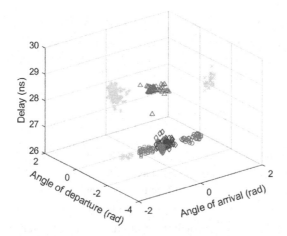

Figure 7.2 Results of KPM clustering. © 2006 IEEE. Reprinted, with permission, from [2].

partition into seven clusters realizes a good trade-off between cluster compactness and separation. It is also found that the pruning algorithm improves the visibility without changing cluster parameters.

7.2.2   Sparsity-Based Clustering

In this subsection, a sparsity-based method [5] is described to cluster channel impulse response (CIR) for single-input–single-output (SISO) channels. The proposed clustering algorithm involves solving a sparsity-based optimization problem. The main idea can be summarized as follows [5, 6]:

1. We assume that CIRs statistically follow the Saleh–Valenzuela (SV) model [18] of eq. (7.3), that is, powers of MPCs generally decrease with delays in terms of $A1$ and $A2$:

$$\left|\overline{\alpha}_{l,k}\right|^2 = \left|\overline{\alpha}_{0,0}\right|^2 \cdot \underbrace{\exp\left(-\frac{T_l}{\Gamma}\right)}_{A1} \cdot \underbrace{\exp\left(-\frac{\tau_{l,k}}{\Lambda_l}\right)}_{A2}, \tag{7.3}$$

where $\alpha_{l,k}$ is the amplitude gain and phase of the kth path within the lth cluster. $\left|\alpha_{0,0}\right|^2$ is the average power of the first MPC in the first cluster. T_l is the arrival delay of the lth cluster. $\tau_{l,k}$ is the excess arrival delay of the kth path within the lth cluster. A_1 and A_2 represent the two terms of MPC power decays with delay for inter- and intra-cluster, respectively. Γ_l and $\Lambda_{l,k}$ are the cluster and MPC power decay constants, respectively.

2. Then, we consider the measured PDP vector $\widehat{P}$ as the given signal and try to recover an original unknown signal vector $\mathbf{P}$, which is close to $\widehat{P}$ and has the formulation of eq. (7.3) using convex optimization, where $\widehat{\boldsymbol{P}}$ and $\mathbf{P}$ are the vectors of $\widehat{P}(\tau)$ and $\widehat{P}(\tau)$, respectively. We use a method to enhance the sparsity of the solution.

3. Finally, we use the curve of $\widehat{P}$ to identify the clusters. As shown in eq. (7.3), the curve of the dB-scaled $\widehat{P}$ generally has a high slope at the first MPC within each cluster, and the slope can thus be used for the cluster identifications. The above idea can be formulated as the following optimization problem [5]:

$$\min_{\widehat{P}} \left\| \widehat{P} - \mathbf{P} \right\|_2^2 + \lambda \left\| \Theta_2 \cdot \Theta_1 \cdot \widehat{P} \right\|_0 \tag{7.4}$$

where $\|\cdot\|_x$ represents ℓ_x norm operation and ℓ_0 norm operation returns the number of nonzero coefficients. $\widehat{P}$ and $\widehat{P}$ have dimension N_p, and λ is a regularization parameter. Θ_1 is the finite-difference operator in the form of eq. (7.5), where $\Delta\tau$ represents the minimum resolvable delay difference of data. Equation (7.5) is used to calculate the slope of $\widehat{P}$. Θ_2 is used to obtain the turning point at which the slope changes significantly and can be expressed as [5]:

$$\Theta_1 = \begin{bmatrix} \frac{\Delta\tau}{|\tau_1-\tau_2|} & -\frac{\Delta\tau}{|\tau_1-\tau_2|} & 0 & \cdots & \cdots & 0 \\ 0 & \frac{\Delta\tau}{|\tau_2-\tau_3|} & -\frac{\Delta\tau}{|\tau_2-\tau_3|} & \cdots & \cdots & 0 \\ \vdots & \ddots & \ddots & \ddots & \ddots & \vdots \\ 0 & 0 & \ddots & \frac{\Delta\tau}{|\tau_{N-2}-\tau_{N-1}|} & -\frac{\Delta\tau}{|\tau_{N-2}-\tau_{N-1}|} & 0 \\ 0 & 0 & \cdots & \cdots & \frac{\Delta\tau}{|\tau_{N-1}-\tau_N|} & -\frac{\Delta\tau}{|\tau_{N-1}-\tau_N|} \end{bmatrix}_{(N-1)\times N}, \tag{7.5}$$

$$\Theta_2 = \begin{bmatrix} 1 & -1 & 0 & \cdots & \cdots & 0 \\ 0 & 1 & -1 & \cdots & \cdots & 0 \\ \vdots & \ddots & \ddots & \ddots & \ddots & \vdots \\ 0 & 0 & \ddots & 1 & -1 & 0 \\ 0 & 0 & \cdots & \cdots & 1 & -1 \end{bmatrix}_{(N-2)\times(N-1)}. \tag{7.6}$$

The term $\lambda \left\| \Theta_2 \cdot \Theta_1 \cdot \widehat{P} \right\|_0$ ensures that the recovered $\widehat{P}$ follows the anticipated behavior of term $A2$ in eq. (7.3). It also implies that the proposed algorithm favors a small number of clusters to avoid over-parameterization. The anticipated behavior of term $A1$ in eq. (7.3) can be incorporated into $\widehat{P}$ using a clustering enhancement approach.

The detailed solutions and implementations of the above optimization problem can be found in [5], and are not repeated in this chapter due to space limitations. To give an example of the performance of this, we compare it with other algorithms using the measurements in [19, 20]. Figure 7.3 shows the example plots of power delay profile (PDP) clustering using different algorithms. It is found that: (1) the clusters identified by the proposed algorithm are distinct, most beginning with a sharp power peak followed by a linear decay. This means the modeling assumption of the SV model is well reflected by the clustering results. (2) For the KMeans and KPM algorithms, we can clearly see that the tail of one PDP cluster is grouped into the next cluster. This may lead to a least-squared regression curve of PDPs with a positive slope (within

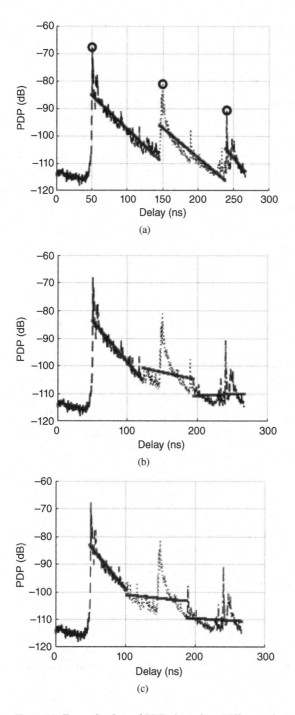

Figure 7.3 Example plots of PDP clustering. Different clusters are plotted with different colors. The black curves represent noise data and the dotted lines represents the least-squared regression of PDPs within clusters. (a) Proposed algorithm. The first peaks in each cluster are marked with black circles for clarity. (b) KMeans algorithm. (c) KPM algorithm. © 2016 IEEE. Reprinted, with permission, from [5].

the cluster), which results in the parameterized intra-cluster PDP model having an (erroneously) large delay spread. Further examples can be found in [5].

7.2.2.1   Kernel-Power-Density-Based Clustering

In this subsection we introduce a kernel-power-density (KPD)-based algorithm [8] for MPC clustering in MIMO channels. The main feature of KPD is threefold [8]: (1) the KPD uses the kernel density and only considers the neighboring points when computing the density; (2) the KPD uses the relative density (i.e., normalized within a local region) and a threshold is used to determine whether two clusters are density-reachable; and (3) the impact of power is incorporated in the clustering. We present KPD in the following steps [7, 8]:

1. Calculating density: For each MPC sample, say x, calculate the density ρ using the K nearest MPCs as follows:

$$
\rho_x = \sum_{y \in K_x} \exp(P_y) \cdot \exp\left(-\frac{|\tau_x - \tau_y|}{(\sigma_\tau)^2}\right)
$$
$$
\cdot \exp\left(-\frac{|\phi_x^T - \phi_y^T|}{(\sigma_{\phi T})}\right) \cdot \exp\left(-\frac{|\phi_x^R - \phi_y^R|}{(\sigma_{\phi R})}\right),
\tag{7.7}
$$

where y is an arbitrary MPC that $y \neq x$. K_x is the set of the K nearest MPCs for the MPC x. $\sigma_{(.)}$ is the standard deviation of MPCs. In eq. (7.7) we use the Laplacian kernel density for the angular domain as it has been widely observed that the angle of MPC follows the Laplacian distribution [21, 22].

2. Calculating relative density: For each MPC sample, calculate the relative density ρ^* using the K nearest MPCs' density, as follows:

$$
\rho_x^* = \frac{\rho_x}{\max\limits_{y \in K_x \cup \{x\}} \{\rho_y\}}.
\tag{7.8}
$$

By using the relative density, it is able to identify the clusters with relatively weak power. It can be seen that ρ^* ranges from 0 to 1.

3. Searching key MPCs: For each MPC x, if ρ^* equals to 1, label it as the key MPC $\hat{x}$. We thus obtain the set of key MPCs as follows:

$$
\hat{\Phi} := \{x | x \in \Phi, \rho_x^* = 1\}.
\tag{7.9}
$$

The key MPCs can be considered as the initial cluster centroids.

4. Clustering: For each MPC x, define its high-density-neighboring MPC as:

$$
\tilde{x} := \arg\min_{y \in \Phi, \rho_y^* > \rho_x^*} \{d(x, y)\},
\tag{7.10}
$$

where d represents the Euclidean distance. Similar to the idea of density-reachable in DBSCAN [23], we connect each MPC to its high-density-neighboring MPC and the connectedness path is defined as

$$p_x = \{x \rightarrow \tilde{x}\}. \tag{7.11}$$

We thus obtain a connectedness map, ζ_1, as follows:

$$\zeta_1 := \{p_x | x \in \Phi\}. \tag{7.12}$$

Note that two MPCs can be connected to each other over multiple paths. Those MPCs that are connected and reachable to the same key MPC in ζ_1 are grouped as one cluster.

5. Cluster merging: For each MPC, connect it to its K nearest MPCs and the connectedness path is defined as

$$q_x := \{x \rightarrow y, y \in K_x\}. \tag{7.13}$$

We thus obtain another connectedness map, ζ_2, as follows:

$$\zeta_2 := \{q_x | x \in \Phi\}. \tag{7.14}$$

If (1) two key MPCs are reachable in ζ_2 and (2) any MPC in any path connecting the two key MPCs has $\rho^* \geq \chi$, where χ is a density threshold, we merge the two key MPCs' clusters as one new cluster. As shown in Figure 7.4(a), clusters 2 and 3, 4 and 5, and 6 and 7 are merged respectively, and we finally obtain the results in Figure 7.4(b). Compared with the raw MPCs from measurement, the resulting cluster in Figure 7.4(b) looks fairly convincing. Detailed analysis of KPD parameter selection can be found in [8].

We describe the performance of the KPD algorithm under different "cluster conditions." Intuitively, a channel with a large cluster number and angular spread would have reduced clustering performance. We use the F measure [24] to evaluate the clustering performance, which is a robust external quality measure and ranges from 0 to 1, and a larger value indicates higher clustering quality. We test the impact of the cluster number and angular spread on the clustering accuracy. We use the SCME MIMO channel model to generate MPCs, and different cluster numbers are used in the simulation. A total of 300 random channels are simulated for each cluster number and spread case. Figure 7.5 shows that the proposed KPD algorithm, having the highest value of the F measure, shows a fairly good performance, and the value of the F measure decreases only slightly for larger cluster numbers. Figure 7.6 shows the impact of cluster angular spread on the F measure. It is found that the F measure generally decreases with the increasing cluster angular spread. The KPD algorithm shows a fairly good performance for arbitrary cluster sizes. Further performance examples can be found in [8]. Note that the above comparisons are limited to the simulated data in [8], and further validations with different measurements and simulations are still necessary.

7.2.2.2   Time-Cluster-Spatial-Lobe-Based Clustering

In this subsection we describe the time-cluster-spatial lobe (TCSL) clustering algorithm proposed by NYU [27], and implemented in the NYUSIM channel simulator [28, 29]. The TCSL scheme was proposed since extensive comparison to measured

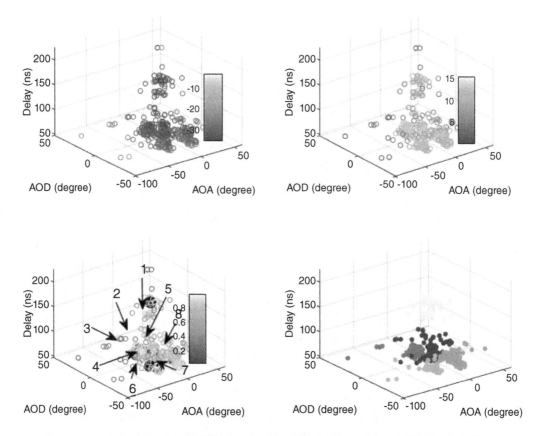

Figure 7.4 Illustration of KPD clustering from [8]. (a) Plots of the relative density estimated from measurement, where the grayscale bar indicates the level of relative density. The eight solid black points are the key MPCs. (b) Clustering results with the KPD algorithm, where the clusters are plotted with different shades of gray.

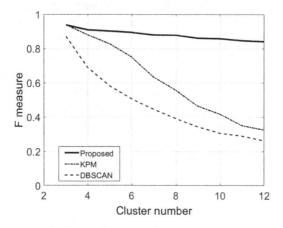

Figure 7.5 Impact of cluster number on the F measure. © 2017 IEEE. Reprinted, with permission, from [8].

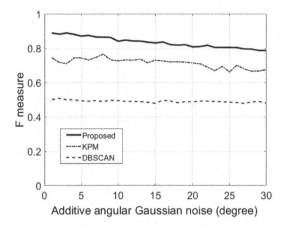

Figure 7.6 Impact of cluster angular spread on the F measure. © 2017 IEEE. Reprinted, with permission, from [8].

field data yielded a fit to 2D and 3D measurements using the TCSL algorithm, as the classical joint time–space modeling did not fit the extensive measurement data from urban NYC at millimeter wave (mmWave) using directional antennas as well as synthesized omnidirectional antenna patterns [25–27, 30, 31]. In the TCSL framework, time clusters (TCs) are composed of MPCs traveling close in time, and that arrive from potentially different directions in a short propagation time window [27]. Spatial mainlobes (SLs), determined by the beamwidth of the antenna (horn or lens or phased array), represent the main directions of arrival (or departure) from which energy arrives (and is measured over several hundred nanoseconds). These definitions decouple the time and space dimensions by extracting temporal and spatial statistics separately.

The time-partitioning methodology is illustrated in Figure 7.7, where the beginning and end times of each TC are extracted utilizing a 25-ns minimum inter-cluster void interval. Sequentially arriving MPCs that occur within 25 ns of each other are assumed to belong to one TC. For instance, in the omnidirectional PDP shown in Figure 7.7 there are two TCs which consist of eight and six subpath components with random delays, amplitudes and AoAs. The total power in one TC is also random, as it is composed of the sum of randomly varying subpath powers, which is borne out by field measurements. In addition, the propagation phases of each MPC can be taken to be *i.i.d.*, uniform between 0 and 2π [18]. This simple clustering method is easily adjustable to resolve temporal statistics over arbitrary time resolutions using a different minimum inter-cluster void interval. The value of 25 ns for the minimum inter-cluster void interval was found to match the measured data in the outdoor UMi scenario, and makes sense from a physical standpoint, since MPCs tend to arrive in clusters at different time delays over many angular directions, most likely due to the free-space air gaps between reflectors such as buildings, lampposts, and streets [27]. Similarly, the value of 6 ns for the minimum inter-cluster void interval was used for

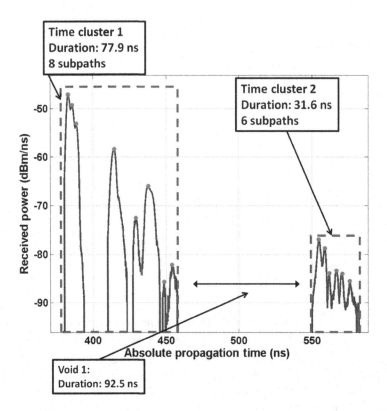

Figure 7.7 An example of measured omnidirectional PDP in the New York City measurements at mmWave frequencies [25–27]. In the PDP there are two time clusters (TCs) consisting of eight and six subpath components. The subpath components are found to have randomly varying AoAs, delays and amplitudes based on a peak detection algorithm. © 2016 IEEE. Reprinted, with permission, from [27].

the indoor office scenario since the width of a typical hallway in the measured indoor office environment is about 1.8 m (i.e., ~6 ns propagation delay) [32]. By counting the number of TCs and intra-cluster subpaths, and extracting TC and subpath delays and power levels from all available measured PDPs, measurement-based statistical distributions are obtained and allow reconstruction of time-varying impulse responses that embody the statistics of the collected data [27].

The concept of SLs is depicted in Figure 7.8, which shows that energy arrives at distinct mean pointing AoAs over a contiguous range of azimuth angles and a −10-dB power threshold with respect to the maximum received angle power [27]. The 3D spatial distribution of received power was reconstructed from the 28- and 73-GHz LoS and NLoS directional received powers [25–27] by linearly interpolating adjacent power level segments in azimuth and elevation with a 1° resolution and extracting 3D spatial angular statistics. A −10-dB threshold below maximum peak power was employed in the 3D power spectrum in both LoS and NLoS environments for the outdoor UMi scenario, where all power segments below this threshold were

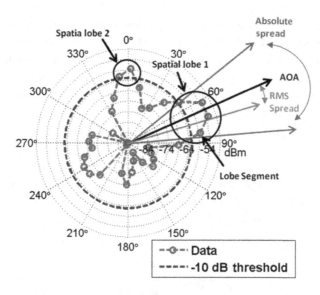

Figure 7.8 An example of measured power azimuth spectrum in the New York City measurements at mmWave frequencies [25–27]. In the power spectrum, there are two spatial lobes (SLs) using a −10-dB power threshold with respect to the maximum received angle power [27]. An SL has well-defined properties, including its mean pointing angle, its absolute angle spread, and its RMS angle spread. © 2016 IEEE. Reprinted, with permission, from [27].

disregarded for further processing [27]. Similarly, a −15-dB threshold was applied for the indoor office scenario [32].

Per the definitions and illustrations given above, a TC contains MPCs that travel close in time, but may arrive from different SL angular directions. Similarly, an SL may contain many MPCs arriving (or departing) in a spatial beam (angular cluster) but with different time delays. These features have been observed in real-world propagation measurements [25, 26, 33], which have shown that MPCs belonging to the same TC can arrive from distinct spatial angles and that energy arriving or departing in a particular spatial direction can span hundreds or thousands of nanoseconds in propagation delay spread, detectable due to high-gain rotatable directional antennas. The TCSL clustering scheme is physically based, and is derived from field observations based on about 1 terabyte of measured data over many years, and can be used to extract TC and SL statistics for any measurement or ray-tracing data sets [27]. The key parameters for the TCSL algorithm are the number of TCs, the number of intra-cluster subpaths, the number of SLs, the TC and subpath delays, the TC and subpath power levels, and the SL power levels.

7.2.2.3   Cluster Initialization Using Improved Subtractive

As described in Section 3.2.1, predefining the number of clusters and their initial positions is critical for this algorithm to work. Therefore, in [34] a density-based initialization algorithm is applied to MPC data sets to find the number of clusters and the initial centroid positions (originally proposed in [35, 36]). Afterwards, one or more runs of the KPM algorithm assigns each MPC to the nearest centroid and updates the power-weighted centroid positions. As a distance measure the BMCD (balanced MCD) is used (proposed in [34]), which introduces additional normalization factors for the angular domains, as known from the delay domain of the MCD. The normalizers are calculated as:

$$\delta_{\text{AoD/AoA}} = 2 \cdot \frac{\text{std}_j(d_{\text{MCD, AoD/AoA}}(\mathbf{x}_j, \bar{\mathbf{x}}))}{\text{max}_j^2(d_{\text{MCD, AoD/AoA}}(\mathbf{x}_j, \bar{\mathbf{x}}))}, \tag{7.15}$$

where $\text{std}_j()$ provides the standard deviation of the MCD between all MPC positions $\mathbf{x}_j$ and the center of the data space $\bar{\mathbf{x}}$ and max_j the corresponding maximum.

The improved subtractive performs the following steps:

1. Calculate the normalization constant β:

$$\beta = \frac{N}{\sum_{j=1}^{N} d_{\text{MPC}}(\mathbf{x}_j, \bar{\mathbf{x}})}, \tag{7.16}$$

where N is the total number of MPCs and $d_{\text{MPC}}(\mathbf{x}_j, \bar{\mathbf{x}})$ is the BMCD between MPC position $\mathbf{x}_j$ and the center of the data space $\bar{\mathbf{x}}$.

2. Calculate a density value for each MPC position $\mathbf{x}_i$:

$$P_i^{\mathbf{m}} = \sum_{j=1}^{N} \exp\left(-\mathbf{m}^{\text{T}} \cdot \beta \cdot \mathbf{d}_{\text{MPC}}(\mathbf{x}_i, \mathbf{x}_j)\right). \tag{7.17}$$

The product $\mathbf{m}^{\text{T}} \cdot \beta$ scales in fact the influence of neighboring MPCs and its inverse is called *neighborhood radius*. For MPC data sets it is good practice to find appropriate radii for direction of arrival (DoA) and direction of departure (DoD) and delay dimensions separately. That is why $\mathbf{m}$ and $\mathbf{d}$ are vectors with three components:

$$\mathbf{d}_{\text{MPC}}(\mathbf{x}_i, \mathbf{x}_j) = [d_{\text{MPC, DoA}}(\mathbf{x}_i, \mathbf{x}_j), d_{\text{MPC, DoD}}(\mathbf{x}_i, \mathbf{x}_j), d_{\text{MPC, Delay}}(\mathbf{x}_i, \mathbf{x}_j)]^{\text{T}}. \tag{7.18}$$

3. Choose the point $\mathbf{x}_k$ with the highest density value as a new cluster centroid if its density value is above a certain threshold. If not, stop the initialization procedure.
4. Subtract the new centroid from the data set by updating the density values:

$$P_i^{\mathbf{m}} = P_i^{\mathbf{m}} - P_k^{\mathbf{m}} \cdot \exp\left(-\eta \cdot \mathbf{m}^{\text{T}} \cdot \beta \cdot \mathbf{d}_{\text{MPC}}(\mathbf{x}_i, \mathbf{x}_k)\right). \tag{7.19}$$

$\eta \in (0, 1]$ scales the density subtraction. Return to step 3.

Afterwards, the KPM algorithm can be initialized with this cluster centroid.

Neighborhood Radius
The neighborhood radius can be found automatically by using the so-called *correlation self-comparison technique* [34, 36]. This technique is applied for each component of $\mathbf{m}$ separately.

1. Calculate the set of density values for all MPCs P^{m_l} for an increasing m_l with $m_l \in \{1, 5, 10, 15, ...\}$. Set the other components of $\mathbf{m}$ to 1 (e.g., $\mathbf{m} = [m_l, 1, 1]^T$).
2. Calculate the correlation between P^{m_l} and $P^{m_{l+1}}$. If the correlation rises above a certain threshold (e.g., 0.99), choose m_l as value for $\mathbf{m}$ in this dimension.

7.2.3   MR-DMS Clustering

MR-DMS (multi-reference detection of maximum separation) is a hierarchical clustering approach that starts with a single cluster and subsequently divides it into smaller ones. It is described in [37, 38]. The separation is done by evaluating the distances between all MPCs of a cluster seen from multiple reference points in the data space to find the biggest gap. As the distance measure the BMCD is used, which is already described for the improved subtractive. The optimum number of clusters can be found either by applying cluster validation indices to the results or by defining a threshold during the separation process.

The MR-DMS performs the following steps:

1. Spread N reference points over the data space. (e.g., $N = 16$).
2. Add all MPCs to one cluster D_1.
3. If the maximum number of clusters is not reached: For each recent cluster D_k iterate over all references $\mathbf{r}_n (n = 1, \ldots, N)$ and calculate the distance between all MPC positions $\mathbf{x}_i$ in the current cluster and the reference point and store it as a vector:

$$\mathbf{d}_n^k(i) = d_{\text{MPC}}(\mathbf{x}_i, \mathbf{r}_n). \tag{7.20}$$

4. Sort all vectors $\mathbf{d}_n^k$ in ascending order.
5. Calculate the derivative $(\mathbf{d}_n^k)'$ which is in fact the distance between neighboring MPCs in cluster k seen from reference n.
6. Find the maximum distance/separation over all clusters and references $\max_{k,n,i}((\mathbf{d}_n^k)')$.
7. Split the found cluster at the position of maximum separation. Return to step 3.

Defining the Threshold

Calculating cluster results for different numbers of clusters and applying validation indices to the results can be computationally very expensive. Therefore, thresholding can be used during the separation process to automatically detect an appropriate number of clusters.

After calculating the derivatives in step 5, only those clusters are considered for the maximization in step 6 whose maximum derivative/separation exceeds a certain threshold for at least one reference. For example, this threshold could be defined dynamically by considering the distribution of $(\mathbf{d}_n^k)'$:

$$th_n^k = \text{mean}((\mathbf{d}_n^k)') + \alpha \cdot \text{std}((\mathbf{d}_n^k)'), \tag{7.21}$$

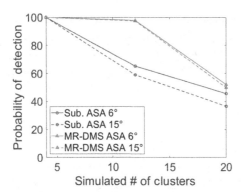

Figure 7.9 Probability of correct estimated number of clusters between improved subtractive and the MR-DMS vs. number of clusters (NCL) and the cluster angular spread of arrival (ASA) are varied (NCL = {4, 12, 20}, ASA = {6°, 15°}).

with a scaling constant α. Only those clusters whose distance between two neighboring MPCs is significantly larger than the other distances in the same cluster are considered (the cluster provides an obvious gap). Stop the algorithm if all clusters are below the defined threshold.

In Figure 7.9 a performance comparison between both algorithms (improved subtractive and the MR-DMS) is shown. The algorithms are validated over 500 drops of the WINNER channel model scenario urban macro cell (C2), similar to [38]. Here, six different scenarios are used where the number of clusters (NCL) and the angular spread of arrival (ASA) are varied (NCL = {4, 12, 20}, ASA = {6°, 15°}). The curves show the probability of correct estimation of the NCL (range: −2/+4) over the simulated NCL. The upper bound for the range is chosen as +4 since the two strongest clusters are split into three subclusters, providing four additional clusters. Figure 7.10 illustrates estimated clusters by the improved subtractive and the MR-DMS algorithm from a MIMO measurement campaign in Bonn (Germany). Details of the campaign settings can be found in [39]. The MPCs are extracted using the RIMAX algorithm. The figures show the azimuth–azimuth–delay dimensions of the data sets.

7.2.3.1   Other Algorithms

In [40], the fuzzy-c-means algorithm is used as an alternative to the KPM. It is found that with random initialization, the fuzzy-c-means algorithm outperforms the KPM. In [9], the density-based spatial clustering for applications with noise (DBSCAN) algorithm is applied to cluster local MPCs. In [41], a fixed inter-cluster void interval, which represents the minimum propagation time between likely reflection or scattering objects, is used to distinguish clusters in the time domain. In [42], a hierarchical agglomerative clustering algorithm is used to search for clusters jointly in the delay–angle-space domain and the performance is validated by ray-tracing simulation. Those algorithms are also used in MPC clustering; however, the details are not presented due to limited space in this chapter.

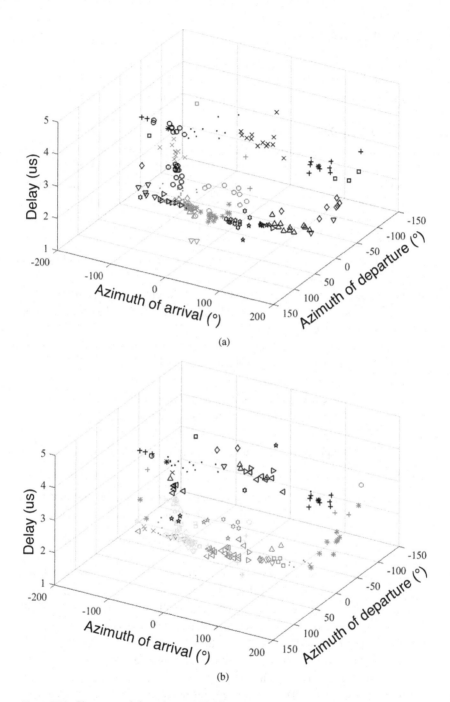

(a)

(b)

Figure 7.10 Cluster result based on a MIMO measurement campaign in Bonn (Germany).

7.2.4   Tracking Algorithms

In this section we present some tracking algorithms used for channel modeling. Note that some of them are actually for tracking MPCs, which can also be used for tracking clusters by applying for the center points of clusters.

7.2.4.1   MCD-Based Tracking

This cluster tracking mechanism is able to capture the movement of clusters with very low complexity. The idea is based on the distance between the clusters' centroids. MCD is chosen as a suitable distance metric to cope with angular periodicity as well as data scaling. The algorithm is as shown in Algorithm 7.1 [3].

Algorithm 7.1

1. Calculate the distance between any old and any new centroid using the MCD.
2. For each new centroid:
   a. calculate the distance and index of the closest old centroid;
   b. If smallest distance > threshold, treat the centroid as a new cluster.
3. For each old centroid:
   a. check the number of close new centroids within the distance threshold;
   b. If number = 1, the old cluster is moved;
   c. If number > 1, the cluster is split:
      – the closest new cluster is treated as if the old cluster moved;
      – other close ones are treated as new clusters.

Two subsequent sets of a number of N_{old} old and N_{new} new cluster centroids $c_i^{(old)}$ and $c_j^{(new)}$, are considered, where $i = 1, \ldots, N_{old}$ and $j = 1, \ldots, N_{new}$.

1. The distances between the centroids are arranged in the distance matrix $\mathbf{D}$ with dimension $N_{old} \times N_{new}$, where each element is calculated as [3]

$$[D]_{i,j} = \text{MCD}\left(c_i^{(old)}, c_j^{(new)}\right), \tag{7.22}$$

that is, the distance between the ith old and jth new centroid. All further evaluations can now easily be done by searching in the distance matrix.

2a. For each column of $\mathbf{D}$, search for the smallest entry in the distance matrix. The indices i and j of this value identifies the closest old cluster.

2b. If the distance $[D]_{i,j}$ between a new cluster and the closest old cluster exceeds a specified threshold, the cluster is treated as a new cluster.

3. We now check for each old cluster, if it has moved.

3a. For each row in $\mathbf{D}$ count the number of elements smaller than the threshold.

3b. If only one new cluster is in the vicinity of the old cluster, the old cluster has moved.

3c. If many new clusters are in the vicinity of the old cluster, the old cluster moved toward the closest new one. The other close ones are treated as new.

To every new cluster a unique cluster-ID (CLID) is assigned. If a movement is identified, the moved cluster inherits the CLID from its predecessor.

7.2.4.2   Two-Way Matching Tracking

This tracking algorithm is proposed in [43, 44] for dynamic MPC tracking and also uses MCD. The main difference compared to [3] is that it requires a two-way matching between two consecutive snapshots in the tracking and improves the accuracy. The main steps are as follows [43]:

1. Calculate MCD between any MPC within time i and any MPC within time $i + 1$ and obtain an MCD matrix $\mathbf{D}$ with dimension $N(i) \times N(i + 1)$, where N indicates the number of MPCs.
2. If conditions

$$
\begin{aligned}
D_{u,v} &\le \varepsilon \\
u &= \arg \min_u \left(D_{u \in N(i), v} \right) \\
v &= \arg \min_v \left(D_{u, v \in N(i+1)} \right)
\end{aligned}
\tag{7.23}
$$

are satisfied, the uth MPC at time i and the vth MPC at time $i + 1$ are considered to be the same MPC. To match them, a unique MPC ID is assigned to them. ε is a specified threshold used to measure the similarity between two MPCs.
3. Examine all other MPCs between time i and $i + 1$ and match all MPCs according to eq. (7.23).
4. Calculate MCD between any MPC within $i + 1$ and any MPC within $i + 2$ and repeating steps 1 and 2. If the wth MPC at time $i + 2$ is found to match the vth MPC at time $i + 1$, the wth MPC (at time $i + 2$) inherits the MPC ID from the vth MPC (at time $i + 1$), and so forth.
5. Repeat the preceding steps for the times after $i + 2$, and do matching in every two consecutive indices of time windows (or quasi-stationary windows [45–47]). Assign MPC IDs for all MPCs.

Figure 7.11 shows an example plot of MPC tracking in vehicle-to-vehicle environments [43]. We can see that only those MPCs with similar evolutions on both angular and delay domains are grouped, and the tracking algorithm of two-way matching generally leads to reasonable results. It is also noteworthy that the evolutions of MPCs in angular and delay domains are independent. Furthermore, the track of evolution (i.e., the slope of the LS fit curves) is generally independent of angle and delay. This follows the physical insight that the scatterers are randomly distributed in the dynamic V2V channels.

7.2.4.3   Kalman Filter-Based Tracking

This method uses a Kalman filter for tracking and predicting cluster positions [4]. Figure 7.12 shows the clustering and tracking framework.

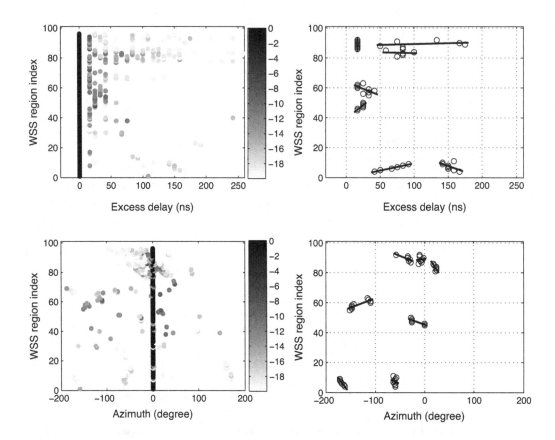

Figure 7.11 Example plot of tracking. (a) Detected MPCs (in dB) on delay domain. (b) MPC tracking on delay domain. (c) Detected MPCs (in dB) on azimuth domain. (d) MPC tracking on azimuth domain. In (b) and (d), MPCs with a lifetime of less than six quasi-stationary windows are not plotted for clarity. © 2015 IEEE. Reprinted with permission, from [43].

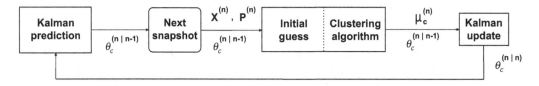

Figure 7.12 Clustering and tracking framework. © 2007 IEEE. Reprinted, with permission, from [4].

Cluster Data Model

Each cluster is determined by the following parameters:

1. A unique cluster-ID c.
2. The cluster power at time i. Denoting the set of path indices belonging to cluster c at time snapshot i by $\Gamma_c^{(i)}$, the cluster power is calculated as $\gamma_c^{(i)} = \sum_{l \in \Gamma_c^{(i)}} P_l^{(i)}$.

3. The number of paths within the clusters $L_C^{(i)} = |\Gamma_c^{(i)}|$, where every path is assumed to belong to one cluster, uniquely.

4. The cluster centroid position in the angle–angle–delay domain $\mu_c^{(i)}$. The cluster centroid position can be calculated as

$$
\mu_c^{(i)} = \begin{bmatrix} \tau_c^{(i)} & \varphi_{Rx,c}^{(i)} & \varphi_{Tx,c}^{(i)} \end{bmatrix}^T = \begin{bmatrix} \frac{1}{\gamma_c^{(i)}} \cdot \sum_{l \in \Gamma_c^{(i)}} P_l^{(i)} \tau_l^{(i)} \\ \text{angle}\left(\sum_{l \in \Gamma_c^{(i)}} P_l^{(i)} \exp\left(j\varphi_{Rx,l}^{(i)} \right) \right) \\ \text{angle}\left(\sum_{l \in \Gamma_c^{(i)}} P_l^{(i)} \exp\left(j\varphi_{Tx,l}^{(i)} \right) \right) \end{bmatrix},
$$
(7.24)

where the mean angle is calculated by averaging angles over their respective complex representation. For tracking, the centroid speed is also of interest, so we combine the position and speed in the cluster tracking parameter vector

$$
\theta_c^{(i)} = \begin{bmatrix} \tau_c^{(i)} & \Delta\tau_c^{(i)} & \varphi_{Rx,c}^{(i)} & \Delta\varphi_{Rx,c}^{(i)} & \varphi_{Tx,c}^{(i)} & \Delta\varphi_{Tx,c}^{(i)} \end{bmatrix}^T.
$$
(7.25)

5. The cluster's joint spread, which is the power-weighted covariance matrix of the path parameters within one cluster at time i. The cluster spread matrix is calculated by

$$
\mathbf{C}_c^{(i)} = \frac{\sum_{l \in \Gamma_c^{(i)}} P_l^{(i)} \left(x_c^{(i)} - \mu_c^{(i)} \right) \left(x_c^{(i)} - \mu_c^{(i)} \right)^T}{\gamma_c^{(i)}}.
$$
(7.26)

Kalman Cluster Model

State-space model. Only the cluster centroid position is used. The following is the state equation:

$$
\theta_c^{(i)} = \phi\theta_c^{(i)} + w^{(i)},
$$
(7.27)

where $w^{(i)}$ denotes the state-noise with covariance matrix $\mathbf{Q}$, and ϕ is the state-transition matrix given by

$$
\phi = \mathbf{I}_3 \otimes \begin{bmatrix} 1 & 1 \\ 0 & 1 \end{bmatrix},
$$
(7.28)

where identity matrices are denoted by $\mathbf{I}_d$ with d denoting the dimension, and $\otimes$ denotes the Kronecker matrix product. Since we can observe only the cluster centroids and not their speed, we use the following observation model

$$
\mu_c^{(i)} = \mathbf{H}\theta_c^{(i)} + v^{(i)},
$$
(7.29)

where $\mu_c^{(i)}$ describes the observed cluster centroid position; thus, $\mathbf{H}$ is given by

$$
\mathbf{H} = \mathbf{I}_3 \otimes \begin{bmatrix} 1 & 0 \end{bmatrix}
$$
(7.30)

and $v^{(i)}$ denotes the observation noise with covariance matrix $\mathbf{R}$.

Tracking equations. The derivation of the Kalman filter is straightforward and leads to the following prediction and update equations:

Prediction:
$$\theta_c^{(i|i-1)} = \phi\theta_c^{(i|i-1)}$$
$$\mathbf{M}^{(i|i-1)} = \phi\mathbf{M}^{(i|i-1)}\phi^T + \mathbf{Q}. \tag{7.31}$$

Update:
$$\mathbf{K}^{(i|i)} = \mathbf{M}^{(i|i-1)}\mathbf{H}^T\left(\mathbf{H}\mathbf{M}^{(i|i-1)}\mathbf{H}^T + \mathbf{R}\right)^{-1}$$
$$\theta_c^{(i|i)} = \theta_c^{(i|i-1)} + \mathbf{K}^{(i|i)}\left(\mu_c - \mathbf{H}\theta_c^{(i|i-1)}\right) \tag{7.32}$$
$$\mathbf{M}^{(i|i)} = \left(\mathbf{I} - \mathbf{K}^{(i|i)}\mathbf{H}\right)\mathbf{M}^{(i|i-1)}.$$

Cluster association. A major problem in multitarget tracking is how to associate the predicted with the identified cluster centroids. Since we are tracking clusters that show a certain extent in parameter space, the Euclidean distance does not provide a good association. Instead, we use the following probability-based method. The distance between a cluster and a cluster centroid is defined by

$$\left(\tilde{\mu}|\mu_c, \mathbf{C}_c\right) = \frac{1}{(2\pi)^{3/2}|\mathbf{C}_c|^{1/2}}\exp\left(-\frac{1}{2}(\tilde{\mu} - \mu_c)^T\mathbf{C}_c^{-1}(\tilde{\mu} - \mu_c)\right). \tag{7.33}$$

Since a small distance between the two centroids now corresponds to a large value of this function, we refer to it as the closeness function.

7.2.4.4   Threshold-Based Tracking

This method was first proposed for tracking delay and amplitude changes of MPCs in [48], and then extended to the angular domain in [49]. Consider that evaluation of all the snapshots provides a delay matrix $T \in R^{M \times L}$, a DoA matrix $\Phi \in R^{M \times L}$ and an amplitude matrix $A \in C^{M \times L}$. Finally, M is the number of measured snapshots. In general, the delay and angle of one MPC change very slowly, so that the following constraint is imposed on the delay change between snapshots:

$$\varepsilon = \left|T_{i,l} - \tilde{\tau}_i\right| \leq \frac{1}{2W}, \tag{7.34}$$

where W is the measurement bandwidth, $T_{i,l}$ is the estimated delay of the lth MPC in the ith snapshot and $\tilde{\tau}_i$ is the predicted value of this MPC in the ith snapshot. We add a similar constraint on the AoA

$$\eta = \left|\cos\left(\Phi_{i,l}\right) - \cos\left(\tilde{\varphi}_i\right)\right| \leq \frac{1}{2N^R}, \tag{7.35}$$

where N^R is the number of the elements of the virtual array, $\Phi_{i,l}$ is the estimated AoA of the lth MPC in the ith snapshot and $\tilde{\varphi}_i$ is the value that is tracked in the $(i-1)$th or the $(i+1)$th snapshot. The tracked MPCs might show "gaps," that is, these are not visible for certain snapshots; details of the algorithm to bridge those gaps, and when to interpret MPCs as "new," are given in [48].

7.2.4.5   Probability-Based Tracking

To model the MPCs in time-variant channels, a probability-based tracking algorithm is proposed in [50]. The proposed algorithm is developed in two steps: (1) recognize the trajectories of MPCs; and (2) cluster MPCs based on the trajectories. Note that the trajectories here represent the moving paths of MPCs in successive snapshots, and one trajectory connects two MPCs in successive snapshots, which means that these two MPCs are the same MPC in different snapshots. In the first stage, a novel probability-based tracking process is proposed, which is conducted by maximizing the total sum probability of all trajectories. In the second stage, a tracking based approach is provided to cluster MPCs.

Let $A_1, \ldots, A_M$ denote the MPCs in snapshot S_i and $B_1, \ldots, B_M$ denote the MPCs in snapshot S_{i+1}. l represents an ordered pair of the MPCs in successive snapshots, that is, l_{A_x, B_y}, where $x, y \in [1, \ldots, M]$, represents the trajectory from A_x in snapshot S_i to B_y in snapshot S_{i+1}. Let $\mathbf{L}$ be the set of all such trajectories and if there are more than one MPC in both snapshots, there could be many possible trajectories between the two snapshots, whereas the ground truth is only a specific subset of trajectories in $\mathbf{L}$. The main idea of [50] is to identify the true trajectories of the MPCs in every two successive snapshots, and trajectory l_{A_x, B_y} is weighted by a moving probability $P(A_x, B_y)$, as shown in Figure 7.13(a).

In [50], to accurately identify the trajectories, the truth trajectories are obtained by maximizing the total probabilities of all selected trajectories, as follows:

$$p^* = \arg\max_{L \subset \mathbf{L}} \sum_{(A_x, B_y) \in L} p'(A_x, B_y), \qquad (7.36)$$

where $L = \{(A_1, B_{y_1}), (A_2, B_{y_2}), \ldots, (A_M, B_{y_M})\}$ subject to $y_1, y_2, \ldots, y_M$ are a permutation of integers $1, 2, \ldots, M$. Let $\mathbf{U}$ and n denote the set of parameters $[\phi^T, \phi^R, \tau, \alpha]$ and the number of parameters, respectively. A normalized Euclidean distance of $\mathbf{U}$ is used to measure the distance between A_x and B_y, as follows:

$$D_{A_x, B_y} = \sqrt{\sum_{i=1}^{n} N[(\mathbf{U}_{A_x}(i) - \mathbf{U}_{B_y}(i))^2]}. \qquad (7.37)$$

Based on D_{A_x, B_y}, an aggregated pairwise probability is proposed to measure the possibility of each trajectory. Generally, a shorter distance between two MPCs leads to a higher probability of the moving path/trajectory between them, and vice versa. Hence, the probability of trajectory $p(A_x, B_y)$ can be obtained by using the reciprocal of D_{A_x, B_y}. Furthermore, for each MPC, in order to ensure the sum of all the possible trajectories' probabilities equals to 1, $\sum_{y=1}^{M} p'(A_x, B_y) = 1$, $p'(A_x, B_y)$ is obtained by normalizing $p(A_x, B_y)$ as follows:

$$p'(A_x, B_y) = \begin{cases} 1 & D_{A_x, B_y} = 0, \\ 0 & D_{A_x, B_z} = 0, y \neq z, \\ \dfrac{1}{D_{A_x, B_y} \sum_{z=1}^{M} D_{A_x, B_z}^{-1}} & \text{others.} \end{cases} \qquad (7.38)$$

Note that $D_{A_x, B_y} = 0$ means MPC A_x is not moving or is remaining relatively static with respect to both transmitter and receiver, respectively, during two snapshots. To solve the problem in eq. (7.36), the Kuhn–Munkres algorithm (K-M) is adopted, which is able to find the maximum weight perfect-matching in a bipartite graph of

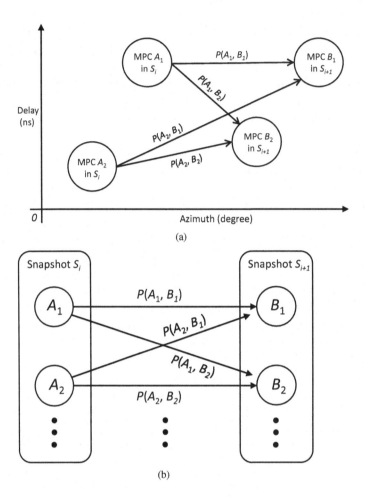

Figure 7.13 Illustration of the trajectories between snapshots S_i and S_{i+1}. (a) Delay and azimuth domain. (b) Bipartite graph domain. © 2017 IEEE. Reprinted with permission, from [50].

a general assignment problem invented by Kuhn and improved by Munkres. In the bipartite graph, every node in two subsets links to each other and every link has its own weight. In the algorithm, the MPCs in two successive snapshots are considered as the two subsets in the bipartite graph, and the trajectories between each snapshot are considered as the links between two subsets, which is weighted by the moving probability, as shown in Figure 7.13(b). Using the K-M algorithm, the best solution can be obtained, which indicates the most possible trajectories of MPCs in successive snapshots.

In this algorithm, a heuristic approach is provided to cluster MPCs with the purpose of comparing the moving probability of the MPCs in the same snapshot with a preset threshold P_T, where $P_T = 0.8$ is suggested based on simulations. For MPC A_x in snapshot S_i, the moving probabilities to the MPCs in snapshot S_{i+1} reflect the similarity of these MPCs in S_{i+1}. For example, if $p'(A_x, B_{y_1})$ and $p'(A_x, B_{y_2})$ are greater than the threshold P_T, it implies that the MPCs B_{y_1} and B_{y_2} are fairly similar and belong to the same one cluster:

$$\mathbf{K}_x = \{B_y | p'(A_x, B_y) > P_T, A \in S_i, B \in S_{i+1}\}. \tag{7.39}$$

As seen in eq. (7.39), different A in S_i may indicate different clustering results; for example, $\mathbf{K}_1 = \{B_1, B_2, B_3\}$, $\mathbf{K}_2 = \{B_1, B_2\}$, $\mathbf{K}_3 = \{B_1, B_2\}$. In this case, the results with the most occurrences can be selected, which is $\mathbf{K} = \{B_1, B_2\}$ in the example above. Note that if some MPCs do not belong to any cluster in the results, these MPCs are considered as individual clusters, although it rarely happens. In this way, MPCs can be clustered based on their relationships during successive snapshots.

7.2.5 Multipath Fading Behavior of the Clusters

In this subsection we will highlight some of the major differences in the fading behavior of the clusters as compared to the legacy narrowband channel models. Supported with realistic small-scale fading measurements, we demonstrate that these differences have a significant impact on the channel modeling methodology.

In contrast to the 3GPP LTE systems [51], where spatial processing is applied to narrowband channels, mmWave systems do spatial processing of wideband channels [52, 53]. Channel models for 3GPP LTE systems (e.g., 3GPP SCM [54], WINNER [55] and COST 2100 [13]) are based on certain narrowband assumptions, for example, uncorrelated scattering (US) of resolvable MPCs corresponding to a cluster of scatterers in the space domain. In these models, fading behavior of a multipath cluster is modeled as a zero-mean Rayleigh distributed random process using the sum-of-sinusoids (SOS) principle. Millimeter-wave systems, on the other hand, are supposed to operate with wider bandwidths and higher beamforming gains. As a result, radio channels for mmWave systems behave quite differently as compared to legacy narrowband channels. Irrespective of the carrier frequency, probing a propagation channel with a wideband signal results in a finer resolution of multipath echoes in the delay domain of the PDP. Different from narrowband channels, increased resolution of MPCs in the delay domain results in reduced signal fading [56, 57]. Additionally, fading behavior of a radio channel converges from Rayleigh toward the Rician fading regime [58]. Channels measurements presented in [59] illuminated different scattering objects inside a classroom with directional antennas emulating beamforming at the TX and RX. Presented results therein and in Figure 7.14 explain that fading behavior of a cluster nearly vanishes with increased system bandwidth. This means that channel fading behavior becomes more deterministic and in the literature it is often termed as *channel hardening* [62]. Similarly, spatial filtering of MPCs with increased beamforming gains also results in reduced fading depths, as demonstrated in [63]. Figure 7.15 shows a summary of results presented in [63], which shows the fading envelope of signal reflected from a wall cluster when illuminated with different TX–RX antenna gains. It is easy to follow that the fading envelope of the

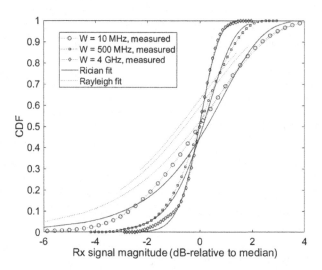

Figure 7.14 Double bounce reflections in a classroom scenario, 34 GHz carrier frequency, directional TX–RX antenna gains 11–12 dBi, cross-polarization setup. © 2017 IEEE. Reprinted with permission, from [60].

RX signal power process becomes more deterministic with increased beamforming gain. From the channel modeling perspectives, results presented in [59, 63] show that increased system bandwidth and increased beamforming gains results in very high Rician K-factors for the clusters. Therefore, the complex channel impulse response does not remain a zero-mean random process. High Rician K-factor also implies that the phase process of the received signal fading envelope also becomes deterministic [64]. Therefore, the phase of the RX signal is not a uniformly distributed *i.i.d.*, random process as considered in the Rayleigh fading random process.

Large-amplitude fading of the RX signal in the narrowband channel results in a considerable randomness in the TX–RX polarization setups inside the 2×2 polarization coupling matrix. Therefore, in narrowband channel models like WINNER II [55], the cross-polarization ratio (XPR) is modeled as a log-normally distributed random variable with considerably high standard deviation. On the other hand, channel hardening due to bandwidth and beamforming gain of the mmWave systems result in reduced randomness in the polarization coupling matrix [65, 66]. Results presented in [65] show that bandwidth has little impact on the average XPR values of the RX signal obtained from a cluster reflection; however, the standard deviation in the XPR drops almost exponentially with increase in system bandwidth. This implies that channel hardening also results in a more deterministic behavior of XPR and polarization coupling matrix.

From the discussion above, one may argue that the modeling of clusters in terms of their stochastic-deterministic fading behavior has to be adopted with frequency, bandwidth and the beamforming gain of the systems. Considering the mmWave

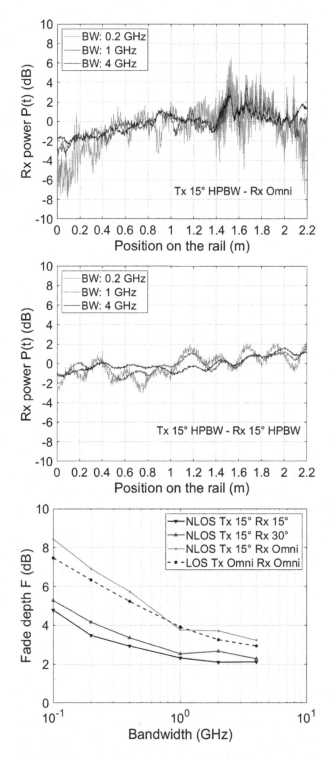

Figure 7.15 Fading behavior analysis of a cluster of scatterers from a wall reflection; the cluster is illuminated with RX antenna HPBWs $(15°, 30°, 360°)$, TX antenna HPBW is $15°$, absolute bandwidth (BW) = $\{0.2, 1, 4\}$ GHz. © 2018 IEEE. Reprinted with permission, from [61]

system aspects (i.e., higher bandwidth and higher beamforming gains), straight-forward parameterization of 3GPP channel models like 3GPP SCM and WINNER II may not be accurate.

7.3   Conclusion

In this chapter we present a brief introduction of clustering and tracking algorithms for MPCs in radio channel modeling. Some existing algorithms and recent progress are summarized. The multipath fading behavior of the clusters is discussed. The results in this chapter can be used to cluster and track real-world measurement data, and can be further used for the cluster-based (dynamic) channel modeling for 4G/5G communications.

References

[1]   N. Czink, P. Cera, J. Salo, E. Bonek, J.-P. Nuutinen and J. Ylitalo, "Improving clustering performance using multipath component distance," *Electronics Letters*, vol. 42, no. 1, pp. 33–35, 2006.

[2]   N. Czink, P. Cera, J. Salo, E. Bonek, J.-P. Nuutinen and J. Ylitalo, "A framework for automatic clustering of parametric MIMO channel data including path powers," *Vehicular Technology Conference, 2006. VTC-2006 Fall. 2006 IEEE 64th*, IEEE, 2006, pp. 1–5.

[3]   N. Czink, C. Mecklenbrauker and G. Del-Galdo, "A novel automatic cluster tracking algorithm," *2006 IEEE 17th International Symposium on Personal, Indoor and Mobile Radio Communications*, IEEE, 2006, pp. 1–5.

[4]   N. Czink, R. Tian, S. Wyne, F. Tufvesson, J.-P. Nuutinen, J. Ylitalo, E. Bonek and A. F. Molisch, "Tracking time-variant cluster parameters in MIMO channel measurements," *Second International Conference on Communications and Networking in China, 2007. CHINACOM'07*, IEEE, 2007, pp. 1147–1151.

[5]   R. He, W. Chen, B. Ai, A. F. Molisch, W. Wang, Z. Zhong, J. Yu and S. Sangodoyin, "On the clustering of radio channel impulse responses using sparsity-based methods," *IEEE Transactions on Antennas and Propagation*, vol. 64, no. 6, pp. 2465–2474, 2016.

[6]   R. He, W. Chen, B. Ai, A. F. Molisch, W. Wang, Z. Zhong, J. Yu and S. Sangodoyin, "A sparsity-based clustering framework for radio channel impulse responses," *Proceedings of IEEE VTC*, IEEE, 2016, pp. 1–5.

[7]   R. He, Q. Li, B. Ai, Y. Geng, A. F. Molisch, K. Vinod, Z. Zhong and J. Yu, "An automatic clustering algorithm for multipath components based on kernel-power-density," *Proceedings of IEEE WCNC*, Mar. 2017, pp. 1–6.

[8]   R. He, Q. Li, B. Ai, Y. L.-A. Geng, A. F. Molisch, K. Vinod, Z. Zhong and J. Yu, "A kernel-power-density based algorithm for channel multipath components clustering," *IEEE Transactions on Wireless Communications*, vol. 16, no. 11, pp. 7138–7151, 2017.

[9]   M. Gan, Z. Xu, C. F. Mecklenbräuker and T. Zemen, "Cluster lifetime characterization for vehicular communication channels," *2015 9th European Conference on Antennas and Propagation (EuCAP)*, IEEE, 2015, pp. 1–5.

[10] C. Gustafson, K. Haneda, S. Wyne and F. Tufvesson, "On mm-wave multipath clustering and channel modeling," *IEEE Transactions on Antennas and Propagation*, vol. 62, no. 3, pp. 1445–1455, 2014.

[11] R. He, B. Ai, A. F. Molisch, G. L. Stuber, Q. Li, Z. Zhong and J. Yu, "Clustering enabled wireless channel modeling using big data algorithms," *IEEE Communications Magazine*, vol. 56, no. 5, pp. 177–183, May 2018.

[12] Y. Li, R. He, S. Lin, K. Guan, D. He, Q. Wang and Z. Zhong, "Cluster-based nonstationary channel modeling for vehicle-to-vehicle communications," *IEEE Antennas and Wireless Propagation Letters*, vol. 16, pp. 408–411, 2017.

[13] L. Liu, C. Oestges, J. Poutanen, K. Haneda, P. Vainikainen, F. Quitin, F. Tufvesson and P. De-Doncker, "The COST 2100 MIMO channel model," *IEEE Wireless Communications*, vol. 19, no. 6, pp. 92–99, 2012.

[14] T. Santos, J. Karedal, P. Almers, F. Tufvesson and A. F. Molisch, "Modeling the ultra-wideband outdoor channel: Measurements and parameter extraction method," *IEEE Transactions on Wireless Communications*, vol. 9, no. 1, pp. 282–290, 2010.

[15] M. Steinbauer, H. Ozcelik, H. Hofstetter, C. F. Mecklenbrauker and E. Bonek, "How to quantify multipath separation," *IEICE Transactions on Electronics*, vol. 85, no. 3, pp. 552–557, 2002.

[16] S. Mota, F. Perez-Fontan and A. Rocha, "Estimation of the number of clusters in multipath radio channel data sets," *IEEE Transactions on Antennas and Propagation*, vol. 61, no. 5, pp. 2879–2883, 2013.

[17] S. Sangodoyin, V. Kristem, C. Bas, M. Käske, J. Lee, C. Schneider, G. Sommerkorn, J. Zhang, R. Thomä and A. F. Molisch, "Cluster-based analysis of 3D MIMO channel measurement in an urban environment," *Military Communications Conference, MILCOM 2015-2015 IEEE*, IEEE, 2015, pp. 744–749.

[18] A. A. M. Saleh and R. Valenzuela, "A statistical model for indoor multipath propagation," *IEEE Journal on Selected Areas in Communications*, vol. 5, pp. 128–137, Feb. 1987.

[19] S. Sangodoyin, R. He, A. F. Molisch, V. Kristem and F. Tufvesson, "Ultrawideband MIMO channel measurements and modeling in a warehouse environment," *2015 IEEE International Conference on Communications (ICC), 2015 IEEE International Conference on*, IEEE, 2015, pp. 2277–2282.

[20] S. Sangodoyin, V. Kristem, A. F. Molisch, R. He, F. Tufvesson and H. M. Behairy, "Statistical modeling of ultrawideband MIMO propagation channel in a warehouse environment," *IEEE Transactions on Antennas and Propagation*, vol. 64, no. 9, pp. 4049–4063, 2016.

[21] 21D. S. Baum, J. Hansen and J. Salo, "An interim channel model for beyond-3G systems: Extending the 3GPP spatial channel model (SCM)," *Vehicular Technology Conference, 2005. VTC 2005-Spring. 2005 IEEE 61st*, vol. 5, IEEE, 2005, pp. 3132–3136.

[22] Q. H. Spencer, B. D. Jeffs, M. A. Jensen and A. L. Swindlehurst, "Modeling the statistical time and angle of arrival characteristics of an indoor multipath channel," *IEEE Journal on Selected Areas in Communications*, vol. 18, no. 3, pp. 347–360, 2000.

[23] M. Ester, H.-P. Kriegel, J. Sander and X. Xu, "A density-based algorithm for discovering clusters in large spatial databases with noise," *Kdd*, vol. 96, pp. 226–231, 1996.

[24] B. Larsen and C. Aone, "Fast and effective text mining using linear-time document clustering," *Proceedings of the Fifth ACM SIGKDD International Conference on Knowledge Discovery and Data Mining*, ACM, 1999, pp. 16–22.

[25] T. S. Rappaport, G. R. MacCartney, M. K. Samimi and S. Sun, "Wideband millimeter-wave propagation measurements and channel models for future wireless communication system design (invited paper)," *IEEE Transactions on Communications*, vol. 63, pp. 3029–3056, Sept. 2015.

[26] T. S. Rappaport, S. Sun, R. Mayzus, H. Zhao, Y. Azar, K. Wang, G. N. Wong, J. K. Schulz, M. Samimi and F. Gutierrez, "Millimeter wave mobile communications for 5G cellular: It will work!," *IEEE Access*, vol. 1, pp. 335–349, May 2013.

[27] M. K. Samimi and T. S. Rappaport, "3-D millimeter-wave statistical channel model for 5G wireless system design," *IEEE Transactions on Microwave Theory and Techniques*, vol. 64, pp. 2207–2225, Jul. 2016.

[28] S. Sun, G. R. MacCartney and T. S. Rappaport, "A novel millimeter-wave channel simulator and applications for 5G wireless communications," *Proceedings of IEEE ICC*, May 2017, pp. 1–7.

[29] S. Ju, O. Kanhere, Y. Xing and T. S. Rappaport, "A millimeter wave channel simulator NYUSIM with spatial consistency and human blockage," *IEEE Global Communications Conference (GLOBECOM)*, Dec. 2019.

[30] G. R. MacCartney, T. S. Rappaport, M. K. Samimi and S. Sun, "Millimeter-wave omnidirectional path loss data for small cell 5G channel modeling," *IEEE Access*, vol. 3, pp. 1573–1580, 2015.

[31] S. Sun, G. R. MacCartney, M. K. Samimi and T. S. Rappaport, "Synthesizing omnidirectional antenna patterns, received power and path loss from directional antennas for 5G millimeter-wave communications," *2015 IEEE Global Communications Conference (GLOBECOM)*, Dec. 2015, pp. 1–7.

[32] S. Ju, Y. Xing, O. Kanhere and T. S. Rappaport, "Millimeter wave and sub-terahertz spatial statistical channel model for an indoor office building," *IEEE Journal on Selected Areas in Communications*, vol. 39, no. 6, pp. 1561–1575, June 2021.

[33] T. S. Rappaport, R. W. Heath, R. C. Daniels and J. N. Murdock, *Millimeter Wave Wireless Communications*. Pearson/Prentice-Hall: Upper Saddle River, NJ, 2015.

[34] A. Yacob, *Clustering of multipath parameters without predefining the number of clusters*, Techn. Univ., Masterarbeit–Ilmenau, (Supervisor C. Schneider), 2015.

[35] S. L. Chiu, "A cluster extension method with extension to fuzzy model identification," *Proceedings of 1994 IEEE 3rd International Fuzzy Systems Conference,* vol. 2, June 1994, pp. 3132–3136.

[36] M.-S. Yang and K.-L. Wu, "A modified mountain clustering algorithm," *Pattern Analysis & Applications*, vol. 8, pp. 125–138, Sept. 2005.

[37] M. Ibraheam, *Clustering of multipath parameters based on multi variate Gaussian-mixture models and alternative approaches in real and model-based multipath environments*. Techn. Univ., Masterarbeit–Ilmenau (Supervisor C. Schneider), 2013.

[38] C. Schneider, M. Ibraheam, S. Häfner, M. Käske, M. Hein and R. Thomä, "On the reliability of multipath cluster estimation in realistic channel data sets," *8th European Conference on Antennas and Propagation (EuCAP)*, Apr. 2014.

[39] G. Sommerkorn, M. Käske, C. Schneider, S. Häfner and R. Thomä, "Full 3D MIMO channel sounding and characterization in an urban macro cell," *2014 XXXIth URSI General Assembly and Scientific Symposium (URSI GASS 2014)*, Aug. 2014.

[40] C. Schneider, M. Bauer, M. Narandzic, W. T. Kotterman and R. S. Thoma, "Clustering of MIMO channel parameters: performance comparison," *IEEE 69th Vehicular Technology Conference, 2009. VTC Spring 2009*, IEEE, 2009, pp. 1–5.

[41] M. K. Samimi and T. S. Rappaport, "3-D statistical channel model for millimeter-wave outdoor mobile broadband communications," *2015 IEEE International Conference on Communications (ICC)*, IEEE, 2015, pp. 2430–2436.

[42] 3GPP TSG RAN WG1 R1-163115, "A hierarchical agglomerative clustering algorithm for channel modelling," technical report, 3GPP, 2016.

[43] R. He, O. Renaudin, V. Kolmonen, K. Haneda, Z. Zhong, B. Ai and C. Oestges, "A dynamic wideband directional channel model for vehicle-to-vehicle communications," *IEEE Transactions on Industrial Electronics*, vol. 62, no. 12, pp. 7870–7882, 2015.

[44] R. He, O. Renaudin, V.-M. Kolmonen, K. Haneda, Z. Zhong, B. Ai and C. Oestges, "Statistical characterization of dynamic multi-path components for vehicle-to-vehicle radio channels," in *Proceedings of the IEEE VTC'15*, 2015, pp. 1–6.

[45] R. He, O. Renaudin, V. Kolmonen, K. Haneda, Z. Zhong, B. Ai and C. Oestges, "Characterization of quasi-stationarity regions for vehicle-to-vehicle radio channels," *IEEE Transactions on Antennas and Propagation*, vol. 63, no. 5, pp. 2237–2251, 2015.

[46] R. He, O. Renaudin, V.-M. Kolmonen, K. Haneda, Z. Zhong, B. Ai and C. Oestges, "Non-stationarity characterization for vehicle-to-vehicle channels using correlation matrix distance and shadow fading correlation," in *Proceedings of the 35th Progress in Electromagnetics Research Symposium*, 2014, pp. 1–5.

[47] A. Ispas, C. Schneider, G. Ascheid and R. Thomä, "Analysis of the local quasi-stationarity of measured dual-polarized MIMO channels," *IEEE Transactions on Vehicular Technology*, vol. 64, no. 8, pp. 3481–3493, 2015.

[48] J. Karedal, F. Tufvesson, N. Czink, A. Paier, C. Dumard, T. Zemen, C. F. Mecklenbrauker and A. F. Molisch, "A geometry-based stochastic MIMO model for vehicle-to-vehicle communications," *IEEE Transactions on Wireless Communications*, vol. 8, no. 7, pp. 3646–3657, 2009.

[49] F. Luan, A. F. Molisch, L. Xiao, F. Tufvesson and S. Zhou, "Geometrical cluster-based scatterer detection method with the movement of mobile terminal," *2015 IEEE 81st Vehicular Technology Conference (VTC Spring)*, IEEE, 2015, pp. 1–6.

[50] C. Huang, R. He, Z. Zhong, Y. L.-A. Geng, Q. Li and Z. Zhong, "A novel tracking based multipath component clustering algorithm," *IEEE Antennas and Wireless Propagation Letters*, vol. 16, no. 1, pp. 2679–2683, 2017.

[51] 3GPP, "Evolved universal terrestrial radio access (E-UTRA) physical channels and modulation," release 10 3GPP TS 36.211, 3rd Generation Partnership Project (3GPP).

[52] A. Alkhateeb, G. Leus and R. W. Heath, "Limited feedback hybrid precoding for multiuser millimeter wave systems," *IEEE Transactions on Wireless Communications*, vol. 14, pp. 6481–6494, Nov. 2015.

[53] M. Iwanow, N. Vucic, M. H. Castaneda, J. Luo, W. Xu and W. Utschick, "Some aspects on hybrid wideband transceiver design for mmwave communication systems," *WSA 2016; 20th International ITG Workshop on Smart Antennas*, Mar. 2016, pp. 1–8.

[54] 3GPP, "Spatial channel model for mimo simulations," technical report, 2003.

[55] P. Kyosti, J. Meinila, L. Hentila, X. Zhao, T. Jamsa, C. Schneider, M. Narandzic, M. Milojevic, A. Hong, J. Ylitalo, V.-M. Holappa, M. Alatossava, R. Bultitude, Y. de-Jong and T. Rautiainen, "Ist-4-027756 winner ii deliverable 1.1.2. v.1.2, winner ii channel models," technical report, ISTWINNERII, 2007.

[56] M. V. Clark and L. J. Greenstein, "The relationship between fading and bandwidth for multipath channels," *IEEE Transactions on Wireless Communications*, vol. 4, pp. 1372–1376, July 2005.

[57] W. Q. Malik, B. Allen and D. J. Edwards, "Bandwidth-dependent modelling of smallscale fade depth in wireless channels," *IET Microwaves, Antennas Propagation*, vol. 2, pp. 519–528, Sept. 2008.

[58] A. F. Molisch, "Ultra-wide-band propagation channels," *Proceedings of the IEEE*, vol. 97, pp. 353–371, Feb. 2009.

[59] N. Iqbal, C. Schneider, J. Luo, D. Dupleich, R. Müller, S. Haefner and R. S. Thomä, "On the stochastic and deterministic behavior of mmwave channels," *The 11th European Conference on Antennas and Propagation (EuCAP 2017)*, Mar. 2017.

[60] N. Iqbal, C. Schneider, J. Luo, D. Dupleich, R. Müller and R. S. Thomä, "Modeling of directional fading channels for millimeter wave systems," *IEEE 86th Vehicular Technology Conference (VTC-Fall)*, 2017.

[61] D. Dupleich, N. Iqbal, C. Schneider, S. Häfner, R. Müller, S. Skoblikov, J. Luo, G. Del Galdo and R. Thomä, "Influence of system aspects on fading at mmwaves," *IET Microwaves, Antennas and Propagation*, vol. 12, no. 4, pp. 516–524, Feb. 2018.

[62] B. M. Hochwald, T. L. Marzetta and V. Tarokh, "Multiple-antenna channel hardening and its implications for rate feedback and scheduling," *IEEE Transactions on Information Theory*, vol. 50, pp. 1893–1909, Sept. 2004.

[63] D. Dupleich, N. Iqbal, C. Schneider, S. Haefner, R. Muller, S. Skoblikov, J. Luo and R. Thoma, "Investigations on fading scaling with bandwidth and directivity at 60 GHz," *2017 11th European Conference on Antennas and Propagation (EUCAP)*, Mar. 2017, pp. 3375–3379.

[64] M. Pätzold, *Mobile Radio Channels*. Wiley: Chichester, 2011.

[65] N. Iqbal, J. Luo, C. Schneider, D. Dupleich, R. Müller, S. Haefner and R. S. Thomä, "Stochastic/deterministic behavior of cross polarization discrimination in mmwave channels," *IEEE ICC Wireless Communications Symposium (ICC'17 WCS)*, May 2017.

[66] W. Q. Malik, "Polarimetric characterization of ultrawideband propagation channels," *IEEE Transactions on Antennas and Propagation*, vol. 56, pp. 532–539, Feb. 2008.

8 Dispersion Characteristics

Alenka Zajić, David Matolak, Theodore S. Rappaport,
George MacCartney, Jr., Yunchou Xing, Shu Sun, Alexander Maltsev,
Andrey Pudeyev, Aki Hekkala, Pekka Kyosti, Anmol Bhardwaj,
Katsuyuki Haneda, Usman Virk, Camillo Gentile, Kate A. Remley
and Andreas F. Molisch

8.1 Introduction

When designing a wireless communication system, it is essential to have a channel model that can quickly and accurately generate the channel impulse response (CIR) needed for system simulations. With this objective in mind, researchers have proposed various channel models for wireless communications [1–3]. These models are often classified as physical or analytical channel models. The physical channel models are further classified as deterministic and stochastic models [4].

Deterministic models such as ray-tracing (RT) offer an accurate model of the propagation environment, but their high computational complexity prohibits the intensive link or system-level simulations required during system design. Hence, the models with lower computational complexity that could emulate a large class of radio-propagation environments are preferred. These requirements have led to stochastic channel models, which are often classified into geometry-based stochastic models (GSCMs) and nongeometrical stochastic models. In this chapter we focus on the GSCM models such as those in [5–10].

In contrast to deterministic models, the GSCM models the physical parameters of plane waves without directly relating them to any particular (or very detailed) radio environment. To achieve that, the channel realizations are determined as realizations of a multidimensional random process that attributes multipath plane waves to physical but imaginary distribution of electromagnetic field scatterers.

The stochastic generation of plane wave parameters can be done in several different ways. Here we distinguish two classes of GSCM models. The first class describes generation of multipath components (MPCs) in the parametric domain, that is, the MPCs are not related to particular scatterers [6, 10, 11]. The second class describes generation of MPCs according to the interaction with scattering objects during the physical model synthesis [12–14].

Typical representatives of the first class of the GSCMs are the 3GPP spatial-channel-model [10], the channel model developed in the WINNER project [6], and the reference model for evaluation of IMT-Advanced radio interface technologies [7]. In this channel modeling approach, the propagation channel is characterized by statistical parameters obtained from the radio channel measurements. This gives a possibility of using the same framework of the model for the simulations in different frequencies and the different number or type of antennas. These GSMs are popular because of their scalability and reasonably low complexity [15].

The typical representative of the second class of the GSCMs is the COST 259 model [12, 16, 17]. In this channel modeling approach, the scattering objects are placed on a particular geometrical structure such as a circle, ellipsoid, etc. or in a 2D/3D coordinate system such as a plane or volume, and their abstraction is performed in the form of multipath clusters. By assigning visibility regions to each of the clusters, a simplified RT engine can be obtained. The randomness in this approach is attained by random selection of visibility regions and the intra-cluster structure. In the remainder of this chapter we focus on this type of GSCM.

8.2   Review of Theory of Geometry-Based Modeling of Stochastic Channel Models

8.2.1   Geometry-Based Modeling of Frequency Flat Multipath Fading Channels

Clarke [18] was the first to propose a channel model for non-line-of-sight (NLoS) frequency flat multipath fading channels based on the statistical characteristics of the electromagnetic fields of the received signal. The model assumes a fixed transmitter with a vertically polarized antenna. The field incident on the mobile antenna is assumed to consist of N azimuth plane waves with arbitrary carrier phases, arbitrary azimuth angles of arrival, and all waves having equal average amplitudes. It should be noted that the equal average amplitude assumption is based on the fact that, in the absence of a direct line-of-sight (LoS) path, the scattered components arriving at a receiver will experience similar attenuation over small-scale distances.

Figure 8.1 illustrates a plane wave incident on a receiver traveling with a velocity v in the x-direction. The angle of arrival is measured in the $x-y$ plane with respect

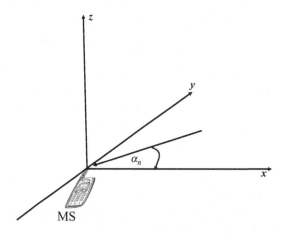

Figure 8.1 Plane waves arriving at the receiver at random angles α_n.

to the direction of motion. Every wave that is incident on the receiver undergoes a Doppler shift due to the motion of the receiver and arrives at the receiver at the same time. Note that no variation in delay due to multipath is assumed in this model – this is a consequence of the frequency flat fading assumption. For the nth wave arriving at an angle α_n to the x-axis, the Doppler shift in hertz is given by

$$v_n = \frac{v}{\lambda} \cos \alpha_n = v_d \cos \alpha_n, \tag{8.1}$$

where v_d is the maximum Doppler frequency in hertz, λ is the carrier wavelength and v is the speed of the receiver. The angle of arrival, α_n, depends on the scattering environment and the antenna radiation pattern.

The vertically polarized plane waves arriving at the mobile receiver have electric and magnetic field components given by

$$E_z = E_0 \sum_{n=1}^{N} C_n \cos(2\pi f_c t + 2\pi v_n t + \phi_n), \tag{8.2}$$

$$H_x = -\frac{E_0}{\eta} \sum_{n=1}^{N} C_n \sin \alpha_n \cos(2\pi f_c t + 2\pi v_n t + \phi_n), \tag{8.3}$$

$$H_y = -\frac{E_0}{\eta} \sum_{n=1}^{N} C_n \cos \alpha_n \cos(2\pi f_c t + 2\pi v_n t + \phi_n), \tag{8.4}$$

where E_0 is the real amplitude of the local average E-field (E_0 is assumed to be constant), C_n is a real random variable representing the amplitude of individual waves, η is the intrinsic impedance of free space (377 Ω) and f_c is the carrier frequency. The random phase of the nth arriving component is denoted by ϕ_n.

The amplitudes of the E- and H-fields are normalized such that the ensemble average of the C_ns is given by

$$\sum_{n=1}^{N} E[C_n^2] = 1. \tag{8.5}$$

Since the Doppler shift is very small when compared to the carrier frequency, the three field components may be modeled as narrowband random processes with mutually independent random variables C_n, α_n and ϕ_n. If N is sufficiently large, the three components E_z, H_x and H_y can be approximated as Gaussian random variables (according to the central limit theorem [19]). The phase angles ϕ_n are assumed to have a uniform probability density function (pdf) on the interval $(0, 2\pi]$. Based on the analysis in [13], the E-field can be expressed in an in-phase and quadrature form as

$$E_z = h_i(t) \cos(2\pi f_c t) - h_q(t) \sin(2\pi f_c t), \tag{8.6}$$

where

$$h_i(t) = E_0 \sum_{n=1}^{N} C_n \cos(2\pi v_n t + \phi_n), \tag{8.7}$$

$$h_q(t) = E_0 \sum_{n=1}^{N} C_n \sin(2\pi v_n t + \phi_n). \tag{8.8}$$

Both $h_i(t)$ and $h_q(t)$ are Gaussian random processes which are denoted as h_i and h_q, respectively, at any time t. Then, h_i and h_q are uncorrelated zero-mean Gaussian random variables with an equal variance given by

$$E[h_i^2] = E[h_q^2] = E[E_z^2] = E_0^2/2, \tag{8.9}$$

where the $E[\cdot]$ denotes the ensemble average.

The envelope of the received E-field is given by

$$|E_z(t)| = \sqrt{h_i^2 + h_q^2} = x(t), \tag{8.10}$$

and has a Rayleigh distribution with the time-average power of the received signal equal to $\sigma^2 = E_0^2/2$.

The MPC is generally modeled as a linear time-variant filter with low-pass impulse response, as shown in eq. (8.6):

$$h(t, \tau) = \sum_{n=1}^{N_p} \tilde{a}_n(t)\delta(\tau - \tau_n), \tag{8.11}$$

where $h(t, \tau)$ is the channel response at time t due to an impulse applied at time $t - \tau$, and $\delta(\cdot)$ is the Dirac delta function. Clarke's model can be related to this general channel model by characterizing flat frequency fading as the transmission of an unmodulated carrier. Then, the MPC has the *complex baseband impulse response*

$$h(t, \tau) = (h_i(t) + jh_q(t))\delta(\tau - \hat{\tau}) = E_0 \sum_{n=1}^{N_p} C_n e^{j(2\pi v_n t + \phi_n)}\delta(\tau - \hat{\tau}), \tag{8.12}$$

where all time delays τ_n are equal to $\hat{\tau}$ due to flat frequency fading.

In the presence of a specular component, (i.e., LoS or a strong reflected path), the complex baseband impulse response becomes a superposition of a strong specular component and scattered components in eq. (8.12). To account for a specular component, Clarke's model needs to be modified as follows:

$$h(t) = \frac{E_0\sqrt{K}}{\sqrt{K+1}}e^{j(2\pi v_d t \cos \alpha_0)} + \frac{E_0}{\sqrt{K+1}} \sum_{n=1}^{N} C_n e^{j(2\pi v_n t + \phi_n)}, \tag{8.13}$$

where K is the ratio of the received specular to scattered power (i.e., Rician factor), and α_0 is the angle of arrival of the specular component. The Rician distribution is named after S.O. Rice of Bell Laboratories fame named after S.O. Rice of Bell Laboratories, who, determined the distribution for the case of noise with a deterministic sinusoid tone added to it, and who first defined the K factor as describing the level of the nonfading (specular) signal to the random scattered components [2].

8.2.2 Correlation Functions and Doppler Power Spectra for Isotropic Scattering Channels

The correlation function and the Doppler spectrum are important statistics that characterize temporal selectivity of a communication system. These statistics enable the system designer to make informed decisions when choosing modulation, interleaving and coding schemes at the transmitting end and the type of channel estimator and decoder at the receiving end.

Assuming that the received pass-band signal $r(t)$ is wide-sense stationary (WSS),[1] the autocorrelation of $r(t)$ is

$$
\begin{aligned}
R_{rr}(\tau) &= \mathrm{E}[r(t)r(t+\tau)] \\
&= \mathrm{E}[h_i(t)h_i(t+\tau)]\cos(2\pi f_c \tau) - \mathrm{E}[h_q(t)h_q(t+\tau)]\sin(2\pi f_c \tau) \\
&= R_{h_i h_i}(\tau)\cos(2\pi f_c \tau) - R_{h_q h_q}(\tau)\sin(2\pi f_c \tau),
\end{aligned}
\tag{8.14}
$$

where

$$
R_{h_i h_i}(\tau) = R_{h_q h_q}(\tau) \tag{8.15}
$$

$$
R_{h_i h_q}(\tau) = -R_{h_q h_i}(\tau). \tag{8.16}
$$

It is reasonable to assume that the phases ϕ_n and ϕ_m are independent for $m \neq n$ since their associated delays and Doppler shifts are independent. Furthermore, since the phases are uniformly distributed over $[0, 2\pi)$, the autocorrelation $R_{h_i h_i}(\tau)$ can be obtained as follows:

$$
R_{h_i h_i}(\tau) = \mathrm{E}_{\tau, \alpha_n}[h_i(t)h_i(t+\tau)] = \frac{E_0^2}{2}\mathrm{E}_{\alpha_n}[\cos(2\pi v_d \tau \cos \alpha_n)]. \tag{8.17}
$$

Similarly, the cross-correlation $R_{h_i h_q}(\tau)$ is

$$
R_{h_i h_q}(\tau) = \mathrm{E}_{\tau, \alpha_n}[h_i(t)h_q(t+\tau)] = \frac{E_0^2}{2}\mathrm{E}_{\alpha_n}[\sin(2\pi v_d \tau \cos \alpha_n)]. \tag{8.18}
$$

To evaluate the expectations in eqs. (8.17) and (8.18), statistical channel models assume a particular distribution of incident power on the receiver antenna, $p(\alpha)$. Clarke's 2D *isotropic scattering* model assumes that the plane waves propagate in the x–y plane and arrive at the receiver from all directions with equal probability, that is, $p(\alpha) = 1/(2\pi)$, for $\alpha \in [0, 2\pi)$, with an isotropic receiver antenna with gain $G(\alpha) = 1$. Then, the expectation in eq. (8.17) becomes:

$$
R_{h_i h_i}(\tau) = \frac{E_0^2}{2}\int_{-\pi}^{\pi}\cos(2\pi v_d \tau \cos \alpha)p(\alpha)G(\alpha)d\alpha = \frac{E_0^2}{2}J_0(2\pi v_d \tau), \tag{8.19}
$$

where $J_0(\cdot)$ is the zeroth-order Bessel function of the first kind.

---

[1] See Section 8.2.5 for details on WSS.

Similarly, the cross-correlation function for 2D isotropic scattering and an isotropic antenna with $G(\alpha) = 1$ becomes

$$R_{h_i h_q}(\tau) = \frac{E_0^2}{2} \frac{1}{2\pi} \int_{-\pi}^{\pi} \sin(2\pi v_d \tau \cos \alpha) d\alpha = 0. \tag{8.20}$$

This result implies that $h_i(t)$ and $h_q(t)$ are uncorrelated functions. Furthermore, because $h_i(t)$ and $h_q(t)$ are Gaussian random variables, it means that they are also independent random processes. However, note that the fact that $h_i(t)$ and $h_q(t)$ are independent is a direct result of the symmetry of the 2D isotropic scattering environment and isotropic antenna gain pattern. In case of a nonisotropic scattering environment, $h_i(t)$ and $h_q(t)$ may not be independent random processes.

The autocorrelation of the complex baseband impulse response $h(t) = h_i(t) + jh_q(t)$ can be written as

$$R_{hh}(\tau) = \frac{1}{2}E[h^*(t)h(t+\tau)] = R_{h_i h_i}(\tau) + jR_{h_i h_q}(\tau). \tag{8.21}$$

The autocorrelation function of the complex baseband impulse response in the presence of both a strong specular component and a scatter component can be obtained by substituting eq. (8.13) into (8.21) and evaluating the expectation; the result is

$$R_{hh}(\tau) = \frac{1}{K+1} \frac{E_0^2}{2} J_0(2\pi v_d \tau) + \frac{K}{K+1} \frac{E_0^2}{2} \cos(2\pi v_d \tau \cos \alpha_0)$$

$$+ j\frac{K}{K+1} \frac{E_0^2}{2} \sin(2\pi v_d \tau \cos \alpha_0). \tag{8.22}$$

The power spectral density (PSD) of $h_i(t)$ and $h_q(t)$ is the Fourier transform of $R_{h_i h_i}(\tau)$ or $R_{h_q h_q}(\tau)$. For autocorrelation in eq. (8.19), the corresponding PSD is

$$S_{h_i h_i}(v) = \mathcal{F}[R_{h_i h_i}(\tau)] = \begin{cases} \frac{E_0^2}{2\pi v_d} \frac{1}{\sqrt{1-(v/v_d)^2}} & |v| \le v_d \\ 0 & \text{otherwise} \end{cases}. \tag{8.23}$$

The PSD of the complex impulse response $h(t) = h_i(t) + jh_q(t)$ is

$$S_{hh}(v) = S_{h_i h_i}(v) + jS_{h_i h_q}(v). \tag{8.24}$$

Often, $S_{hh}(v)$ is called the **Doppler power spectrum**. For 2D isotropic scattering and an isotropic antenna, $R_{h_i h_q}(\tau) = 0$ and the Doppler power spectrum is $S_{hh}(v) = S_{h_i h_i}(v)$ (which is real and even). If both a specular component and a scattered component are present, the Doppler power spectrum becomes

$$S_{hh}(v) = \begin{cases} \frac{1}{K+1} \frac{E_0^2}{2\pi v_d} \frac{1}{\sqrt{1-(v/v_d)^2}} + \frac{K}{K+1} \frac{E_0^2}{2} \delta(v - v_d \cos \alpha_0) & |v| \le v_d \\ 0 & \text{otherwise} \end{cases}. \tag{8.25}$$

The pass-band Doppler power spectrum of the received pass-band signal $r(t)$ can be obtained from the baseband Doppler power spectrum as follows:

$$S_{rr}(\nu) = \frac{1}{2}[S_{hh}(\nu - f_c) + S_{hh}(-\nu - f_c)]. \tag{8.26}$$

For 2D isotropic scattering and an isotropic antenna, the pass-band Doppler power spectrum becomes

$$S_{rr}(\nu) = \begin{cases} \dfrac{E_0^2}{4\pi\nu_d} \dfrac{1}{\sqrt{1-((\nu-f_c)/\nu_d)^2}} & |\nu - f_c| \leq \nu_d \\ 0 & \text{otherwise} \end{cases}. \tag{8.27}$$

8.2.3   Correlation Functions and Doppler Power Spectra for Nonisotropic Scattering Channels

One of the main assumptions in Clarke's channel model is that scattering is isotropic, that is, uniform distribution of angles of arrival of MPCs at the receiver. However, it has been argued [20–24], and experimentally demonstrated [25–34] that scattering encountered in many environments, such as suburban areas, urban areas between buildings, street canyons, highways, etc., is nonisotropic, resulting in a nonuniform probability density function (pdf) for angles of arrival (AoA) at the receiver. As has been discussed in [30], the assumption of a uniform pdf for the AoAs introduces small errors on the first-order statistics of the received signal, but a significant error on the second-order statistics, such as correlation functions and level crossing rates.

In the literature, several different scatterer distributions, such as quadratic pdf [35], cosine pdf [36], Laplace and Gaussian pdfs [37] and von Mises pdf [38], have been used to characterize nonuniform scattering environments. In addition to showing good fit to measurements [39], the von Mises pdf approximates many of the other distributions (e.g., uniform, sinusoid, and Gaussian) and, in contrast to most other distributions, leads to closed-form solutions for many useful statistics. Hence, this distribution is often used to characterize nonisotropic scattering [38].

The von Mises pdf was introduced by R. von Mises in 1918 to study the deviations of measured atomic weights from integral values [40]. This pdf plays a prominent role in statistical modeling and analysis of angular variables [36, pp. 57–68].

The von Mises pdf is defined as [40]

$$p(\alpha) \overset{\triangle}{=} \frac{1}{2\pi I_0(k)} \exp\left[k \cos(\alpha - \mu)\right], \tag{8.28}$$

where $\alpha \in [-\pi, \pi)$ represents AoAs, $I_0(\cdot)$ is the zeroth-order modified Bessel function of the first kind, $\mu \in [-\pi, \pi)$ is the mean angle at which the scatterers are distributed and k controls the spread of scatterers around the mean. When $k = 0$, $p(\alpha) = 1/(2\pi)$ is a uniform distribution yielding 2D isotropic scattering. As k increases, the scatterers become more clustered around angle μ and the scattering becomes increasingly nonisotropic. For small k, this function approximates the cardioid pdf [41, p. 60], which is similar to the cosine pdf [36], while for large k it resembles a

Gaussian pdf with mean μ and standard deviation $1/\sqrt{k}$ [41, p. 60]. In general, the von Mises pdf can approximate the wrapped Gaussian pdf [41, p. 66]. It is interesting to note that the von Mises pdf appears in a number of other applications in telecommunications. For example, this pdf is referred to as the Tikhonov pdf in partially coherent communications [42, p. 406] and has been used in phase-lock-loop-related problems [43]. It also has been shown that the phase of a sine wave with Gaussian noise for large signal to noise ratios has a von Mises pdf [44].

Assuming that the received bandpass signal $r(t)$ is WSS, the autocorrelation of $r(t)$ can be evaluated starting from eq. (8.14), as in Section 8.2.2. By assuming that the phases ϕ_n and ϕ_m are independent for $m \neq n$ and uniformly distributed over $[0, 2\pi)$, the autocorrelation $R_{h_i h_i}(\tau)$ and the cross-correlation $R_{h_i h_q}(\tau)$ can be written as in eqs. (8.17) and (8.18), respectively. To evaluate the expectations in eqs. (8.17) and (8.18), the nonisotropic scattering model assumes that the plane waves propagate in the $x-y$ plane and arrive at the receiver from all directions with nonequal probability, that is, the von Mises pdf describes the angles of arrival. To simplify calculations, the nonisotropic scattering model still assumes an isotropic receiver antenna with gain $G(\alpha) = 1$. Then, the autocorrelation function in eq. (8.17) becomes

$$R_{h_i h_i}(\tau) = \frac{E_0^2}{2} \Re \left\{ \frac{I_0 \left(\sqrt{k^2 - 4\pi^2 v_d^2 \tau^2 + j4\pi k v_d \tau \cos \mu} \right)}{I_0(k)} \right\}, \tag{8.29}$$

where $\Re\{\cdot\}$ denotes the real part operation. For $k = 0$, the von Mises pdf becomes the uniform pdf and the autocorrelation function in eq. (8.29) becomes identical to the autocorrelation function in eq. (8.17).

Similarly, the cross-correlation function for 2D nonisotropic scattering and an isotropic antenna with $G(\alpha) = 1$ is

$$R_{h_i h_q}(\tau) = \frac{E_0^2}{2} \Im \left\{ \frac{I_0 \left(\sqrt{k^2 - 4\pi^2 v_d^2 \tau^2 + j4\pi k v_d \tau \cos \mu} \right)}{I_0(k)} \right\}, \tag{8.30}$$

where $\Im\{\cdot\}$ denotes the imaginary part operation. Note that the cross-correlation function is not zero in nonisotropic scattering channels.

The autocorrelation of the complex impulse response $h(t) = h_i(t) + jh_q(t)$ can be obtained as

$$R_{hh}(\tau) = \frac{1}{2}E[h^*(t)h(t+\tau)] = R_{h_i h_i}(\tau) + j R_{h_i h_q}(\tau)$$

$$= \frac{E_0^2}{2} \frac{I_0 \left(\sqrt{k^2 - 4\pi^2 v_d^2 \tau^2 + j4\pi k v_d \tau \cos(\mu)} \right)}{I_0(k)}. \tag{8.31}$$

The autocorrelation function of the complex impulse response in the presence of both specular and scattered components can be obtained by substituting eq. (8.13) into (8.31) and evaluating the expectation; the result is [38]

$$R_{hh}(\tau) = \frac{1}{K+1} \frac{E_0^2}{2} \frac{I_0\left(\sqrt{k^2 - 4\pi^2 v_d^2 \tau^2 + j4\pi k v_d \tau \cos(\mu)}\right)}{I_0(k)}$$

$$+ \frac{K}{K+1} \frac{E_0^2}{2} \exp(j2\pi v_d \tau \cos(\alpha_0)). \tag{8.32}$$

The Doppler power spectrum of the complex impulse response $h(t) = h_i(t) + jh_q(t)$ can be obtained by calculating Fourier transform of the autocorrelation function of the complex impulse response in eq. (8.32). The result of this Fourier transform is [38]

$$S_{hh}(v) = \frac{1}{K+1} \frac{E_0^2}{2\pi v_d \sqrt{1 - (v/v_d)^2}} \frac{\exp(kv \cos\mu/v_d) \cosh(k \sin\mu \sqrt{1 - (v/v_d)^2})}{I_0(k)}$$

$$+ \frac{K}{K+1} \frac{E_0^2}{2} \delta(v - v_d \cos\alpha_0), |v| \le v_d, \tag{8.33}$$

where $\cosh(\cdot)$ is the hyperbolic cosine. For $k = 0$, this Doppler power spectrum reduces to eq. (8.25), that is, the Doppler power spectrum for Clarke's 2D isotropic scattering.

8.2.4   Geometry-Based Modeling of Selective Multipath Fading Channels

Up to this point we have considered channel models that are appropriate for narrowband transmission, where the inverse signal bandwidth is much greater than the time spread of the propagation path delays. Such a channel introduces very little or no distortion into the received signal and is said to exhibit flat fading. For digital communication systems, this means that the duration of a modulated symbol is much greater than the time spread of the propagation path delays. Under this condition, all frequencies in the transmitted signal will experience the same random attenuation and phase shift due to multipath fading.

In contrast, in frequency-selective fading channels, frequency components in the transmitted signal experience different phase shifts along the different paths. As the differences in path delays become larger, even closely separated frequencies in the transmitted signal can experience significantly different phase shifts. This type of propagation can be characterized by the geometrical channel model illustrated in Figure 8.2. Considering only single reflections, all scatterers that are associated with a particular path length are located on an ellipse, with the transmitter and receiver located at the foci. Different delays correspond to different co-focal ellipses. Frequency-selective channels have strong scatterers that are located on several ellipses that represent significantly different delays (compared to a symbol duration). This means that if the bandwidth of the signal is W, we can differentiate between signals arriving at least $\Delta\tau = 1/W$ s apart. As we move outward, energy reflected from successive ellipses takes longer to get to the receiver. In this way, at the receiver, we observe multiple copies of the transmitted signal, each copy taking progressively longer to reach the receiver. This model does not consider second- or higher-order

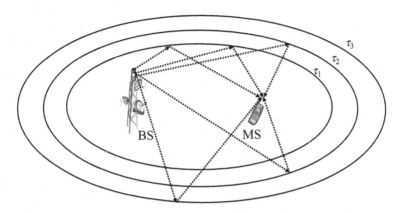

Figure 8.2 Geometrical model for frequency-selective multipath fading channels.

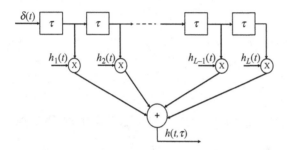

Figure 8.3 Frequency-selective multipath fading channel model with discrete MPCs.

reflections, where energy is reflected between scatterers. Indeed, most RT models (such as Jake's model) are *single-bounced scattering* models – they assume that most of the scattered energy arriving at the receiver is due to single-bounced scattering, though [45] and the COST 273/2100 models consider multiple-scattering processes [45–47].

This geometrical model can be described using a tapped-delay line with number of taps at different delays. Each tap is the result of a large number of MPCs and, therefore, the taps will experience multipath fading. If we use $\hat{s}(t)$ to denote the complex envelope of the transmitted signal, the complex envelope of the received signal can be written as

$$\hat{r}(t) = \sum_{l=1}^{L} h_l(t)\hat{s}(t - \tau_l), \tag{8.34}$$

where L is the number of taps, and $h_l(t)$ and τ_l are the complex gains and path delays associated with each tap. Sometimes it is convenient if the tap delays are multiples of some small number τ, leading to the τ-*spaced tapped delay line channel model* shown in Figure 8.3. Many of the tap coefficients in the tapped delay line are zero, that is, no energy is received at these delays. The time-variant channel tap coefficients $\{h_l(t)\}$ can be generated using the channel models described in Section 8.2.1.

In a typical digital communication system, data symbols are sent to the transmitter every T seconds and T-spaced samples are taken at the output of the receiver, where T is the band duration. Hence, it is desirable to have a channel model with T-spaced impulse response. The *T-spaced channel model* is similar to the τ-spaced channel model, except that the channel taps are T-spaced. This discrete system model is by far the most popular in the literature (e.g., [19]). This is largely due to the prevalence of digital hardware for measuring and/or computing the CIR. Most channel sounding devices operate in the digital domain and thus sample the PDP at discrete intervals. Each delay bin is the sum of all energy received over the duration of that bin. This is analogous to discrete frequency spectra, in which each frequency bin is the sum of all energy over the bandwidth that bin occupies. Details on how the τ-spaced channel model can be converted into a T-spaced channel model can be found in [19].

8.2.5 Correlation Functions for F-to-M Frequency-Selective Multipath Fading Channels

For frequency-selective fading, multipath-fading channels are modeled as *time-varying* in addition to being time-dispersive. The *input delay-spread* function $h(t, \tau)$ is defined as the response of the channel at time t to a unit impulse at time $t - \tau$. This function is useful to analyze how rapidly the channel varies in time and quantify how dispersive the channel is.

As defined in eq. (8.6), if we take the Fourier transform of $h(t, \tau)$ with respect to time t, we can quantify how much the channel varies over time (delay), that is,

$$S(v, \tau) = \mathcal{F}_t[h(t, \tau)] = \int_{-\infty}^{\infty} h(t, \tau) e^{-j2\pi v t} dt, \qquad (8.35)$$

where v is the Doppler frequency in hertz.

On the other hand, if we take the Fourier transform of $h(t, \tau)$ with respect to delay τ, we can quantify how dispersive the channel is at a given time. Analytically, the *time-variant transfer function* $H(t, f)$ is defined as

$$H(t, f) = \mathcal{F}_\tau[h(t, \tau)] = \int_{-\infty}^{\infty} h(t, \tau) e^{-j2\pi f \tau} d\tau. \qquad (8.36)$$

Finally, if we take the double Fourier transform of both time and delay simultaneously, the *Doppler variant transfer function* is obtained as

$$B(v, f) = \mathcal{F}_{t,\tau}[h(t, \tau)] = \int_{-\infty}^{\infty} \int_{-\infty}^{\infty} h(t, \tau) e^{-j2\pi v t} e^{-j2\pi f \tau} dt d\tau. \qquad (8.37)$$

Here, we note that the CIR $h(t, \tau) = h_i(t, \tau) + jh_q(t, \tau)$ can be modeled as a complex random process, where the in-phase and quadrature components, that is, $h_i(t, \tau)$ and $h_q(t, \tau)$, are correlated random processes. Hence, all of the transmission functions defined here are random processes. A thorough characterization of a channel requires knowledge of the joint pdf of all the transmission functions. However, this is a difficult task and a more reasonable approach is to obtain statistical correlation functions for the individual transmission functions. If the underlying process is Gaussian, then a complete statistical description is provided by the means and autocorrelation functions. Here, we assume zero-mean Gaussian random processes so that only the autocorrelation functions are of interest. Since there are four transmission functions, there are four autocorrelation functions that can be defined as follows [19, 48]:

$$R_h(t,s;\tau,\eta) = \frac{1}{2}E[h(t,\tau)h^*(s,\eta)], \tag{8.38}$$

$$R_S(\tau,\eta;v,\mu) = \frac{1}{2}E[S(\tau,v)S^*(\eta,\mu)], \tag{8.39}$$

$$R_H(f,m;t,s) = \frac{1}{2}E[H(f,t)H^*(m,s)] \tag{8.40}$$

$$R_B(f,m;v,\mu) = \frac{1}{2}E[B(f,v)B^*(m,\mu)], \tag{8.41}$$

where $(\cdot)^*$ denotes the complex conjugate operation.

The channel is said to be **WSS** if the fading statistics remain constant over short periods of time. This implies that the channel correlation functions depend on the time variables t and s only through the time difference $\Delta t = s - t$. Hence, for WSS channels, the correlation functions become

$$R_h(t,t+\Delta t;\tau,\eta) = R_h(\Delta t;\tau,\eta), \tag{8.42}$$

$$R_S(\tau,\eta;v,\mu) = \psi_S(\tau,\eta;v)\delta(v-\mu), \tag{8.43}$$

$$R_H(f,m;t,t+\Delta t) = R_H(f,m;\Delta t), \tag{8.44}$$

$$R_B(f,m;v,\mu) = \psi_B(f,m;v)\delta(v-\mu), \tag{8.45}$$

where

$$\psi_S(\tau,\eta;v) = \int_{-\infty}^{\infty} R_h(\Delta t;\tau,\eta)e^{-j2\pi v\Delta t}d\Delta t, \tag{8.46}$$

$$\psi_B(f,m;v) = \int_{-\infty}^{\infty} R_H(f,m;\Delta t)e^{-j2\pi v\Delta t}d\Delta t, \tag{8.47}$$

are Fourier transform pairs. It can be shown that WSS channels give rise to scattering with uncorrelated Doppler shifts. This behavior suggests that signal components with different Doppler shifts have uncorrelated attenuations and phase shifts.

Please note that channel can be wide sense stationary uncorrelated scattering (WSSUS) without having Gaussian statistics. By adding the requirement for Gaussian statistics, we impose the Gaussian WSSUS model for which second-order statistics provide a complete description.

The channel is said to exhibit *uncorrelated scattering* (US), if the contributions from elemental scatterers corresponding to different delays are uncorrelated. This implies that the channel correlation functions depend on the frequency variables f and m only through the frequency difference $\Delta f = m - f$ [48]. Hence, for US channels, the correlation functions become

$$R_h(t,s;\tau,\eta) = \psi_h(t,s;\tau)\delta(\eta - \tau), \tag{8.48}$$

$$R_S(\tau,\eta;v,\mu) = \psi_S(\tau;v,\mu)\delta(\eta - \tau), \tag{8.49}$$

$$R_H(f,f+\Delta f;t,s) = R_H(\Delta f;t,s), \tag{8.50}$$

$$R_B(f,f+\Delta f;v,\mu) = R_B(\Delta f;v,\mu), \tag{8.51}$$

where

$$\psi_h(t,s;\tau) = \int_{-\infty}^{\infty} R_H(\Delta f;t,s)e^{j2\pi\Delta f\tau}d\Delta f, \tag{8.52}$$

$$\psi_S(\tau;v,\mu) = \int_{-\infty}^{\infty} R_B(\Delta f;v,\mu)e^{j2\pi\Delta f\tau}d\Delta f. \tag{8.53}$$

Wide-sense stationary uncorrelated scattering channels are a very special type of multipath-fading channel. These channels display US in both the time-delay and Doppler shift, and hence reduce to the following simple forms:

$$R_h(t,t+\Delta t;\tau,\eta) = \psi_h(\Delta t;\tau)\delta(\eta - \tau), \tag{8.54}$$

$$R_S(\tau,\eta;v,\mu) = \psi_S(\tau,v)\delta(\eta - \tau)\delta(v - \mu), \tag{8.55}$$

$$R_H(f,f+\Delta f;t,t+\Delta t) = R_H(\Delta f;\Delta t), \tag{8.56}$$

$$R_B(f,f+\Delta f;v,\mu) = \psi_B(\Delta f;v)\delta(v - \mu). \tag{8.57}$$

Fortunately, many radio channels can be accurately modeled as WSSUS channels.

It should be noted that stationarity is never fulfilled over arbitrarily large regions (for WSS) or bandwidth (for US). This fact has been recognized already in [48]. The *region of stationarity* and *stationarity bandwidth* define the spatial (temporal) and bandwidth size within which stationarity is approximately valid [49]. A variety of tests for the size of stationarity regions have been introduced, and it has been found that different quantities, such as power, PDP and angular spectrum, have different-sized regions of stationarity. Finally, the concept of WSSUS can also be generalized to the spatial domain [50].

8.2.6 Channel Characteristic Parameters: Delay Spread, PDP and Channel Coherence

In order to evaluate the characteristics of a propagation channel, a set of measured impulse responses need to be analyzed. This section presents a number of parameters describing different aspects of the radio channel. In general, their computation requires the WSSUS assumption. As it is not practically feasible to collect a set of statistical channel realizations, we generally assume that the channel system functions are ergodic. Hence, it is for instance possible to approximate the statistical expectation using a temporal (or spatial) average over a set of successive measurements.

A channel's frequency selectivity is quantified by its span in the delay domain, that is, its *delay spread*. As the channel's frequency response becomes more varied, an interacting signal will exhibit a broader envelope in the delay domain. In practice, delay spread is the result of different propagation delays of received signals from reflectors at large distances compared to the LoS path between the transmitter and receiver.

To introduce the PDP, we observe the channel for a short-period over which it is time-invariant, that is, $h(t, \tau)$ does not depend on t and can be expressed as $h(\tau)$. The PDP is defined as the squared envelope of $h(\tau)$ [2]:

$$P(\tau) = |h(\tau)|^2. \tag{8.58}$$

The *root mean square* (RMS) delay spread is related to the statistics of the PDP. Specifically, the RMS delay spread σ_τ^2 is defined as the second central moment of the PDP [2]:

$$\sigma_\tau^2 = E[\bar{\tau}^2] - E[\bar{\tau}]^2, \tag{8.59}$$

where

$$E[\bar{\tau}^n] = \frac{\int \tau^n P(\tau) d\tau}{\int P(\tau) d\tau}. \tag{8.60}$$

The RMS delay spread is a significant parameter for the analysis of inter-symbol interference.

Closely related to the RMS delay spread is the concept of *coherence bandwidth* W_c. Coherence bandwidth is inversely proportional to delay spread. The coherence bandwidth is a statistical measure of the bandwidth over which the channel exhibits approximately equal gain. In other words, the coherence bandwidth is the frequency bandwidth $W_c = f_1 - f_2$ within which the frequency autocorrelation function crosses a given threshold. For the 90% threshold, the coherence bandwidth can be approximated from the RMS delay spread as

$$W_c \approx \frac{1}{50\sigma_\tau}. \tag{8.61}$$

In the literature, the coherence bandwidth is most often defined for the 50% correlation threshold, in which case

$$W_c \approx \frac{1}{5\sigma_\tau}, \tag{8.62}$$

though strictly speaking the relationship is an *uncertainty relationship* [51]. Furthermore, the particular numerical values of the proportionality constants depend on the shape of the PDP as well as the correlation level.

A channel's time variance is qualified by the *Doppler spread*, that is, by the span of the channel's Doppler spectrum, because channels that vary faster exhibit a broader frequency range in the Doppler domain. Similar to the RMS delay spread, the *RMS Doppler spread* σ_ν^2 is the second central moment of the Doppler spread:

$$\sigma_\nu^2 = E[\bar{\nu}^2] - E[\bar{\nu}]^2, \tag{8.63}$$

where

$$E[\bar{v^n}] = \frac{\int v^n P(v)dv}{\int P(v)dv}. \tag{8.64}$$

Closely related to the Doppler spread is the *coherence time* T_c. The coherence time is a statistical measure of the time period over which the channel does not change appreciably. Put differently, the channel will affect two signals differently if the period between their arrival times exceeds T_c. The coherence time is inversely proportional to RMS Doppler spread. In the literature, the following ratio is used to estimate T_c:

$$T_c \approx \frac{1}{5\sigma_v}. \tag{8.65}$$

In mobile channels where time variation can be mainly attributed to movement at the transmitter or receiver, the coherence time is often estimated using the maximum Doppler shift f_d:

$$T_c \approx \frac{0.423}{f_d}. \tag{8.66}$$

8.2.7   COST 2100 MIMO Channel Model

The COST 2100 multiple-input–multiple-output (MIMO) channel model [46] is a GSCM that was built on the framework of the earlier COST 259 [16] and COST 273 [47] models and can be extended to millimeter-wave (mmWave) frequencies. The channel model [17] was the first GSCM considering multi-antenna base stations, while full MIMO systems were later targeted by the COST 273 model. The COST 2100 channel model extends the COST 273 model to cover MIMO systems at large, including multiuser, multicellular and cooperative aspects, without requiring a fundamental shift in the original modeling philosophy.

The COST 2100 channel model was originally proposed for simulating the radio channel between a static multiple-antenna base station (BS) and a multiple-antenna mobile station (MS). In most cases, the MPCs are mapped to the corresponding scatterers, and are characterized by their delay, azimuth of departure (AoD), elevation of departure (EoD), azimuth of arrival (AoA) and elevation of arrival (EoA). Clusters are formed by grouping scatterers that generate MPCs with similar delays and directions (azimuth and elevation). Figure 8.4 depicts the scattering mechanisms from the BS to the MS. There are three types of clusters in the COST 2100 model [52], as illustrated in Figure 8.4. Local clusters are located around the MS or the BS, and those are characterized by single-bounce scatterers only. Far clusters are divided into single-bounce and multiple-bounce clusters. They are distributed throughout the simulation area, with the number of clusters following a Poisson distribution. Given the geometrical cluster distribution, the large-scale parameters (LSPs) of a channel are actually controlled by the number of clusters that are active (i.e., visible to the MS) and thus contributing to the channel. While local clusters are always visible, the visibility of a far cluster is determined by the concept of visibility region, which confines the cluster activity

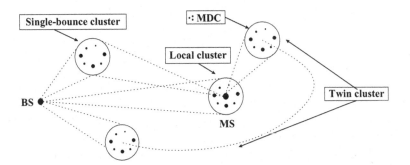

Figure 8.4 General structure of the COST 2100 channel model. © 2012 IEEE. Reprinted, with permission, from [46].

within a limited geographical area. The far clusters include clusters with single-bounce scatterers and clusters with multiple-bounce scatterers. Single-bounce clusters can explicitly be mapped to a certain position by matching their delay and angles through a geometric approach. On the contrary, the multiple-bounce clusters are described by two representations, as viewed from the BS and MS sides, respectively, and called twin clusters. Visually, a twin cluster contains two identical images of one cluster, appearing at both sides (Figure 8.4). In a specific environment, the ratio of twin to single-bounce clusters is set to be constant [52].

The CIR is obtained by the superposition of the MPCs from all active clusters determined by the position of the MS. The amplitude of each MPC is jointly determined by the path loss, the large-scale properties of the cluster to which it belongs, and its own small-scale properties.

Key Modeling Concepts in COST 2100

Visibility regions. A visibility region (VR) is a circular region given fixed size in the simulation area. It determines the visibility of only one cluster. When the MS enters a VR, the related cluster smoothly increases its visibility, as shown in Figure 8.5. This is accounted for mathematically by a VR gain, which grows from 0 to 1 upon entry to the VR. Furthermore, when the MS is located in an area where multiple VRs overlap, multiple clusters are visible simultaneously. In the COST 2100 model, the VRs are uniformly distributed in the simulation area, the VR density is related to the average number of visible clusters determined experimentally [16, 17, 52].

Clusters. A cluster is depicted as an ellipsoid in space as viewed from the BS and from the MS, as illustrated in Figure 8.6. The local cluster and the far clusters are characterized with specific positions and orientations toward the BS and MS, respectively, so their spatial spreads match their corresponding delay and angular spreads. The geometric correspondence between the cluster spatial spread and the cluster delay and angular spreads is simple. For instance, the length a_C, width b_C and height h_C of the single-bounce cluster in Figure 8.6 correspond to the cluster delay, azimuth and elevation spreads, respectively. It is important to note that the COST framework does

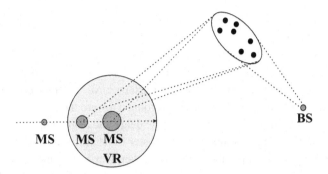

Figure 8.5 Illustration of the visibility region concept. The size of the circle around the MS represents the visibility level of the cluster to the BS–MS channel; when the MS moves outside the cluster visibility region, the related cluster becomes totally inactive in the transmission. © 2012 IEEE. Reprinted, with permission, from [46].

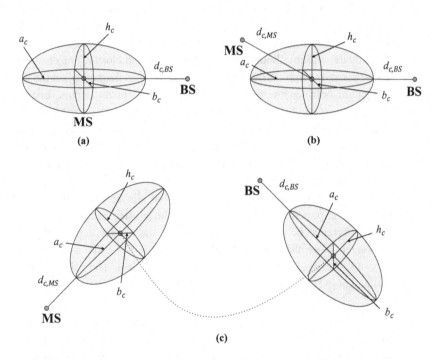

Figure 8.6 Spatial description of (a) local, (b) single-bounce, (c) twin clusters, respectively. © 2012 IEEE. Reprinted, with permission, from [46].

not prescribe the specific way in which the clusters are simulated – they could be based on geometrical clusters, or delay/angle combinations that follow the prescribed cluster ADPSs and fading statistics. The specific implementation is up to the user, and just has to be suitable for the bandwidth at hand. This is an important difference to the 3GPP model, which prescribes precisely the way in which the simulation is to be performed, and where the implementation recipe was designed with a specific maximum bandwidth in mind.

(1) Local cluster(s): A local cluster has an omnidirectional spread in the azimuth plane. Its spatial spread is only determined by its delay and elevation spread. (2) Single-bounce clusters: Single-bounce clusters have independent delay and azimuth spreads. Each single-bounce cluster is rotated toward the BS so that its spatial spreads along its different axes adequately fit the delay and angular spreads as viewed from the BS. The position of a single-bounce cluster is determined by a random vector originating from the BS and rotated with a Gaussian distributed angle relative to the imaginary line between the BS and the center of its corresponding VR. The length of the vector follows a lower-bounded nonnegative distribution. (3) Twin clusters [53]: In this case, the cluster ellipsoids at the BS and MS sides are rotated toward the BS and MS, respectively, similar to the single-bounce clusters. To determine the position of a twin cluster, the method applied for a single-bounce cluster is performed twice: first from the BS side and then from the VR side. This approach is used to control the delay and angles of the twin cluster once the MS is located inside the related VR. A cluster-link delay was introduced in the COST 273 model to compensate for the extra delay caused by the multiple-bounce propagation via a twin cluster. The cluster-link delay is a nonnegative random variable. Its minimum value is defined when the single-bounce propagation between the two centers of the twin cluster occurs. Local and single-bounce clusters can be treated as special twin clusters with a cluster-link delay always equal to zero. Since the cluster-link delay is a large-scale property, it should be applied to all MPCs belonging to the corresponding cluster. (4) Cluster parameterizations: The clustering of paths enables the large-scale properties of the channel to be characterized (i.e., the delay and angular spreads of the MPCs within each cluster, the cluster-link delay, the random shadowing level S_n and the cluster attenuation L_n). The cluster attenuation exponentially increases when the cluster excess delay increases, that is, the difference between the total delay of the cluster and the delay of the LoS component. Note that uncorrelated clustering is normally assumed, meaning that LSPs of different clusters are statistically independent. The values of these LSPs are tabulated in [52] for macro-, micro-, and pico-cellular scenarios.

Time evolution. The COST 2100 framework enables a time-varying channel description using a single realization of the clusters as long as the environment remains static. Note that the environment (i.e., the clusters and the VRs) is generated independent of the MS position. This is actually very similar to the generation of virtual environments. While virtual environments reproduce the exact location and shape of scatterers (buildings, obstacles, etc.), clusters and their visibility regions stochastically represent a typical environment. As mentioned, a whole different approach is followed from WINNER II, where small (stationary) pieces of MS motion are connected by correlating the LSPs between these pieces, thereby enabling explicitly nonstationary channels to be simulated. In the COST family of models, the whole environment is first generated and the movement of the MS in this simulation area causes the visibility of different clusters to change as the MS enters and leaves different VRs, resulting implicitly in nonstationary channel simulations. This also implies that the COST 2100 model structure and parameterization are independent of the MS speed: the higher the speed, the faster the MS moves in and out of VRs , decreasing the stationarity length of the channel. Thereby, scenarios involving high-speed MSs can readily be simulated using the COST 2100 approach.

8.3 Overview of Geometry-Based Stochastic Channel Models for mmWave and THz Communications

During the last few years there has been increased research activity in investigating the properties of radio channels at the frequency bands above 6 GHz. The GSCM has gained popularity as a channel modeling approach at the microwave frequencies because of their reasonable compromise between accuracy and complexity.

The seminal work in outdoor GSCM modeling investigated fixed point-to-point wireless links at 38 GHz [54]. The geometrical model is shown in Figure 8.7. The estimated parameters of outdoor GSCMs at 28 and 73 GHz are presented in [55, 56].

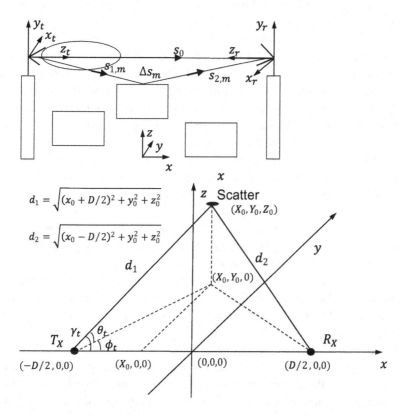

Figure 8.7 Fixed radio link propagation geometry [57] and geometrical model. © 2000 IEEE. Reprinted, with permission, from [54].

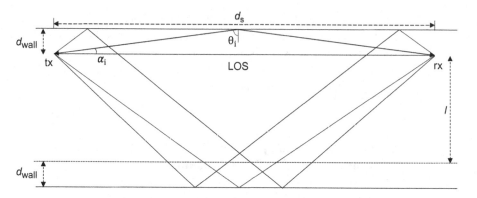

Figure 8.8 Street-canyon GSCM [61].

An extension of this model for urban scenarios at 20 GHz was proposed in [57, 58]. In these models, geometrical properties of wave-scattering objects such as a height distribution of rooftops, tilt and orientation of building surfaces and their reflectivity are modeled as random variables. More recently, work in [59–61] proposed GSCMs for street-canyon scenarios, as illustrated in Figure 8.8. Work in [62–64] expands urban canyon modeling by including over-the-rooftop propagation effects into the GSCM by characterizing buildings as rectangular and cylindrical obstacles. A hybrid uniform theory of diffraction-physical optics approach is derived for accurate prediction of the path loss in the vertically and horizontally polarized fields. Similar to the street-canyon scenario, the scatterer distribution in the indoor GSCM can reflect a specific layout of physical scatterers that characterize the propagation channel. An example is found in the GSCM for inter-vehicular links [65], where the body and roof of a link-shadowing car and the ground are identified as major scattering points when simulating a channel between two vehicles. The car shape is simplified when modeled for ease of determining the scattering points. Using the uniform theory of diffraction to calculate the diffraction losses, the model provides good agreement of path loss in comparison with measurements.

It is worth noting that the GSCM is not limited only to outdoor scenarios, but can be used for indoor scenarios as well. The estimation of indoor GSCM parameters for indoor 60 GHz channels is reported in [66].

Wireless traffic volume is expected to expand tremendously in the next few years and wireless data rates exceeding 10 Gbit/s will be required in the near future [67]. The opening up of carrier frequencies in the terahertz range (THz) is the most promising approach to providing sufficient bandwidth required for ultrafast and ultra-broadband data transmissions. Suitable frequency windows can be found around 110–170 GHz (D-band) and 300–350 GHz. The large bandwidths paired with higher-speed wireless links can open the door to a large number of novel applications such as ultrahigh-speed pico-cell cellular links, wireless short-range communications, secure wireless communication for military and defense applications and on-body communication for health monitoring systems. All these potential applications have motivated researchers to

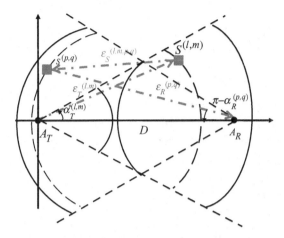

Figure 8.9 GSCM for THz point-to-point communications. © 2015 IEEE. Reprinted, with permission, from [75].

model propagation beyond 100 GHz. The key difference between THz and mmWave propagation effects is that the high antenna directivity gives rise to a scattering pattern that is somewhat different from other indoor (GHz or mmWave) channels observed in [68–74] and needs additional modeling.

The first RT and path-loss models for THz communications were reported in [76–83]. Furthermore, the first statistical model for THz channels has been reported in [84]. The proposed model adapts the frequency-dependent path gains model [85] and the indoor Saleh–Valenzuela model [86] for THz frequencies by running a large number of RT models to extract statistical parameters needed for the model. The first THz GSCM was proposed in [75, 87] for fixed point-to-point communications on desktop. The geometry of the model is shown in Figure 8.9. In addition to scattering mechanisms that are common to all indoor channels in sub-THz channels, signals may reflect off the objects that are behind the receive (RX) antenna, travel back to the objects near the transmit (TX) antenna and reflect back to be received by the RX antenna. This essentially produces the second arriving path, even without any scatterers between the TX and RX. This propagation effect has been captured in [75]. Furthermore, as frequency increases, rough scattering effects are more pronounced, the effects of which have been captured in [88]. Additionally, the work in [75, 87] derives relevant correlation functions needed for system design.

8.4 Stochastic-Based Tap Delay-Line Model

8.4.1 Background

The stochastic-based tapped-delay line (TDL) models for communication channels evolved with the rise of digital communications that began in the 1960s [48]. For guided wave channels, such as the public switched telephone network, the channel

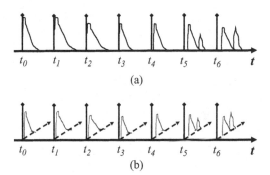

Figure 8.10 Time-variant CIR.

is essentially deterministic (once the switch connections are made), and this yields a deterministic LTI channel, which when sampled can be modeled as a deterministic TDL [89]. As commercial wireless communication systems developed, more investigators applied TDL models [86, 90–93]. In these early cellular systems, mobility and Doppler shift were the cause of time variation, and because of the complexity of deterministic modeling when many time-varying MPCs exist, stochastic TDLs were employed. Thus stochastic-based (SB) TDLs arise from the sampling required in digital communications and the apparent randomness that results from a large number of time-varying MPCs, and can be written as:

$$y(t) = \int x(\tau)h(t, t - \tau)d\tau. \tag{8.67}$$

Worth noting here is that this equation typically represents the complex baseband responses, and not bandpass signals and responses [2, 94]. In principle, $h(t, \tau)$ means a CIR that can change instantaneously, for *any* value of t. Yet, as previously noted, in nearly all practical wireless systems, the CIR changes only slowly over time as compared to a single transmission symbol – this is known as slow fading [2]. Figure 8.10 shows an example timeline in which ideal impulses are input periodically at times t_i, with period $\Delta t = t_{i+1} - t_i$, yielding CIRs that change over time. All CIRs have duration less than Δt. Parts (a) and (b) of Figure 8.10 show a conceptual separation into the two axes t and τ, which yields another conceptual interpretation shown in Figure 8.11. The interpretation here can be useful for measurement design, specifically for the relationship between the maximum expected MPC delay τ_{max} and the CIR measurement repetition period Δt. For unambiguous CIR estimates, τ_{max} must be less than Δt. This requires that one know (or assume) the maximum value of CIR delay. The measurement repetition rate is often called the snapshot rate.

Implicit in Figure 8.11 is the periodic *sampling* of the continuous-time CIRs of Figure 8.10. This yields a discrete time (complex baseband) CIR that can be described by [2]:

$$h(t, \tau) = \sum_{n=1}^{N_p} a_n(t)e^{j\phi_n(t)}\delta(\tau - \tau_n(t)). \tag{8.68}$$

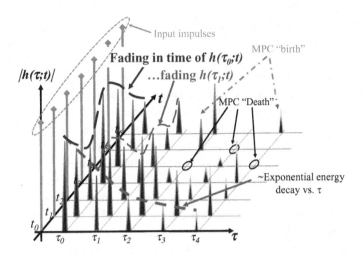

Figure 8.11 Time- and delay-variant CIR and MPCs. © 2008 IEEE. Reprinted with permission.

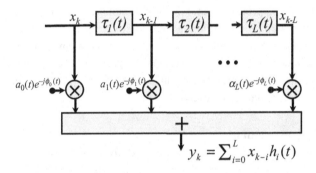

Figure 8.12 Tapped-delay line model. © 2008 IEEE. Reprinted with permission.

In this version of the CIR we have made the complex nature explicit by including the phase term $\phi_n(t)$, and have also made the time variation of the propagation delays explicit. Ideal sampling (infinite bandwidth) is still assumed. This equation yields the model in Figure 8.12, where the input is the discrete time sequence $\ldots x_k, x_{k+1}, \ldots$ and the output is the discrete time sequence $\ldots y_k, y_{k+1} \ldots$. As previously noted, for many practical cases, for signals of moderate bandwidth each of the impulses here is composed of *multiple* components within the delay resolution $\Delta = 1/B$, and the variation of these components (primarily their phases) yields variation of the resolved impulses, and this variation is termed small-scale fading. Worth pointing out is that as signal bandwidth increases and delay resolution correspondingly decreases, the MPCs represented by this equation and in Figure 8.11 become truly resolvable and represent individual MPCs, and may be nonfading over time and space, as in some ultra-wideband channel [93, 95].

The TDL embodied by eq. (8.68) and the representation in Figure 8.11 represent a finite impulse response (FIR) filter, linear but time-varying. The coefficients

$a_i(t)e^{j\phi_i(t)}$ are commonly termed taps. Appealing features of such models include their ease of use in computer simulations, and in realizing hardware channel emulators. Depending upon the random models used for the MPCs, such SB TDLs also provide repeatability in terms of channel statistics. The most common statistical model for the taps is complex Gaussian: This yields an amplitude $a_i(t)$with a Rician distribution and phase $\phi_i(t)$ with a uniform distribution (on $[0,2\pi)$). The complex Gaussian model arises from the central limit theorem, in which a large number of MPCs of nearly equal amplitude and uniform random phases combine.

The delay blocks in Figure 8.12 can take any value in reality, and some established models, such as, COST 207 [92], do provide explicit real number delay values that are separated by arbitrary values of delays. For most TDL models, though, the assumption is that each delay is identical, and equal to the delay resolution Δ. Use of an equal delay for each block simplifies analyses and simulations, but via a fine enough quantization of the delay axis, arbitrary delays are easily accommodated.

A primary virtue of SB TDLs is their efficiency. Once the TDL length and tap statistics are specified, computer implementation of the SB TDL amounts to the generation of sequences of appropriate random variable samples, and this is very easy and efficient with modern computer simulation tools. A slight complexity increase accrues if the taps are correlated, but the implementation is far faster than for more physically based models such as GBSCMs [9, 14, 96]. As an example, execution time and the number of operations required to implement a GBSCM may be a factor of tens to hundreds of times larger than that required for an SB TDL [97].

8.4.1.1   SB TDLs and Measurements

Stochastic-based TDL channel models developed from measurements have been widely employed. There are generally four types of signals that are used to probe channels for estimation of CIRs: (1) short-duration pulses; (2) direct-sequence spread spectrum; (3) multitone; and (4) chirp. Details of these measurement signals are discussed in [2, 98, 99]. Each technique can provide an estimate of the CIR, or equivalently, its Fourier transform the time-variant transfer function $H(f,t)$. Technique (1) produces a CIR estimate directly, whereas technique (2) does so via pulse compression (correlation). Techniques (3) and (4) produce estimates of the transfer function. A PDP, or the magnitude-squared response of the CIR, may also be measured or modeled in order to estimate MPC power decay and time delays. While a PDP does not offer multipath phase information, it is traditional practice to assume that phase of each MPC be drawn from an independent and uniform probability distribution ranging from $-\pi$ to π [2, 100].

Measurements should employ a signal of bandwidth at least as large as that planned for actual signal transmission over the channel. Given a wideband CIR estimate based on a measurement signal bandwidth of B, CIRs appropriate for smaller values of bandwidth can be obtained by combining the complex MPCs. For example, if a model for bandwidth $B/2$ is desired, one can combine two adjacent complex components of eq. (8.68), weighted by the response of the time-domain filters, and this will yield a CIR of length $L/2$ or $L/2 + 1$. Alternatively, for multicarrier signals, sub-bands of $H(f,t)$ could be used for narrower bandwidth channels. For actual transmission system analysis, such as, in time-domain system simulations of single-carrier signals, symbol-spaced taps are usually used.

The channel measurement snapshot period Δt should be larger than the expected maximum channel MPC delay τ_{max} but smaller than – preferably much smaller than – the channel's coherence time t_c. The coherence time is most easily estimated as the reciprocal of the maximum expected Doppler spread f_D, with $f_D = v/\lambda$ where v is relative velocity and λ wavelength. To satisfy the Nyquist criterion Δt should be less than $1/(2f_D)$. Most wireless channels, at least those that do not involve very high-speed platforms and/or extremely long multipath delays, are *underspread*, such that $\tau_{max} f_D \ll 1$. In this case, it is not difficult to select a snapshot period Δt that is both larger than τ_{max} and smaller than t_c.

As with other models, such as RT, interacting objects in the environment *can* be associated with specific TDL model taps, although this connection is not often made in the case of TDL models, and when it is, it is typically only done during actual measurements in a specific environment. Once measurements are taken and the set of stored CIRs is processed for the purpose of computing channel statistics and model development, specific connections between the TDL models and the measurement environment are absent. The most common (possibly only) exception to this is the case where the environment has an LoS component and a strong earth surface reflection as the first two TDL model taps.

Often, individual MPCs, or components associated with a particular propagation path from a particular physical environment, may undergo diffraction around a building corner or may incur penetration loss due to large obstructions. In such instances, it is useful to incorporate diffraction or partition loss modeling as a deterministic attenuation factor on the MPCs in a TDL model. Diffraction measurements at cmWave and mmWave bands have demonstrated that diffraction loss can be predicted well using simple models of the physical environment [101–103], and penetration loss measurements for common indoor partitions at 73 GHz are given in [104]. These models, such as the knife-edge diffraction (KED) and the creeping-wave linear model or partition loss attenuation factors, may be applied to adjust individual MPC amplitudes at cmWave and mmWave frequencies for indoor and outdoor environments [101, 103].

Diffuse scattering from rough surfaces may introduce large signal variations over very short distances (just a few centimeters), which must be considered in TDL channel modeling [105]. Such rapid variations of the channel must be anticipated for proper design of channel state feedback algorithms, link adaptation schemes, beamforming/beamtracking algorithms, and for ensuring efficient design of MAC and network layer transmission control protocols (TCPs) that induce retransmissions [106, 107]. Measurement of diffuse scattering at 60 GHz on several rough and smooth wall surfaces [107, 108] has demonstrated large signal-level variations in the first-order specular and in the nonspecular scattered components (with fade depths up to 20 dB) as a user moved over a few centimeters.

In environments with a large number of MPCs, for modeling purposes, one may choose to retain only the strongest components, up to some fraction of the total channel response energy, for example, 95%. Other alternatives for thresholding include discarding all MPCs that are below some level x dB relative to the strongest MPC, and discarding MPCs whose signal-to-noise ratio (SNR) is below some minimum value, such as 5 dB [109]. This latter approach implies that estimation of the (measurement receiver) noise average power is also required, such that the noise figure and operating bandwidth are considered for interpretation of results.

8.4.2   Statistically Nonstationary TDLs

Current channel modeling research does not end with "steady state" models. In the quest for greater model fidelity to support higher link reliability and an expanding variety of wireless services and higher service qualities, channel models are now more than ever taking into account transients and rapid dynamics. These effects cause changes in local area channel statistics, and hence models that take such effects into account are termed statistically nonstationary (NS). Traditional SB TDLs did not account for these effects, but they can be modified to some degree to do so.

Physical mechanisms that cause so-called NS effects include the time variation ("drift") of MPC delays, and the appearance and disappearance of MPCs. The MPC delay drift is typically slow with respect to signaling durations, but if needed, it can be accounted for in SB TDLs by adjusting delays over time. Conceptually this would appear in Figure 8.11 as having the lines associated with each MPC delay τ_i no longer run parallel to the time t axis, but instead run at some moderate angle(s). If one is using a TDL model with multiple components within each tap, these individual components can be made to transition from one tap to another. In models with a large number of taps, tracking and moving these subcomponents may considerably add to model complexity.

Similarly, the appearance/disappearance of MPCs can be incorporated by switching on or off MPCs at appropriate values of delay [110]. For stochastic realizations, discrete Markov chains are convenient for these switching processes. If these "switched" MPCs have sufficient energy with respect to existing taps, one may wish to smooth the on–off transitions in some way, such as via interpolation.

As discussed in greater detail in Chapter 10, attenuation caused by human blockage will greatly impact cellphone link performance, and a useful statistical model for the effects of human blockage can be represented in a four-state Markov model, where the four states of the model correspond to the four regions of a blockage event – the unshadowed, shadowed, rising and decaying regions [100, 103, 111, 112]. The double KED (DKED) model assumes a human blocker to be represented as a screen with four sides, or as an infinitely long vertical screen with two sides [113]. A DKED antenna gain (DKED-AG) model that incorporates directional antenna patterns to accurately predict the upper and lower envelopes of measured received power during a blockage better agrees with real-world measurements when compared to 3GPP/METIS blockage models [100, 103, 112] that underpredict attenuation when a blocker is close to either the TX or RX. Measurements have shown that when a human blocker is within 0.5 m of either the TX or RX, rapid fades as deep as 40 dB can occur when using narrowbeam and directional antennas [100, 103]. Therefore, channel models at mmWave frequencies need to represent the channel dynamics where human blockage causes severe signal fades when a signal path is blocked and when diffraction is no longer viable. The large fades due to diffraction loss have implications for the design of physical-layer protocols and frame structures that incorporate phased-array antennas to maintain a link, while finding other spatial paths [103].

8.4.3   Clustering and Mixed SB TDL and GBSCMs

As described in the previous chapter, MPCs often occur in groups or clusters that are localized in delay. Such localizing also occurs in the spatial angular domain, and application of combined delay–angle clustering is becoming increasingly popular. The earliest model that described clustering in delay was developed by Saleh and Valenzuela (SV) [86]. The SV model was developed from measurements for an indoor office environment, but the structural features of this model have been observed in other environments as well. The CIR for the SV model is

$$h(t, \tau) = \sum_{k=1}^{M} \sum_{i=1}^{L_k} a_{i,k}(t) e^{j\phi_{i,k}(t)} \delta(\tau - \tau_{i,k}(t) - T_k(t)). \tag{8.69}$$

where the kth cluster of MPCs has delay T_k, and the ith MPC of the kth cluster has additional relative delay $\tau_{i,k}$. These MPC "inter-arrival" times are exponentially distributed, as the initial cluster delays and intra-cluster delays are selected according to a Poisson point process random model. In the original SV model, all parameters were time-invariant. Amplitudes of the MPCs within a cluster are exponentially decaying, and the amplitude of the envelope of clusters also takes an exponentially decaying form versus delay. Work at mmWave frequencies [103, 114, 115] showed Rician fading of voltage amplitudes of MPCs, as well as total received power (or received energy under the PDP), and exponential or sinusoidally exponential decaying trends for spatial autocorrelation (the autocorrelation coefficient function is defined as equation (1) in [115]) at 28 GHz and 73 GHz with rapid decorrelation at approximately 0.67–33.3 wavelengths of spacing, depending on the receiver orientation with respect to the environment and the transmitter [103, 115]. Therefore, the spatial correlation can be modeled by a "damped oscillation" function of:

$$f(\Delta X) = \cos(a\Delta X) e^{-b\Delta X}, \tag{8.70}$$

where ΔX denotes the space between antenna positions, a is an oscillation distance with units of radians/λ (wavelength), $T = 2\pi/a$ can be defined as the spatial oscillation period with units of λ or cm, and b is a constant with units of λ^{-1} whose inverse $d = 1/b$ is the spatial decay constant with units of λ [103]. Model parameters a and b in eq. (8.70) are obtained using the MMSE method to find the best fit between the empirical spatial autocorrelation curve and theoretical exponential model given by eq. (8.70). The "damped oscillation" pattern can be explained by superposition of MPCs with different phases at different linear track positions. As the separation distance of linear track positions increases, the phase differences among individual MPCs will oscillate as the separation distance of track positions increases due to alternating constructive and destructive combining of the multipath phases. This "damped oscillation" pattern is obvious in LoS environments, where phase difference among individual MPCs is not affected by shadowing effects that occurred in NLoS environments [103]. Worth pointing out here is that small-scale fading and spatial correlation statistics for directional measurements in [115] are focused on the amplitude of the total received signal (i.e., by taking the square root of the square of eq. (8.69)), whereas the spatial correlation and small-scale fading of individual MPCs are presented in [114]. The small correlation distance in most cases is favorable for spatial multiplexing in MIMO since it allows for uncorrelated spatial data streams to be transmitted from closely spaced (a fraction to several wavelengths) antennas. However, it is worth noting that in a lightly clustered environment, for example in a rural macro-cell (RMa) scenario, measurements have shown that there is only one cluster/spatial lobe due to lack of scattering objects, which will lead to lack of angular diversity in the RMa channel [116, 117].

The 3GPP models utilize various forms of the TDL modeling approach and take several forms, with the simplest being the "link level" models. These link level models dispense with much of the geometric environment setup, to enable users to create delay line models for several settings (e.g., urban micro, indoor office, etc.). The link level models can be either cluster delay line (CDL) models or TDLs. For the latter, generally applicable to SISO links, taps are specified in one of two ways: (1) via a table format listing tap delay, fading amplitude distribution and relative power; or (2) via spatial filtering of the more general CDL. The table format is a traditional one, similar to the models where all taps – other than a possible LoS tap – have Clarke's Doppler spectrum. The assumption is Rayleigh fading on all taps (except the LoS tap) in two of the TDL models (TDL-D and TDL-E). Hence, a large number of MPCs per delay bin is assumed, even though the 3GPP models claim applicability up to a signal bandwidth of 2 GHz. In the 3GPP CDL channel model, clusters are characterized by a joint delay–angle pdf, such that a group of traveling multipaths depart and arrive from a unique angle of departure (AoD)–angle of arrival (AoA) combination centered around a mean propagation delay [117, 120, 121]. High-resolution parameter extraction algorithms, such as, SAGE, and clustering, such as KPowerMeans algorithms [118, 119] that have high computational complexity, are often employed to obtain cluster characteristics.

The 3GPP CDL creates MPCs in both the (azimuth) angular and temporal delay domains via the following sequence of steps (see [120, section 7.7.1] for details: (1) generate random AoAs and AoDs for each of the M rays for each of the N clusters; (2) couple the AoAs and AoDs; (3) generate (cross-polarization ratios) XPRs for all clusters; (4) generate random phases for each MPC of each cluster; and (5) generate geometry-based complex amplitude coefficients.

Parameters for all the CDL generation steps are setting-specific, and are provided in [120, section 7.7.1]. Doppler effects are incorporated via basic linear mobility models for the user terminal (UT). The TDL can also be extended for MIMO applications via introduction of a spatial correlation matrix.

The most general and most comprehensive model in the 3GPP set is the "system-level" model. After scenario selection, this model begins with definition of coordinate systems, both global and local. The local coordinate systems allow for specification of antenna array geometries for MIMO systems. Following this is specific antenna modeling, including modeling of multiple polarizations. Then comes the modeling of large-scale effects, such as determination of LoS or NLoS, and specification of path loss and shadowing. In this large-scale modeling, if an indoor–outdoor link is encountered, attenuation due to building materials ("penetration loss") is also modeled. Subsequent to the large-scale propagation modeling is small-scale parameter modeling. This consists of generation of random delays, then generation of cluster powers, then essentially the five steps described in CDL modeling, with the additional specification of elevation angle parameters as well as azimuth angle parameters.

In the 3GPP channel model, the directional antenna element radiation pattern is defined in [120, Table 7.3.1] and the channel matrix **H** is defined by eqs. (8.71) and (8.72) [120, eqs. 7.5–7.29]).

$$
H^{NLoS}_{x^R_j, x^T_i, n, m}(t) = \sqrt{\frac{P_n}{M}} \begin{bmatrix} F_{R,x^R_j,\theta}(\phi^{R,E}_{n,m}, \phi^{R,A}_{n,m}) \\ F_{R,x^R_j,\phi}(\phi^{R,E}_{n,m}, \phi^{R,A}_{n,m}) \end{bmatrix}^T
$$

$$
\begin{bmatrix} \exp(j\Phi^{\theta,\theta}_{n,m}) & \sqrt{\kappa^{-1}_{n,m}}\exp(j\Phi^{\theta\phi}_{n,m}) \\ \sqrt{\kappa^{-1}_{n,m}}\exp(j\Phi^{\phi\theta}_{n,m}) & \exp(j\Phi^{\phi\phi}_{n,m}) \end{bmatrix},
$$

$$
\begin{bmatrix} F_{T,x^T_i,\theta}(\phi^{T,E}_{n,m}, \phi^{T,A}_{n,m}) \\ F_{T,x^T_i,\phi}(\phi^{T,E}_{n,m}, \phi^{T,A}_{n,m}) \end{bmatrix} \exp\left(j2\pi\frac{\hat{r}^T_{R,n,m}\cdot\bar{d}_{R,x^R_j}}{\lambda_0}\right)
$$

$$
\exp\left(j2\pi\frac{\hat{r}^T_{T,n,m}\cdot\bar{d}_{T,x^T_i}}{\lambda_0}\right) \exp\left(j2\pi\frac{\hat{r}^T_{R,n,m}\cdot\bar{v}}{\lambda_0}t\right) \tag{8.71}
$$

$$
H^{LoS}_{x^R_j, x^T_i, l}(t) = \begin{bmatrix} F_{R,x^R_j,\theta}(\theta_{LoS,ZoA}, \phi_{LoS,AoA}) \\ F_{R,x^R_j,\phi}(\theta_{LoS,ZoA}, \phi_{LoS,AoA}) \end{bmatrix}^T
$$

$$
\begin{bmatrix} 1 & 0 \\ 0 & -1 \end{bmatrix} \begin{bmatrix} F_{T,x^T_i,\theta}(\theta_{LoS,ZoD}, \phi_{LoS,AoD}) \\ F_{T,x^T_i,\phi}(\theta_{LoS,ZoD}, \phi_{LoS,AoD}) \end{bmatrix}
$$

$$
\exp\left(-j2\pi\frac{d_{3D}}{\lambda_0}\right) \exp\left(j2\pi\frac{\hat{r}^T_{R,LoS}\cdot\bar{d}_{R,x^R_j}}{\lambda_0}\right), \tag{8.72}
$$

$$
\exp\left(j2\pi\frac{\hat{r}^T_{T,LoS}\cdot\bar{d}_{T,x^T_i}}{\lambda_0}\right) \exp\left(j2\pi\frac{\hat{r}^T_{R,LoS}\cdot\bar{v}}{\lambda_0}t\right)
$$

where n and m are the cluster index and the ray index (i.e., the mth ray within cluster n), respectively, P_n is the power of the nth cluster, x^T_i and x^R_j are transmit and receive antenna element indices, κ is the XPR, F is the antenna pattern, ϕ^E and ϕ^A are zenith and azimuth angles, $\bar{d}$ is the location vector, $\hat{r}$ is the spherical unit vector, $\bar{v}$ is the velocity vector, Φ is the initial phase, which is uniformly distributed in $(-\pi, \pi)$ for each ray m of each cluster n for four different polarization combinations ($\theta\theta$, $\theta\phi$, $\phi\theta$, $\phi\phi$) and λ_0 is the wavelength. The CIR is then given by adding the LoS channel coefficient to the NLoS CIR and scaling both terms according to the desired K-factor as in [120, eqs. 7.5–7.30]. All of the channel model parameters such as the angular spread, delay spread, and Rician K-factor can be found in [120, Tables 7.5–7.6, parts I and II].

The NYUSIM channel model is also an SB TDL model [11, 121], and in contrast to the definition for "cluster" used by 3GPP, the NYUSIM model uses the definition of a *time cluster* (TC) and *spatial lobe* (SL) to describe the multipath behavior in omnidirectional CIRs, without forcing the requirement of a cluster to have a one-to-one joint linkage between delay time and spatial angle. The approach in NYUSIM defines separate clusters for time and space, such that all channel energy is fully represented both over time and space at a particular instant. Time clusters are composed of MPCs that are bunched together in time, that is, traveling closely in time delay, and arriving or departing from potentially different directions over a short propagation time window [121]. Spatial lobes denote primary directions (i.e., spans) of departure (or arrival) angles where energy arrives over the entire time delay axis, usually several hundred nanoseconds or more [121]. Per the definitions in [121], a TC contains MPCs traveling close in time, but may arrive from different SL angular directions, such that the temporal and spatial statistics are decoupled and can be recovered separately [109, 117, 121]. Similarly, an SL may contain many MPCs arriving (or departing) in a space (angular cluster) but with different time delays. This distinguishing feature was inspired from real-world propagation measurements [109, 121, 122], which showed MPCs belonging to the same TC can arrive at distinct spatial pointing angles and that energy arriving or departing in a particular pointing direction can span hundreds or thousands of nanoseconds in propagation delay, detectable due to high-gain steerable directional antennas. The TCSL clustering scheme models the directionality of mmWave channels via separate TCs that have time-delay statistics, and via SLs that represent the strongest directions of multipath arrival and departure [121]. Another difference between the NYUSIM model and the 3GPP model is that the 3GPP model has large numbers of clusters (e.g., 12 and 19 clusters for the LoS and NLoS environments in the UMi street-canyon scenario, respectively, and 20 rays per cluster) which may overpredict the diversity of mmWave channels, but NYUSIM yields only up to six TCs and five SLs that are borne out by extensive mmWave field measurements, and this difference has a significant impact on spectral efficiency evaluation [117].

Behavior of wideband mmWave signals as a mobile user moves around a local area is vital for the design of handoff mechanisms and beam-steering algorithms. Studies on properties of local area channel transition and inter-site correlation of shadow fading were conducted in [103, 123] at 73 GHz, as summarized in Table 8.1. These results show that wideband 73 GHz signals have a relatively stationary mean power over slight movements but that average received power can change by more than 25 dB as a mobile transitions around a building corner from LoS to NLoS in a

Table 8.1 Spatial correlation model parameters in eq. (8.70) for 73 GHz, 1 GHz RF bandwidth ($\lambda = 0.41$ cm).

Condition	$a\ (rad/\lambda)$	$T = 2\pi/a$	$b\ (\lambda^{-1})$	$d = 1/b$
LoS omnidirectional	0.45	14.0λ (5.71 cm)	0.10	10.0λ (4.08 cm)
NLoS omnidirectional	0	Not used	0.26	3.85λ (1.57 cm)
LoS directional	0.33–0.50	12.6–19.0λ (5.14–7.76 cm)	0.03–0.15	6.67–33.3λ (2.72–13.6 cm)
NLoS directional	0	Not used	0.04–1.49	0.67–25.0λ (0.27–10.2 cm)

UMi environment [123, 124]. The generic model for local stationarity shows the local mean received power varies with a log Gaussian distribution over a 5×10 m area, with a 2.2 dB standard deviation for NLoS and 4.3 dB for LoS [103].

8.5   Quasi-Deterministic Channel Modeling Approach

8.5.1   Experimental Measurements and Rays Classification

The state-of-the-art for mobile communications channel characterization includes separate descriptions of the path loss models and spatiotemporal channel characteristics, typically comprised of the clustered CIRs and angular spread statistics [125]. Latest works for mmWave channel models also follow such an approach [126, 127] and use different cluster analysis techniques to the experimental data processing. However, such approaches work well for NLoS conditions in rich multipath environments, which is not the main usage case for the mmWave communication system. At the same time, the propagation loss features of mmWave signals lead to the weakness of distant reflections and the domination of direct path and a few strongest reflected rays. This allows developing new approaches to characterization of the mmWave channels in the nonstationary environment. To provide adequate modeling of the channel propagation aspects mentioned above, the quasi-deterministic (QD) approach for channel modeling is proposed and was developed during the MiWEBA FP7 project [128] and IEEE 802.11ay standardization process [129].

   The experimental results obtained for different outdoor environments in the MiWEBA project [128, 130, 132–134] show that mmWave channel for complex large area indoor and outdoor environments may not be completely described by the deterministic RT approach. Detailed analysis of the experimental results leads to the conclusion that realistic mmWave channel models can consist of deterministic components, defined by the scenario and random components, representing unpredictable factors or random objects appeared in this environment. Such an approach, called quasi-deterministic, was offered for modeling access and backhaul mmWave

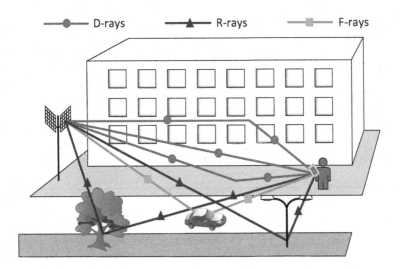

Figure 8.13 D-, R- and F-rays in the QD model illustration

channels at 60 GHz [130, 135–137]. A similar basic idea was used previously for the modeling of ultrawideband (<10 GHz) channels in [138] and sub-6 GHz microcell channels in [139].

The approach builds on the representation of the mmWave CIR comprised of a few QD strong rays (D-rays), a number of relatively weak random rays (R-rays, originating from static surface reflections) and flashing rays (F-rays, originating by reflections from moving cars, buses and other dynamic objects) (see Figure 8.13). D-rays make the major contribution to the signal power, are present all the time and usually can be clearly identified as reflection from scenario-important macro objects. It is logical to include them in the channel model as deterministic, explicitly calculated values. The element of randomness, important for statistical channel modeling, may be introduced at the intra-cluster level by adding a random exponentially decaying cluster to the main D-ray.

R-rays are the reflections from random objects or objects that are not mandatory in the scenario environment. Such rays may be included in the model in a classical statistical way, as rays with parameters (AoD, AoA, PDP) selected randomly in accordance with the predefined distributions.

F-rays may be introduced in the channel model for special nonstationary environments. These rays can appear for a short periods of time, for example, as a reflection from moving cars and other objects. F-rays can be described in the same way as R-rays, but taking into account the statistics of their appearance in time.

All types of rays are then combined in the single clustered CIR, schematically shown in Figure 8.14. Here, cluster refers to MPCs with similar delay, AoD, and AoA parameters. All of these parameters should be similar for all of these MPCs. Physically this means that the paths belonging to the same cluster should have the same physical propagation mechanisms (e.g., produced by one physical reflection surface) [6].

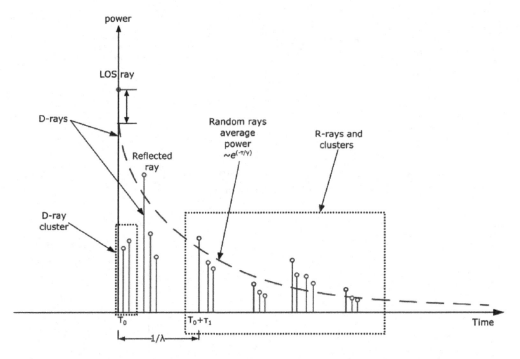

Figure 8.14 QD channel model CIR structure. © 2014 IEEE. Reprinted, with permission, from [140].

For each of the channel propagation scenarios, the strongest propagation paths are determined and associated to rays which produce the substantial part of the received useful signal power. Then the signal propagation over these paths is calculated based on the geometry of the deployment and the locations of the transmitter and receiver, calculating the ray parameters, such as AoD, AoA, power and polarization characteristics. The signal power conveyed over each of the rays is calculated in accordance with theoretical formulas taking into account free-space losses, reflections, antenna polarization and receiver mobility effects like Doppler shift. Some of the parameters in these calculations may be considered as random values, such as reflection coefficients, or as random processes, such as receiver motion. The number of D-rays taken into account is scenario-dependent, and is chosen to be in line with the channel measurement results. Additionally to the D-rays, a number of other reflected waves are received from different directions, coming, for example, from cars, trees, lampposts, benches, houses, etc. (for outdoor scenarios) or from room furniture and other objects (for indoor scenarios). These rays are modeled as R-rays. These rays are defined as random clusters with specified statistical parameters extracted from available experimental data or RT modeling. For a given environmental scenario, the process of the definition of D-rays, R-rays and F-rays, and their parameters, is based both on the experimental measurements and RT reconstruction of the environment. The experimental measurements processing includes a peak detection algorithm with further accumulation of the peak statistics over time, identifying the percentage of

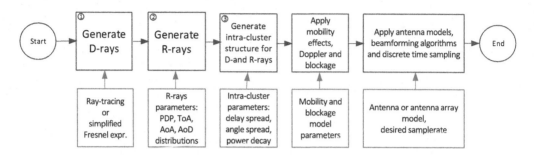

Figure 8.15 Process of CIR generation for the QD approach. Reprinted, with permission, from [130] VDE VERLAG, Berlin, Offenbach, Germany, 2016.

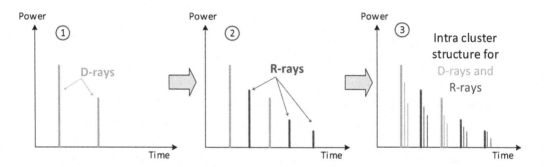

Figure 8.16 Base steps of CIR generation. Reprinted, with permission, from [130] VDE VERLAG, Berlin, Offenbach, Germany, 2016.

the selected ray activity during the observation period. For example, based on the analysis of available experimental data [128], the rays with an activity percentage above 80–90% may be classified as D-rays: strong and always present, if not blocked. The blockage percentage for D-rays may be estimated at around 2–4%. The rays with an activity percentage of about 40–70% are the R-rays, the reflections from far-away static objects, weaker and more susceptible to blockage due to longer travel distances. And finally, the rays with an activity percentage below 30% are the F-rays: the flashing reflections from random moving objects. Such rays are not "blocked," they actually "appear" only for a short time. Figure 8.15 illustrates the CIR generation process.

The core of the algorithm consists of the three major steps of D-ray generation (Section 8.5.2), R-ray generation (Section 8.5.3) and adding the thin intra-cluster structure to the generated D- and R-rays (Section 8.5.4). These three steps are illustrated in Figure 8.16.

8.5.2   D-Rays Modeling

The QD rays are explicitly calculated in accordance with scenario parameters, geometry and propagation conditions. The propagation loss is calculated by the Friis equation, taking into account additional losses from oxygen absorption (Table 8.2, second row). An important part of the proposed QD approach to the channel modeling

Table 8.2 Direct ray parameters.

Parameter	Value
Delay	Direct ray delay is calculated from the model geometry: $\tau_D = d_D/c$ $d_D = \sqrt{L^2 + (H_T - H_R)^2},$ where H_T and H_R are transmitter and receiver heights respectively, and L is the horizontal distance between them.
Power	Direct ray power calculated as free-space pathloss with oxygen absorption: $P_D = 20 \log_{10}\left(\frac{\lambda}{4\pi d_D}\right) - A_0 d_D$ in dB, where λ is the wavelength, and A_0 is the oxygen absorption coefficient (15 dB/km for 60 GHz)
Channel matrix	$H = \begin{bmatrix} 10^{\frac{P_D}{20}} & 0 \\ 0 & 10^{\frac{P_D}{20}} \end{bmatrix} \exp\left(\frac{j2\pi d_D}{\lambda}\right)$
AoA	$0°$ azimuth and elevation
AoD	$0°$ azimuth and elevation

is the calculation of the reflected ray parameters. The calculations are based on the Fresnel equations, additionally taking into account losses due to surface roughness (Table 8.3, second row). The feasibility of the proposed approach to the prediction of the signal power is demonstrated in [141] for outdoor microcell environments and in [142] and [65] for inter-vehicle communication modeling. In general, problems of the signal power prediction are considered in [143]. The D-rays are strictly scenario-dependent, but in all considered outdoor scenarios two basic D-rays are present: the direct LoS ray and the ground-reflected ray. The calculation of those two basic ray parameters will be the same for all scenarios.

8.5.2.1   Direct Ray
A direct LoS ray is a straight ray between TX and RX.

8.5.2.2   Ground-Reflected Ray
A ground-reflected ray presents in all considered scenarios. Its parameters are calculated based on the Friis free-space path-loss equation and the Fresnel equation to take into account reflection and rough surface scattering factor F. Note that the horizontally and vertically polarized components of the transmitted signal will be differently reflected and, thus, the channel matrix should have different diagonal elements.

8.5.2.3   Additional D-Rays
For the open-area scenario, with no significant reflection objects other than the ground, only two D-rays considered. However, in scenarios with richer environments, such as a large square, or for example, a street-canyon scenario, reflection from one or more walls should be taken into account. The principle of calculation of these additional D-rays is the same, a detailed description may be found in [128]. The closest wall can

Table 8.3 Ground-reflected ray parameters.

Parameter	Value
Delay	Ground-reflected ray delay is calculated from the model geometry: $\tau_G = d_G/c$ $$d_G = \sqrt{L^2 + (H_T + H_R)^2}$$
Power	Ground-reflected power calculated as free-space path loss with oxygen absorption, with additional reflection loss calculated on the base of Fresnel equations. Reflection loss R is different for vertical and horizontal polarization $$P_\perp = 20\log_{10}\left(\frac{\lambda}{4\pi d_D}\right) - A_0 d_D + R_\perp + F$$ $$P_\parallel = 20\log_{10}\left(\frac{\lambda}{4\pi d_D}\right) - A_0 d_D + R_\parallel + F$$ $$F = \frac{80}{\ln 10}\left(\pi\sigma_h\sin\phi/\lambda\right)^2 \quad R_\perp = 20\log_{10}\left(\frac{\sin\phi - \sqrt{B_\perp}}{\sin\phi + \sqrt{B_\perp}}\right)$$ $$R_\parallel = 20\log_{10}\left(\frac{\sin\phi - \sqrt{B_\parallel}}{\sin\phi + \sqrt{B_\parallel}}\right)$$ $$B_\parallel = \varepsilon_r - \cos^2\phi \quad B_\perp = \left(\varepsilon_r - \cos^2\phi\right)/\varepsilon_r^2 \text{ in dB,}$$ where $\tan\phi = \frac{H_T + H_R}{L}$, and σ_h is a surface roughness
Channel matrix	$$H = \begin{bmatrix} 10^{\frac{P_\parallel}{20}} & \xi \\ \xi & 10^{\frac{P_\perp}{20}} \end{bmatrix} \exp\left(\frac{j2\pi d_G}{\lambda}\right)$$
AoD	Azimuth: $0°$ Elevation: $\theta_{AoD} = \arctan\left(\frac{L}{H_T - H_R}\right) - \arctan\left(\frac{L}{H_T + H_R}\right)$
AoA	Azimuth: $0°$ Elevation: $\theta_{AoA} = \arctan\left(\frac{H_T + H_R}{L}\right) - \arctan\left(\frac{H_T - H_R}{L}\right)$

be calculated using the geometry and positions of the transmitter and receiver. The calculation of the path properties is similar to the ground ray reflection considered in the previous section, taking into account material properties for the specific environments.

8.5.3   R-Ray Modeling

In order to take into account a number of rays that cannot be explicitly described deterministically (reflections from objects that are not fully specified in the scenario, objects with random or unknown placement, objects with complex geometry, higher-order reflections, etc.) R-rays are introduced in the QD modeling methodology. R-rays may be generated in two different ways: statistically in accordance with the predefined PDP or as deterministic reflections from random objects.

8.5.3.1   Statistical R-Ray Definitions
A statistical approach is the basic means of R-ray generation in the QD channel modeling methodology. The clusters (see Figure 8.14) arrive at moments τ_k according

Table 8.4 R-rays parameters for open area model.

Parameter	Value
Number of rays, N	3
Poisson arrival rate, λ	0.05 ns^{-1}
Power-decay constant, γ	15 ns
K-factor	6 dB
AoA	Elevation: U[−20:20°] Azimuth: U[−180:180°]
AoD	Elevation: U[−20:20°] Azimuth: U[−180:180°]

to a Poisson process and have inter-arrival times that are exponentially distributed. The cluster amplitudes $A(\tau_k)$ are independent Rayleigh random variables and the corresponding phase angles θ_k are independent random variables uniformly distributed over $[0, 2\pi]$.

The random ray contributions to the CIR are given by:

$$h_{\text{cluster}}(t) = \sum_{k=1}^{N_{\text{cluster}}} A(\tau_k) \exp\{j\theta_k\} \delta(t - \tau_k), \qquad (8.73)$$

where τ_k is the arrival time of the kth cluster measured from the arrival time of the LoS ray, $A(\tau_k)$, $P(\tau_k)$ and θ_k are the amplitude, power and phase of the kth cluster. The R-rays are random, with Rayleigh-distributed amplitudes and random phases, with exponentially decaying PDPs. The total power is determined by the K-factor with respect to the direct LoS path:

$$\overline{P(\tau_k)} = P_0 \exp\left\{-\frac{\tau_k}{\gamma}\right\}, \qquad (8.74)$$

$$K = \frac{P_{LOS}}{\sum P(\tau_k)}. \qquad (8.75)$$

Table 8.4 summarizes the R-ray parameters for the open area/large square models. The PDP parameters are derived based on the available experimental data and corresponding RT simulations. The AoA and AoD ranges illustrate the fact that random reflectors can be found anywhere around the receiver, but are limited in height. Uniform distributions are selected for simplicity and can be further enhanced on the base of more extensive measurements.

In the 802.11ad channel model [144], a set of approximations was proposed for diagonal and off-diagonal elements of the channel matrix H for the first- and second-order reflections in typical indoor environments (conference room, cubicle and living room) as combinations of log-normal and uniform distributions on the base of experimental studies [145]. In the QD model the ray amplitude is approximated by the Rayleigh distribution (which is close to log-normal) so the simple fixed polarization matrix H_p may be used for introducing polarization properties to R-rays (matrix H is

obtained by multiplication of the scalar amplitudes A to the polarization matrix H_p).
The polarization matrix H_p for R-rays is defined by:

$$H_p = \begin{bmatrix} a^{V,V} & a^{V,H} \\ a^{H,V} & a^{H,H} \end{bmatrix} = \begin{bmatrix} 1 & \pm 0.1 \\ 0.1 & \pm 1 \end{bmatrix}. \tag{8.76}$$

The values with the $\pm$ sign are assumed to have random sign, ($+1$ or -1, for instance) with equal probability, independently from other values. The polarization matrix is identical for all rays comprising the cluster. Flashing rays, or F-rays, are intended to describe the reflections from fast-moving objects like vehicles and are short in duration. Their properties require additional investigations and analysis, thus the F-rays are not included in the considered QD modeling approach application example.

8.5.3.2   Random Object Reflection R-Rays

The synthetic aperture processing of the experimental results [130] have shown that the reflections from various environmental objects such as trees, lampposts, bus stops, etc. can be clearly identified (with exact estimation of the reflector position) from the experimental data. Such rays should be taken into account along with D-rays, which originate from large-scale objects, but the definition of the position of each reflector makes scenario description complex and very specific. Thus, it is proposed to generate such type of rays (R-rays or F-rays) as reflections from the randomly placed spherical objects, that (unlike the flat objects) can create a specular reflection path between any two points in the 3D space. For now, based on the experimental measurements, the R-rays as reflections from random objects are introduced for the street-canyon scenario only, in addition to statistically generated R-rays described above. Also, the F-rays are generated in this way, with the only difference being the path existence period in the applications where the longer periods of time are analyzed.

8.5.4   Intra-Cluster Structure Modeling

The surface roughness and presence of various irregular objects on the considered reflecting surfaces (bricks, windows, borders, manholes, advertisement boards on the walls, etc.) lead to the separation of a specular reflection ray into a number of additional rays with close delays and angles: a cluster. The intra-cluster structure is introduced in the QD model in the same way as R-rays: as Poisson-distributed in time, exponentially decaying Rayleigh components, dependent on the main ray. The identification of rays inside the cluster in the angular domain requires very high angular resolution. The "virtual antenna array" technique, where a low directional antenna element can be used to perform measurements in multiple positions along the virtual antenna array to form an effective antenna aperture, was used in the MEDIAN project [131]. Note that it is reasonable to assume that different types of clusters may have distinctive intra-cluster structures. For example, properties of the clusters reflected from the road surface are different from the properties of the clusters reflected from brick walls because of the different materials of the surface structure. Also, one may assume the properties of the first- and second-order reflected clusters to be different, with the second-order reflected clusters having larger spreads in temporal and angular domains. All these effects are understood to be reasonable. However, since the number of available experimental results was limited, a common intra-cluster model for all types of clusters was developed. Modifications with different intra-cluster models for different types of clusters may be a subject of future channel model enhancements. In the 802.11ay channel model the intra-cluster structure is added to the D-ray and R-ray base structure (Figure 8.16, step 3). For every base ray, the intra-cluster structure is given by:

$$h_{\text{intra}}(t) = \sum_{m=1}^{N_{\text{intra}}} A(\tau_m) \exp\{j\theta_m\} \delta(t - \tau_m),$$ (8.77)

where τ_m is the arrival time of the mth intra-cluster component measured from the arrival time of the base D-ray or R-ray, $A(\tau_m)$, $P(\tau_m)$ and θ_m are the amplitude, power and phase of the kth intra-cluster component. The intra-cluster components are random, with Rayleigh-distributed amplitudes and random phases, with exponentially decaying PDP. The total power is determined by the K-factor with respect to the base D-ray or R-ray power:

$$\overline{P(\tau_m)} = P_0 \exp\left\{-\frac{\tau_m}{\gamma}\right\},$$ (8.78)

$$K_{\text{intra}} = \frac{P_{\text{base ray}}}{\sum P(\tau_m)}.$$ (8.79)

Generally, the intra-cluster structure generation is very similar to R-ray generation, except that for R-ray generation the LoS rays are used as a timing and power base, and for intra-cluster structure generation the cluster-base D-ray or R-ray is used for that purpose. Combining all D-rays, R-rays and their respective intra-cluster structure components will give the final channel impulse response in the form of eq. (8.73).

8.5.5   Mobility Effects

The mobility effects in the QD channel model are described by directly introducing the velocity vector for each MS. In the multipath channel the MS movement leads to additional phase rotation for each propagation path. For the purposes of the channel modeling, the motion effect can be introduced for D-rays and R-rays in the same way. The additional phase rotation for the kth ray caused by Doppler frequency shift is calculated as:

$$\Delta\varphi_k(t) = 2\pi f_k^D t,$$ (8.80)

$$f_k^D = (\mathbf{v}, \mathbf{r}_k)\frac{f_c}{c},$$

where f_k^D is the Doppler shift for the kth ray, $\mathbf{v}$ is the instantaneous vector of MS velocity (see Figure 8.17), $\mathbf{r}_k$ is the unity vector of the kth ray direction of arrival, f_c is the carrier frequency and $(,)$ denotes the scalar product.

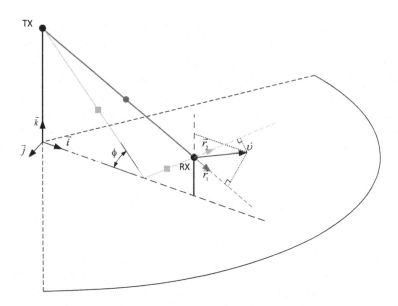

Figure 8.17 Model for mobility effects in 3D channel model for ray r_1 (circle) and r_2 (square). © 2016 IEEE. Reprinted with permission, from [144].

The velocity vector **v** can be represented as the sum of its scalar components ($\mathbf{v} = v_x\boldsymbol{i} + v_y\boldsymbol{j} + v_z\boldsymbol{k}$). For scenarios without preferred direction of motion, such as open areas, the horizontal component of velocity may have uniformly distributed direction and random or fixed value. For example, they may by described by 2D zero-mean Gaussian pdf with appropriate standard deviations σ_x, σ_y:

$$P\left(v_{x,y}\right) = \frac{1}{\sigma_{x,y}\sqrt{2\pi}} \exp\left\{-\frac{v_{x,y}^2}{2\sigma_{x,y}^2}\right\}. \tag{8.81}$$

As was shown in measurements [28], the vertical movement of the pedestrian MS has significant impact on the channel and also should be taken into account. In the important case when the MS is held by a human, the different models of human gait can be applied for vertical motion expressed through variable $z(t)$. In accordance with the QD methodology, the vertical motion is introduced as a stationary Gaussian random process. For the considered case of human gait, the following correlation function of $z(t)$ can be applied:

$$K_z\left[\tau\right] = \sigma_z^2 \exp\left\{-\frac{\tau^2}{\tau_z^2}\right\} \cos\left(2\pi f_0\tau\right), \tag{8.82}$$

with parameters adjusted to the real pedestrian motion at a speed of 3–5 km/h and f_0 being maximum Doppler frequency. The vertical component v_z of the velocity vector **v** can be defined through the user's vertical motion $z(t)$ as the first derivative. With the knowledge of the velocity vector and ray AoA, the values of the phase rotations can be calculated from eq. (8.80) and added to the corresponding D-ray and R-ray phases.

8.5.6   Channel Impulse Response Postprocessing

Channel impulse response postprocessing may include application of the antenna pattern, beamsteering algorithms and sampling the CIR to the desired discrete rate (see Figure 8.15 and also Chapter 1). These steps are the same for 802.11ad and 802.11ay models and are presented in [144]. The MIMO processing in the case of two or more phased antenna arrays is discussed in [129].

8.6   Map-Based Models

8.6.1   Background

The so-called map-based model developed in the METIS project is summarized in this section. The European METIS project was initiated in late 2012. It was one of the first larger-scale actions aiming to investigate components for the next generation, now called 5G, mobile and wireless communication systems. This section briefly summarizes the model; for more details the reader is referred to the references cited in the text.

8.6.2   Introduction

One of the key activities in the METIS project was to develop a radio channel model for the evaluation of technology components. The assumptions, use cases and scenarios, including environment descriptions, are specified in [146].

The first step in the development of a channel model for evaluation purposes is to identify requirements for the channel model. It is not feasible to define a model for all imaginable purposes; thus, it is crucial to understand first the need and the usage of the model. Generally speaking, the main goal of a channel model, as understood in this section, is to generate the wave propagation coefficients in time–space–frequency domains as accurately as possible for the applications of link-level and system-level simulations under acceptable computation complexity. Additionally, it is desirable that the model is straightforward and easily adapted by the engineers. The practical model requirements were identified in the METIS project based on different technology components of the expected 5G systems. The model requirements are defined in detail in [147] and the main requirements are summarized as follows:

1. spatial consistency and mobility;
2. diffuse vs. specular scattering;
3. support for very large arrays;
4. mmWave frequencies (frequency consistency); and
5. flexibility for a variety of simulation needs.

The key target set by the requirements was the consistency. The channel model to be developed had to be consistent and scalable for both spatially distributed antennas and devices, as well as across a wide range of frequency bands to be utilized.

The original intention was to extend the long line of GSCMs for 5G purposes. GSCMs represent a spatial radio channel with randomly drawn directional and propagation parameters, without any particular definition of the environment. However, during the process it was found that it is very difficult, or even impossible, to specify such extensions that would meet the requirements. Therefore, two different channel modeling approaches were selected: stochastic and map-based. This section focuses on the map-based model specified in [148] by the METIS project and more recently with some corrections and complements in [149]. This map-based model is deterministic in the sense that it uses a map or a layout in defining the particular environment and it uses image-based RT techniques on this map or layout to calculate propagation paths from the transmitters to the receivers. This is the difference when comparing the map-based model to the QD model [161, 162]. Furthermore, the map-based model is more general because it is used for both LoS and NLoS scenarios. It is used both for front-haul and back-haul use cases; the number of paths to resolve for a transmitter–receiver link is decided by the user.

8.6.3   Benefits of the Deterministic Model

There are several reasons why a deterministic model is considered to be suitable for 5G wireless communications:

Accuracy: real-world correspondence is better. In the deterministic model, the real-world propagation mechanisms involving LoS, reflection and penetration on smooth surfaces/slabs, diffraction on wedges and diffuse scattering on rough surfaces can be accurately emulated via a RT approach that is based on the proven propagation theories such as geometric optics and uniform theory of diffraction, with acceptable computation complexity.

Site-specific simulations: inherently guaranteed by the maps imported. In the map-based model, which is a deterministic procedure to emulate channels in a specific and deterministic scenario, the properties of each ray, such as path loss, propagation delay, AoA and AoD, are directly calculated by applying the RT principle to a well-defined deployment/environment layout map. The effects of shadowing and blockage are also inherently guaranteed.

Spatial and temporal consistency: inherently guaranteed in map-based models. According to the propagation theories, the channel coefficient of each path is calculated mainly based on the path length, incident angle, reflected/diffracted/scattered angle, as well as the electromagnetic property of the corresponding material. When the TX/RX node or the surrounding object moves, the smooth transition of the parameters leads to a consistent change on the channel coefficients, that is, spatial and temporal consistency can be achieved naturally.

Mesh networks and D2D, massive/distributed MIMO and CoMP: correlation between links inherently guaranteed in map-based models. Similar to the ability to fulfill spatial consistency, a map-based model can simulate the correlation between any two communication links whose transmitters or receivers are either the same or nearby, where the correlation exists in both large-scale fading and small-scale fading. This feature is crucial to support some of the 5G use cases such as mesh networks, D2D, massive/distributed MIMO or CoMP.

Frequency dependency and large bandwidth: interaction formulas are physics based. There is no need to have additional bandwidth handling in a RT tool as long as the bandwidth is less than 10% of the carrier frequency. The calculation based on center frequency can be representative for the whole bandwidth. If extremely large channel bandwidth ($>10\%$) is needed, the bandwidth can be partitioned into several bins and RT is applied separately to each bandwidth bin to obtain the frequency response, that is, channel transfer function. A discrete Fourier transformation can be used to get the CIR from the channel transfer function.

Spherical wave and large antenna arrays beyond consistency interval: inherently guaranteed in map-based models. Under the assumption of plane waves, the same modeling principle and procedure as in the stochastic modeling for the normal antenna array can be reused. In addition, for large antenna arrays beyond the consistency interval, spherical wave calculation should be taken into consideration, where the RT can support it by calculating the propagation parameters such as deterministic delays and arrival/departure angles per each pair of TX–RX antenna elements.

8.6.4   Model Overview

The map-based model is deterministic, utilizing well-known RT methods on simplified environmental maps (see [148] and references therein). The reason for deterministic modeling is to achieve spatial consistency. With maps and a discovery process of pathways it is possible to guarantee location-dependent radio channel coefficients between each TX and RX antenna array (or even antenna elements in very large array cases). Moreover, consistency across frequencies is obtained with textbook formulas for determining reflection, diffraction, scattering, etc. coefficients. In the map-based model the map is mainly a tool to achieve consistency.

As the map-based model is deterministic, the locations including the movement of TX and RX and the links between the TX and RX are specific. The modeling work starts by drawing or importing the environment, which is the map or, as an example, the layout in the case of indoor hotspot. The METIS project defines a set of simplified maps to be used in the analysis work, where maybe the Madrid grid is most often used, describing a 3D irregular urban city scenario with varying building heights and street widths. The other predefined maps are specified for indoor office and shopping malls, stadia and open air festival environments. The simplified map, whether Madrid grid or user-defined, can be used to simulate statistical parameters as is the case with all mainstream channel modeling methodologies. In other words, even though the environment is defined and the calculation method, that is, RT, is deterministic, the output changes (is "statistical") when the parameters in the environment are varied and multiple runs are performed. For example, the location is varied by moving the receiver(s). It corresponds then to the drop simulations of GSCMs. It should be noted

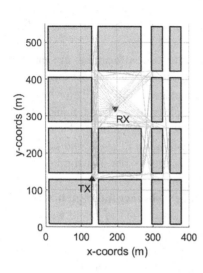 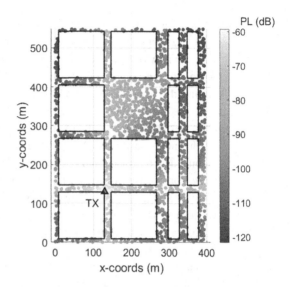

Figure 8.18 Madrid grid.

that the outputs for single runs are also possible. The Madrid grid is presented in Figure 8.18 along with an exemplary simulation case. The paths between a TX and a RX are shown.

The map can also be site-specific. If not purely statistical parameters are searched for but, for example, an installation in the field is studied, the map of that installation site is imported and used as the environment. This enables, for example, network optimization. This section does not cover how the map importing should be done; it is only stated that to build an importer is a straightforward task and there are already multiple softwares on the market for that purpose. The computational complexity increases with site-specific maps as the level of detail normally increases from that of the Madrid grid. The upper limit of the complexity is defined by the available computational resources and the implementation. A snapshot of a tool is shown in Figure 8.19. In that figure the buildings are simple 3D polygons, but they could be modeled in more detail whenever necessary – with the cost of increased computational effort. Part of this complexity definition is also the division of the surfaces into tiles. The tile centers act as the point sources for diffuse scattering. The transceiver locations are defined here and in that figure a BS and the user equipment are positioned.

Once the macroscopic environment is specified, one may add random objects such as pedestrians and vehicles – moving or static – or other blocking objects. The function of random objects is to act as both shadowing and scattering items. Some other QD models identify the main paths deterministically and draw randomly a set of additional paths. The procedure to generate variation and stochastic elements of the map-based model is different. Instead of random paths, the environment is partially defined in a random way by introducing random objects. The concept of random objects is described in more detail in Section 8.6.5.

Figure 8.19 Screenshot of a map-importing tool; BS and user equipment positionings are also shown.

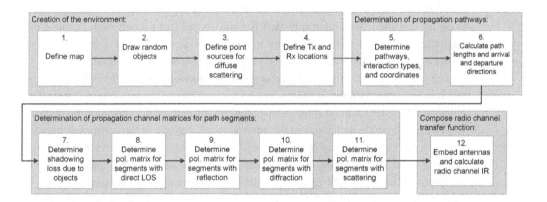

Figure 8.20 Block diagram of the map-based model. In the labels the abbreviation pol. means polarization. © 2016 IEEE. Reprinted, with permission, from [151].

The above process is the first part of the block diagram of the METIS map-based model shown in Figure 8.20. The next parts are described below.

The determination of **propagation pathways** is usually the highest computational workload. It starts by determining the coordinates of interaction points for parameter vectors using mathematical tools of analytical geometry. The principles of this part are intuitively easily understood, although writing an algorithmic description is complicated.

The process starts either from the TX or RX locations (see Figures 8.18 and 8.19). They are the first nodes. All possible second nodes are searched for; this is all nodes that are visible either with a LoS path or via a single specular reflection. Possible pathways are identified by checking whether any obstacles are present. This procedure is repeated to achieve any number of diffraction and specular reflection interactions. The desired accuracy of the computational effort may be used as the limiting factor. After the paths are determined the corresponding path lengths and arrival and departure angles are calculated. The directions are used in a later phase with the radiation patterns of the TX and RX antennas.

Steps 7–11 in Figure 8.20 are for the determination of channel matrices for the **path segments** found (through the nodes explained above). Each blocking or obstructing object is approximated by a simple equivalent block. This can be a rectangular screen or any other suitable model. The original METIS model used the screen always perpendicular to the path blocking to avoid the use of multiple screens. This is, however, implementation-dependent. Next, the propagation matrices are defined for the interactions. These are complex 2×2 matrices describing the gains of polarization components. The examples to calculate them are through the Fresnel reflection coefficient for the specular reflections and through the uniform theory of diffraction (UTD) in case of diffraction.

The theory for the interactions is strictly based on the physics, they are well-known frequency-dependent formulas. However, if for example if for diffuse scattering at mmWave a different formulation is found to be more appropriate, the implementation is easily modified. A similar approach is valid for all steps mentioned; because the computational complexity is high any improved and computationally more effective formula could be used instead of those proposed in the METIS deliverable.

The last step is to compose the **radio channel transfer function** by embedding the antenna radiation patterns in the losses and the composite propagation matrices. The composed transfer function is time-dependent because of the possible motion of TX or RX or time-dependent changes in the environment. The use of the composed transfer function works similarly as in the case of GSCMs. The transfer function (compared to the general format in eq. 8.14) is presented below to give an impression of the intuitiveness of the model as the transfer function contains the antenna patterns $\mathbf{g}$, steering vectors (space angles) $\mathbf{a}$, propagation transfer matrices $\mathbf{h}$ and divergence factors F for the corresponding path segments. The transmitter and receiver subscripts are left out of the formula below for simplicity; the formula is from an RX element u to a TX element s.

$$H(t, \tau) = \Sigma_{k=1}^{K} \mathbf{g}^{Rx}(-\mathbf{a}^{Rx}) e^{\frac{j 2 \pi d_k(t)}{\lambda}} \left(\Pi_{i=1}^{l_k} \mathbf{h}_{k,i}(t) F_{k,i} \right) \mathbf{g}^{Tx}(\mathbf{a}^{Tx}) \frac{\lambda}{4\pi} \delta(\tau - \tau_k). \quad (8.83)$$

8.6.5   Advanced Material on Model Overview

Random objects add variation to environments defined by uncomplicated maps. The motivation to introduce them is partially due to the severity of the blockage effect at mmWave, partially due to identified scenarios of open-air festivals and ultradense urban environments. A high number of people and their impact on the radio channel has to be modeled somehow. Random objects are either dropped randomly on the map with a predefined density or placed in a certain pattern (e.g., spectators in a sport stadium). Objects cause both scattering and blockage (shadowing) of a propagation path. A random object has a scenario-dependent size. The scattering effect is computed by treating an object as a conductive sphere with a certain radar cross-section. For the blockage, the object is interpreted as a rectangular screen and the shadowing is determined by KED across four edges of the screen.

Low complexity approximation of path losses utilizing Berg's recursive model together with identified specular reflections and LoS paths is also to be noticed. A characteristic of the map-based model is that it does not contain a separate path-loss model. This may be a drawback for quick field strength evaluations. On the other hand, it was not seen as feasible to first determine the absolute path contributions and then to normalize the gains and to utilize an empirical path-loss model on top of that. Partially to support faster field strength predictions, the map-based model has an option to utilize principles of Berg's recursive model [152] for urban street canyons and similar. It is a simpler way to approximate contributions of paths compared to, for example, the UTD.

The map-based model has not gained a similar position as geometry-based stochastic channel models, though the first formulas for the interactions were proposed already decades ago, and they have been also verified multiple times by measurement. When looking at reasons for not being widely used, the computational cost is probably the first item to tackle. The path finding may be very time-consuming if an intelligent algorithm is not used or the environment contains many details or the number of paths to find is high (for high accuracy). The algorithm question is an implementation issue; when writing the code any possible way to speed up the calculation should be used. For example, there might be ways to deduce that not all the surfaces or scatterers are electromagnetically visible to the transceivers, or that their electromagnetic effect is negligible. A large set of examples are found in the literature [153–156].

The number of details to take into account defines also strongly the computation time. The surfaces are divided into tiles, with tile centers acting as point sources for specular or diffuse scattering. The raster of the surfaces is a factor here.

The third factor for the computational complexity is the number of paths to find. This means how many reflections and/or diffractions should be searched for. The reflections are calculated much faster than the diffractions. Furthermore, the diffractions are most often more lossy. Therefore, depending on the simulation application, the appropriate number of reflections and diffractions should be defined a priori. The path loss for the diffractions is easily so high that no multiple diffractions can be reasonably included.

An option to speed up the simulation is to accelerate it using hardware. Graphics card-based so-called accelerator cards are already less expensive than a laptop and they offer significant improvements in calculation performance.

The latest 3GPP mmWave channel model TR 38.901 [120, section 8] contains an alternative model called map-based hybrid. This model is similar to the one described here but with differences in the random object handling and, more importantly, with the addition of stochastic cluster creation. The idea of the hybrid is that the deterministic modeling suffers in certain cases from the lack of diffuse paths, and a way to overcome this is to add random paths (clusters). Random paths are added also in cases of missing information about the environment and when the computational complexity is reduced through the stochastic add-ons without sacrificing accuracy. The 3GPP hybrid model contains rules about how to create those clusters very similar to the primary stochastic model in the same TR 38.901. This kind of addition of components to the model means the model is flexible for future editions. The hybrid model is not calibrated, which means the implementations are not compared against a specific test case; this is the reason why the use of the model is on a per-company basis whenever "the system performance is desired to be evaluated or predicted with the use of digital map to take into account the impacts from environmental structures and materials."

Another way to handle the complexity of diffuse scattering is through the use of geometric stochastic propagation graphs (GSPGs) [157–160]. The diffuse tail of the CIR is computationally intractable and in many cases it is not calculated because the energy level in the tail is significantly lower compared to the beginning of the CIR. The main MPCs like the LoS path and first reflections are modeled by the map-based model. However, if the dense MPC of the CIR (higher-order interactions) is desired, GSPG is computationally a very attractive option. Note that the discussion for R-rays does not use the geometry of the environment, that is, it is a random process. Here, the creation of the channel transfer function is divided into two parts: the deterministic part with, for example, a map-based model for the dominant paths; and the stochastic part with GSPG for the scattering on walls (geometry-dependent, but with unknown structures on wall surfaces). The diffuse tail may be with an unbounded number of components.

References

[1] A. Molisch, *Wireless Communications*, 2nd ed., Wiley: Chichester, 2010.

[2] T. Rappaport, *Wireless Communications: Principles and Practice*, 2nd ed., Prentice Hall: Upper Saddle River, NJ, 2002.

[3] A. Zajic, *Mobile-to-Mobile Wireless Channels*, Artech House: Boston, MA, 2013.

[4] P. Almers, E. Bonek, A. Burr, N. Czink, M. Debbah, V. Degli-Esposti, H. Hofstetter, P. Kyosti, D. Laurenson, G. Matz, A. F. Molisch, C. Oestges and H. Ozcelik, "Survey of channel and radio propagation models for wireless MIMO systems," *EURASIP Journal of Wireless Communications and Networks*, vol. 2007, pp. 1–19, Feb. 2007.

[5] R. B. Ertel, P. Cardieri, K. W. Sowerby, T. S. Rappaport and J. H. Reed, "Overview of spatial channel models for antenna array communication systems," *IEEE Personal Communications*, vol. 5, pp. 10–22, Feb. 1998.

[6] P. Kyosti, J. Meinila, L. Hentila, X. Zhao, T. Jamsa, C. Schneider, M. Narandzic, M. Milojevic, A. Hong, J. Ylitalo, V.-M. Holappa, M. Alatossava, R. Bultitude, Y. de Jong and T. Rautiainen, "Ist-4-027756 winner II deliverable 1.1.2. v.1.2, winner II channel models," technical report, 2007.

[7] ITU-R, "Guidelines for evaluation of radio interface technologies for IMT-advanced," technical report, 2008.

[8] P. Petrus, J. H. Reed and T. S. Rappaport, "Geometrically based statistical channel model for macrocellular mobile environments," *Global Telecommunications Conference, 1996. GLOBECOM '96*, vol. 2, Nov. 1996, pp. 1197–1201.

[9] P. Petrus, J. H. Reed and T. S. Rappaport, "Geometrical-based statistical macrocell channel model for mobile environments," *IEEE Transactions on Communications*, vol. 50, pp. 495–502, Mar. 2002.

[10] V6.1.0, G. T., "Spatial channel model for mimo simulations," technical report, 3GPP, 2003.

[11] S. Sun, G. R. MacCartney and T. S. Rappaport, "A novel millimeter-wave channel simulator and applications for 5G wireless communications," in *2017 IEEE International Conference on Communications (ICC)*, May 2017, pp. 1–7.

[12] H. Asplund, A. A Glazunov, A. F. Molisch, K. I. Pedersen and M. Steinbauer, "The COST 259 directional channel model – part II: Macrocells," *IEEE Transactions on Wireless Communications*, pp. 3434–3450, Dec. 2006.

[13] W. C. Jakes, *Microwave Mobile Communications*, 2nd ed., Wiley and IEEE Press: Piscataway, NJ, 1994.

[14] J. C. Liberti and T. S. Rappaport, "A geometrically based model for line-of-sight multipath radio channels," *Proceedings of Vehicular Technology Conference – VTC*, vol. 2, Apr. 1996, pp. 844–848.

[15] J. Medbo, K. Borner, K. Haneda, V. Hovinen, T. Imai, J. Jarvelainen, T. Jamsa, A. Karttunen, K. Kusume, J. Kyrolainen, P. Kyosti, J. Meinila, V. Nurmela, L. Raschkowski, A. Roivainen and J. Ylitalo, "Channel modelling for the fifth generation mobile communications," in *Proceedings of the 8th European Conference on Antennas and Propagation*, 2014, pp. 219–223.

[16] L. M. Correia, *Wireless Flexible Personalized Communications*, Wiley: Chichester, 2001.

[17] A. F. Molisch, H. Asplund, R. Heddergott, M. Steinbauer and T. Zwick, "The COST259 directional channel model – part I: Overview and methodology," *IEEE Transactions on Wireless Communications*, vol. 5, pp. 3421–3433, Dec. 2006.

[18] R. H. Clarke, "A statistical theory of mobile-radio reception," *Bell Systems Technical Journal*, pp. 957–1000, 1968.

[19] G. L. Stüber, *Principles of Mobile Communication*, 2nd ed., Kluwer: Norwell, MA, 2001.

[20] W. R. Braun and U. Dersch, "A physical mobile radio channel model," *IEEE Transactions on Vehicular Technology*, vol. 40, pp. 472–482, May 1991.

[21] F. P. Fontan, M. A. V. Castro, J. Kunisch, J. Pamp, E. Zollinger, S. Buonomo, P. Baptista and B. Arbesser, "A versatile framework for a narrow- and wide-band statistical propagation model for the LMS channel," *IEEE Transactions on Broadcasting*, vol. 43, pp. 431–458, Dec. 1997.

[22] M. Patzold, U. Killat, Y. Li and F. Laue, "Modeling, analysis, and simulation of nonfrequency-selective mobile radio channels with asymmetrical Doppler power spectral density shapes," *IEEE Transactions on Vehicular Technology*, vol. 46, pp. 494–507, May 1997.

[23] M. Patzold, Y. Li and F. Laue, "A study of a land mobile satellite channel model with asymmetrical Doppler power spectrum and lognormally distributed line-of-sight component," *IEEE Transactions on Vehicular Technology*, vol. 47, pp. 297–310, Feb. 1998.

[24] J. S. Sadowsky and V. Kafedziski, "On the correlation and scattering functions of the WSSUS channel for mobile communications," *IEEE Transactions on Vehicular Technology*, vol. 47, pp. 270–282, Feb. 1998.

[25] P. C. Fannin and A. Molina, "Analysis of mobile radio channel sounding measurements in inner city Dublin at 1.808 GHz," *IEE Proceedings – Communications*, vol. 143, pp. 311–316, Oct. 1996.

[26] J. Fuhl, J. P. Rossi and E. Bonek, "High-resolution 3-D direction-of-arrival determination for urban mobile radio," *IEEE Transactions on Antennas and Propagation*, vol. 45, pp. 672–682, Apr. 1997.

[27] K. Kalliola, K. Sulonen, H. Laitinen, O. Kivekas, J. Krogerus and P. Vainikainen, "Angular power distribution and mean effective gain of mobile antenna in different propagation environments," *IEEE Transactions on Vehicular Technology*, vol. 51, no. 5, pp. 823–838, 2002.

[28] J. Karedal, F. Tufvesson, N. Czink, A. Paier, C. Dumard, T. Zemen, C. F. Mecklenbrauker and A. F. Molisch, "A geometry-based stochastic MIMO model for vehicle-to-vehicle communications," *IEEE Transactions on Wireless Communications*, vol. 8, no. 7, pp. 3646–3657, 2009.

[29] A. E. Kuchar, A. Aparicio, J. P. Rossi and E. Bonek, *Wireless Personal Communications: Emerging Technologies for Enhanced Communications,* Kluwer: Boston, MA, 1999.

[30] W. Lee, "Finding the approximate angular probability density function of wave arrival by using a directional antenna," *IEEE Transactions on Antennas and Propagation*, vol. 21, pp. 328–334, May 1973.

[31] P. Petrus, J. H. Reed and T. S. Rappaport, "Effects of directional antennas at the base station on the Doppler spectrum," *IEEE Communications Letters*, vol. 1, pp. 40–42, Mar. 1997.

[32] J. P. Rossi, J. P. Barbot and A. J. Levy, "Theory and measurement of the angle of arrival and time delay of UHF radiowaves using a ring array," *IEEE Transactions on Antennas and Propagation*, vol. 45, pp. 876–884, May 1997.

[33] S. Wyne, A. F. Molisch, P. Almers, G. Eriksson, J. Karedal and F. Tufvesson, "Outdoor-to-indoor office MIMO measurements and analysis at 5.2 GHz," *IEEE Transactions on Vehicular Technology*, vol. 57, no. 3, pp. 1374–1386, 2008.

[34] A. G. Zajic, G. L. Stuber, T. G. Pratt and S. T. Nguyen, "Wideband MIMO mobile-to-mobile channels: Geometry-based statistical modeling with experimental verification," *IEEE Transactions on Vehicular Technology*, vol. 58, pp. 517–534, Feb. 2009.

[35] K. Anim-Appiah, "Complex envelope correlations for non-isotropic scattering," *Electronics Letters*, vol. 34, pp. 918–919, Apr. 1998.

[36] M. D. Austin and G. L. Stuber, "Velocity adaptive handoff algorithms for microcellular systems," *IEEE Transactions on Vehicular Technology*, vol. 43, pp. 549–561, Aug. 1994.

[37] K. I. Pedersen, P. E. Mogensen and B. H. Fleury, "Power azimuth spectrum in outdoor environments," *Electronics Letters*, vol. 33, pp. 1583–1584, Aug. 1997.

[38] A. Abdi, J. A. Barger and M. Kaveh, "A parametric model for the distribution of the angle of arrival and the associated correlation function and power spectrum at the mobile station," *IEEE Transactions on Vehicular Technology*, vol. 51, pp. 425–434, May 2002.

[39] K. I. Pedersen, P. E. Mogensen and B. H. Fleury, "A stochastic model of the temporal and azimuthal dispersion seen at the base station in outdoor propagation environments," *IEEE Transactions on Vehicular Technology*, vol. 49, pp. 437–447, Mar. 2000.

[40] R. von Mises, "Über die ganzzahligkeit der atomgewicht und verwandte fragen," *Physikal. Z.*, vol. 19, pp. 490–500, 1918.

[41] K. V. Mardia, *Statistics of Directional Data*, Academic Press: London, 1972.

[42] M. K. Simon, S. M. Hinedi and W. C. Lindsey, *Digital Communication Techniques: Signal Design and Detection*, Prentice-Hall: Englewood Cliffs, NJ, 1995.

[43] A. J. Viterbi, "Phase-locked loop dynamics in the presence of noise by Fokker-Planck techniques," *Proceedings of the IEEE*, vol. 51, pp. 1737–1753, Dec. 1963.

[44] H. Leib and S. Pasupathy, "The phase of a vector perturbed by Gaussian noise and differentially coherent receivers," *IEEE Transactions on Information Theory*, vol. 34, pp. 1491–1501, Nov. 1988.

[45] A. F. Molisch, "A generic model for MIMO wireless propagation channels in macro and microcells," *IEEE Transactions on Signal Processing*, vol. 52, no. 1, pp. 61–71, 2004.

[46] L. Liu, C. Oestges, J. Poutanen, K. Haneda, P. Vainikainen, F. Quitin, F. Tufvesson and P. D. Doncker, "The COST 2100 MIMO channel model," *IEEE Wireless Communications*, vol. 19, pp. 92–99, Dec. 2012.

[47] A. F. Molisch and H. Hofstetter, "The COST 273 MIMO channel model," in L. Correia (ed.), *Mobile Broadband Multimedia Networks*, Academic Press, 2006.

[48] P. Bello, "Characterization of random time-variant linear channels," *IEEE Transactions on Communications*, vol. 11, pp. 360–393, Dec. 1963.

[49] A. Gehring, M. Steinbauer, I. Gaspard and M. Grigat, "Empirical channel stationarity in urban environments," *Proceedings of the European Personal Mobile Communication Conference*, 2001.

[50] R. Kattenbach, "Statistical modeling of small-scale fading in directional radio channels," *IEEE Journal on Selected Areas in Communications*, vol. 20, no. 3, pp. 584–592, 2002.

[51] B. H. Fleury, "An uncertainty relation for WSS processes and its application to WSSUS systems," *IEEE Transactions on Communications*, vol. 44, no. 12, pp. 1632–1634, 1996.

[52] L. M. Correia, *Mobile Broadband Multimedia Networks*, Academic Press: New York, 2006.

[53] H. Hofstetter, A. F. Molisch and N. Czink, "A twin-cluster MIMO channel model," *First European Conference on Antennas and Propagation, 2006. EuCAP 2006*, IEEE, 2006, pp. 1–8.

[54] H. Xu, T. S. Rappaport, R. J. Boyle and J. H. Schaffner, "Measurements and models for 38 GHz point-to-multipoint radiowave propagation," *IEEE Journal on Selected Areas in Communications*, vol. 18, pp. 310–321, Mar. 2000.

[55] M. R. Akdeniz, Y. Liu, M. K. Samimi, S. Sun, S. Rangan, T. S. Rappaport and E. Erkip, "Millimeter wave channel modeling and cellular capacity evaluation," *IEEE Journal on Selected Areas in Communications*, vol. 32, pp. 1164–1179, June 2014.

[56] M. K. Samimi and T. S. Rappaport, "Ultra-wideband statistical channel model for non line of sight millimeter-wave urban channels," *2014 IEEE Global Communications Conference*, Dec. 2014, pp. 3483–3489.

[57] J. Zhang, J. Richter, L. P. Ivrissimtzis and M. O. Al-Nuaimi, "Measurements and modelling for fixed wireless access systems at millimeter wave frequencies," *Twelfth International Conference on Antennas and Propagation, 2003 (ICAP 2003)*, vol. 1, Mar. 2003, pp. 433–436.

[58] Z. Muhi-Eldeen, L. P. Ivrissimtzis and M. Al-Nuaimi, "Modelling and measurements of millimetre wavelength propagation in urban environments," *IET Microwaves, Antennas Propagation*, vol. 4, pp. 1300–1309, Sept. 2010.

[59] M. Kyro, S. Ranvier, V. M. Kolmonen, K. Haneda and P. Vainikainen, "Long range wideband channel measurements at 81–86 GHz frequency range," *Proceedings of the Fourth European Conference on Antennas and Propagation*, Apr. 2010, pp. 1–5.

[60] M. Kyro, V. M. Kolmonen and P. Vainikainen, "Experimental propagation channel characterization of mm-wave radio links in urban scenarios," *IEEE Antennas and Wireless Propagation Letters*, vol. 11, pp. 865–868, 2012.

[61] E. Torkildson, H. Zhang and U. Madhow, "Channel modeling for millimeter wave MIMO," in *2010 Information Theory and Applications Workshop (ITA)*, Jan. 2010, pp. 1–8.

[62] M. T. Martinez-Ingles, J. Pascual-Garcia, J. V. Rodriguez, R. Lopez-Moya, J. M. Molina Garcia-Pardo and L. Juan-Llacer, "UTD PO solution for estimating the propagation loss due to the diffraction at the top of a rectangular obstacle when illuminated from a low source," *IEEE Transactions on Antennas and Propagation*, vol. 61, pp. 6247–6250, Dec. 2013.

[63] M. T. Martinez-Ingles, J. V. Rodriguez, J. M. Molina-Garcia-Pardo, J. Pascual-Garcia and L. Juan-Llacer, "Parametric study and validation of a UTD-PO multiple-cylinder diffraction solution through measurements at 60 GHz," *IEEE Transactions on Antennas and Propagation*, vol. 61, pp. 4397–4400, Aug. 2013.

[64] M. T. Martinez-Ingles, J. V. Rodriguez, J. M. Molina-Garcia-Pardo, J. Pascual-Garcia and L. Juan-Llacer, "Comparison of a UTD-PO formulation for multiple-plateau diffraction with measurements at 62 GHz," *IEEE Transactions on Antennas and Propagation*, vol. 61, pp. 1000–1003, Feb. 2013.

[65] A. Yamamoto, K. Ogawa, T. Horimatsu, A. Kato and M. Fujise, "Path-loss prediction models for intervehicle communication at 60 GHz," *IEEE Transactions on Vehicular Technology*, vol. 57, pp. 65–78, Jan. 2008.

[66] T. Jamsa, P. K, Anite and K. Kusume, "Deliverable d1.2 initial channel models based on measurements," www.metis2020.com/wpcontent/uploads/deliverables/METIS D1.2 v1.pdf, 2014.

[67] S. Cherry, "Edholm's law of bandwidth," *IEEE Spectrum*, vol. 41, pp. 58–60, July 2004.

[68] S. Geng, J. Kivinen, X. Zhao and P. Vainikainen, "Millimeter-wave propagation channel characterization for short-range wireless communications," *IEEE Transactions on Vehicular Technology*, vol. 58, pp. 3–13, Jan. 2009.

[69] N. Moraitis and P. Constantinou, "Indoor channel measurements and characterization at 60 GHz for wireless local area network applications," *IEEE Transactions on Antennas and Propagation*, vol. 52, pp. 3180–3189, Dec. 2004.

[70] M. Peter and W. Keusgen, "Analysis and comparison of indoor wideband radio channels at 5 and 60 GHz," *2009 3rd European Conference on Antennas and Propagation*, Mar. 2009, pp. 3830–3834.

[71] P. Smulders, "Exploiting the 60 GHz band for local wireless multimedia access: Prospects and future directions," *IEEE Communications Magazine*, vol. 40, pp. 140–147, Jan. 2002.

[72] H. Xu, V. Kukshya and T. S. Rappaport, "Spatial and temporal characteristics of 60 GHz indoor channels," *IEEE Journal on Selected Areas in Communications*, vol. 20, pp. 620–630, Apr. 2002.

[73] H. Yang, P. F. M. Smulders and M. H. A. J. Herben, "Indoor channel measurements and analysis in the frequency bands 2 GHz and 60 GHz," *2005 IEEE 16th International Symposium on Personal, Indoor and Mobile Radio Communications*, vol. 1, pp. 579–583, Sept. 2005.

[74] T. Zwick, T. J. Beukema and H. Nam, "Wideband channel sounder with measurements and model for the 60 GHz indoor radio channel," *IEEE Transactions on Vehicular Technology*, vol. 54, pp. 1266–1277, July 2005.

[75] S. Kim and A. Zajic, "Statistical modeling of THz scatter channels," *Proceedings of IEEE European Conference on Antennas and Propagation EuCAP15*, Apr. 2015, pp. 1–5.

[76] M. Jacob, S. Priebe, R. Dickhoff, T. Kleine-Ostmann, T. Schrader and T. Kurner, "Diffraction in mm and sub-mm wave indoor propagation channels," *IEEE Transactions on Microwave Theory and Techniques*, vol. 60, pp. 833–844, Mar. 2012.

[77] C. Jansen, R. Piesiewicz, D. Mittleman, T. Kurner and M. Koch, "The impact of reflections from stratified building materials on the wave propagation in future indoor terahertz communication systems," *IEEE Transactions on Antennas and Propagation*, vol. 56, pp. 1413–1419, May 2008.

[78] J. M. Jornet and I. F. Akyildiz, "Channel modeling and capacity analysis for electromagnetic wireless nanonetworks in the terahertz band," *IEEE Transactions on Wireless Communications*, vol. 10, pp. 3211–3221, Oct. 2011.

[79] D. W. Matolak, S. Kaya and A. Kodi, "Channel modeling for wireless networks-on-chips," *IEEE Communications Magazine*, vol. 51, pp. 180–186, June 2013.

[80] R. Piesiewicz, C. Jansen, D. Mittleman, T. Kleine-Ostmann, M. Koch and T. Kurner, "Scattering analysis for the modeling of THz communication systems," *IEEE Transactions on Antennas and Propagation*, vol. 55, pp. 3002–3009, Nov. 2007.

[81] R. Piesiewicz, T. Kleine-Ostmann, N. Krumbholz, D. Mittleman, M. Koch and T. Kurner, "Terahertz characterisation of building materials," *Electronics Letters*, vol. 41, pp. 1002–1004, Sept. 2005.

[82] S. Priebe, M. Jacob, C. Jastrow, T. Kleine-Ostmann, T. Schrader and T. Kurner, "A comparison of indoor channel measurements and ray tracing simulations at 300 GHz," *35th International Conference on Infrared, Millimeter, and Terahertz Waves*, Sept. 2010, pp. 1197–1201.

[83] Y. Yang, M. Mandehgar and D. R. Grischkowsky, "Broadband THz pulse transmission through the atmosphere," *IEEE Transactions on Terahertz Science and Technology*, vol. 1, pp. 264–273, Sept. 2011.

[84] S. Priebe and T. Kurner, "Stochastic modeling of THz indoor radio channels," *IEEE Transactions on Wireless Communications*, vol. 12, Sept. 2013, pp. 4445–4455.

[85] A. F. Molisch, "Ultrawideband propagation channels: Theory, measurement, and modeling," *IEEE Transactions on Vehicular Technology*, vol. 54, pp. 1528–1545, Sept. 2005.

[86] A. A. M. Saleh and R. Valenzuela, "A statistical model for indoor multipath propagation," *IEEE Journal on Selected Areas in Communications*, vol. 5, pp. 128–137, Feb. 1987.

[87] S. Kim and A. Zajic, "Statistical modeling and simulation of short-range device-to-device communication channels at sub-THz frequencies," *IEEE Transactions on Wireless Communications*, vol. 15, no. 9, pp. 6423–6433, Sept. 2016.

[88] C. Han, A. O. Bicen and I. F. Akyildiz, "Multi-ray channel modeling and wideband characterization for wireless communications in the terahertz band," *IEEE Transactions on Wireless Communications*, vol. 14, pp. 2402–2412, May 2015.

[89] R. W. Lucky, *Principles of Data Communication*, McGraw-Hill: New York, 1985.

[90] G. L. Turin, F. D. Clapp, T. L. Johnston, S. B. Fine and D. Lavry, "A statistical model of urban multipath propagation," *IEEE Transactions on Vehicular Technology*, vol. 21, pp. 1–9, Feb. 1972.

[91] D. C. Cox, "910 MHz urban mobile radio propagation: Multipath characteristics in New York City," *IEEE Transactions on Vehicular Technology*, vol. 22, pp. 104–110, Nov. 1973.

[92] M. Failli, *Digital Land Mobile Radio Communications. COST 207.* EC: Brussels, 1989.

[93] T. S. Rappaport, S. Y. Seidel and K. Takamizawa, "Statistical channel impulse response models for factory and open plan building radio communicate system design," *IEEE Transactions on Communications*, vol. 39, pp. 794–807, May 1991.

[94] J. G. Proakis and M. Salehi, *Digital Communications*, 5th ed., McGraw-Hill: New York, 2007.

[95] N. Michelusi, U. Mitra, A. F. Molisch and M. Zorzi, "UWB sparse/diffuse channels, part I: Channel models and Bayesian estimators," *IEEE Transactions on Signal Processing*, vol. 60, pp. 5307–5319, Oct. 2012.

[96] R. B. Ertel and J. H. Reed, "Angle and time of arrival statistics for circular and elliptical scattering models," *IEEE Journal on Selected Areas in Communications*, vol. 17, pp. 1829–1840, Nov. 1999.

[97] D. W. Matolak and Q. Wu, "Channel models for v2v communications: A comparison of different approaches," *Proceedings of the 5th European Conference on Antennas and Propagation (EUCAP)*, Apr. 2011, pp. 2891–2895.

[98] A. F. Molisch, *Wireless Communications*, Wiley: Chichester, 2012.

[99] J. D. Parsons, *The Mobile Radio Propagation Channel*, Wiley: Chichester, 2000.

[100] G. R. MacCartney and T. S. Rappaport, "A flexible millimeter-wave channel sounder with absolute timing," *IEEE Journal on Selected Areas in Communications*, vol. 35, pp. 1402–1418, June 2017.

[101] S. Deng, G. R. MacCartney and T. S. Rappaport, "Indoor and outdoor 5G diffraction measurements and models at 10, 20, and 26 GHz," *2016 IEEE Global Communications Conference (GLOBECOM)*, Dec. 2016, pp. 1–7.

[102] G. Liang and H. L. Bertoni, "A new approach to 3-D ray tracing for propagation prediction in cities," *IEEE Transactions on Antennas and Propagation*, vol. 46, pp. 853–863, June 1998.

[103] T. S. Rappaport, G. R. MacCartney, S. Sun, H. Yan and S. Deng, "Small-scale, local area, and transitional millimeter wave propagation for 5G communications," *IEEE Transactions on Antennas and Propagation*, vol. 65, pp. 6474–6490, Dec. 2017.

[104] J. Ryan, G. R. MacCartney and T. S. Rappaport, "Indoor office wideband penetration loss measurements at 73 GHz," *2017 IEEE International Conference on Communications Workshops (ICC Workshops)*, May 2017, pp. 228–233.

[105] V. Degli-Esposti and H. L. Bertoni, "Evaluation of the role of diffuse scattering in urban microcellular propagation," *IEEE VTS 50th Vehicular Technology Conference, 1999. VTC 1999-Fall*, vol. 3, IEEE, 1999, pp. 1392–1396.

[106] T. S. Rappaport, Y. Xing, G. R. MacCartney, A. F. Molisch, E. Mellios and J. Zhang, "Overview of millimeter wave communications for fifth-generation (5G) wireless networks: With a focus on propagation models," *IEEE Transactions on Antennas and Propagation*, vol. 65, pp. 6213–6230, Dec. 2017.

[107] M. Rumney, "Testing 5G: Time to throw away the cables," *Microwave Journal*, Nov. 2016.

[108] mmMAGIC, "Measurement results and final mmMagic channel models," technical report H2020-ICT-671650-mmMAGIC/D2.2, May 2017.

[109] T. S. Rappaport, G. R. MacCartney, M. K. Samimi and S. Sun, "Wideband millimeter-wave propagation measurements and channel models for future wireless communication system design (invited paper)," *IEEE Transactions on Communications*, vol. 63, pp. 3029–3056, Sept. 2015.

[110] D. W. Matolak, I. Sen and W. Xiong, "The 5 GHz airport surface area channel – part I: Measurement and modeling results for large airports," *IEEE Transactions on Vehicular Technology*, vol. 57, pp. 2014–2026, July 2008.

[111] G. R. MacCartney, T. S. Rappaport and S. Rangan, "Rapid fading due to human blockage in pedestrian crowds at 5G millimeter-wave frequencies," *GLOBECOM 2017: 2017 IEEE Global Communications Conference*, Dec. 2017, pp. 1–7.

[112] G. R. MacCartney, S. Deng, S. Sun, T. S. Rappaport, "Millimeter-wave human blockage at 73 GHz with a simple double knife-edge diffraction model and extension for directional antennas," *IEEE 84th Vehicular Technology Conference Fall (VTC 2016-Fall)*, Sept. 2016, pp. 1–6.

[113] J. Kunisch and J. Pamp, "Ultra-wideband double vertical knife-edge model for obstruction of a ray by a person," *2008 IEEE International Conference on Ultra-Wideband*, vol. 2, Sept. 2008, pp. 17–20.

[114] M. K. Samimi, G. R. MacCartney, S. Sun and T. S. Rappaport, "28 GHz millimeter-wave ultrawideband small-scale fading models in wireless channels," *2016 IEEE 83rd Vehicular Technology Conference (VTC 2016-Spring)*, May 2016, pp. 1–6.

[115] S. Sun, H. Yan, G. R. MacCartney and T. S. Rappaport, "Millimeter wave small-scale spatial statistics in an urban microcell scenario," *2017 IEEE International Conference on Communications (ICC)*, May 2017, pp. 1–7.

[116] G. R. MacCartney and T. S. Rappaport, "Rural macrocell path loss models for millimeter wave wireless communications," *IEEE Journal on Selected Areas in Communications*, vol. 35, pp. 1663–1677, July 2017.

[117] T. S. Rappaport, S. Sun and M. Shafi, "Investigation and comparison of 3GPP and NYUSIM channel models for 5G wireless communications," *2017 IEEE 86th Vehicular Technology Conference (VTC Fall)*, Sept. 2017.

[118] N. Czink, P. Cera, J. Salo, E. Bonek, J. Nuutinen and J. Ylitalo, "A framework for automatic clustering of parametric MIMO channel data including path powers," *IEEE Vehicular Technology Conference*, Sept. 2006, pp. 1–5.

[119] B. H. Fleury, M. Tschudin, R. Heddergott, D. Dahlhaus and K. I. Pedersen, "Channel parameter estimation in mobile radio environments using the SAGE algorithm," *IEEE Journal on Selected Areas in Communications*, vol. 17, pp. 434–450, Mar. 1999.

[120] TR 38.901, "Channel model for frequencies from 0.5 to 100 GHz," technical report Rel. 14, V14.0.0, 3GPP, Mar. 2017.

[121] M. K. Samimi and T. S. Rappaport, "3-D millimeter-wave statistical channel model for 5G wireless system design," *IEEE Transactions on Microwave Theory and Techniques*, vol. 64, pp. 2207–2225, July 2016.

[122] T. S. Rappaport, S. Sun, R. Mayzus, H. Zhao, Y. Azar, K. Wang, G. N. Wong, J. K. Schulz, M. Samimi and F. Gutierrez, "Millimeter wave mobile communications for 5G cellular: It will work!," *IEEE Access*, vol. 1, pp. 335–349, May 2013.

[123] G. R. MacCartney and T. S. Rappaport, "Study on 3GPP rural macrocell path loss models for millimeter wave wireless communications," *2017 IEEE International Conference on Communications (ICC)*, May 2017, pp. 1–7.

[124] M. K. Samimi and T. S. Rappaport, "Local multipath model parameters for generating 5G millimeter-wave 3GPP-like channel impulse response," *2016 10th European Conference on Antennas and Propagation (EuCAP)*, Apr. 2016, pp. 1–5.

[125] WINNER, "Spatial channel model for multiple input multiple output (MIMO) simulations, 3GPP TR 25.996 v6.1.0," Sept. 2003.

[126] F. Khan and Z. Pi, "Millimeter-wave Mobile Broadband (MMB): Unleashing 3-300 GHz spectrum," *34th IEEE Sarnoff Symposium*, May 2011, pp. 1–6.

[127] T. S. Rappaport, F. Gutierrez, E. Ben-Dor, J. N. Murdock, Q. Yijun and J. I. Tamir, "Broadband millimeter-wave propagation measurements and models using adaptive-beam antennas for outdoor urban cellular communications," *IEEE Transactions on Antennas and Propagation*, vol. 61, pp. 1850–1859, 2013.

[128] Project 608637 MIWEBA51, I. F. M., "Deliverable d5.1, channel modelling and characterization," 2014.

[129] "Channel models for IEEE 802.11ay," technical report, IEEE doc.: 802.11-15/1150r9, 2017.

[130] A. Maltsev, A. Pudeyev, A. Lomayev and I. Bolotin, "Channel modeling in the next generation mmwave WI-Fi: IEEE 802.11ay standard," *European Wireless 2016; 22th European Wireless Conference*, May 2016, pp. 1–8.

[131] A. Maltsev, A. Pudeyev, I. Bolotin, Y. Gagiev, A. Lomayev, K. Johnsson, T. Sakamoto, H. Motozuka, C. Gentile, N. Golmie, J. Luo, Y. Xin, K. Zeng, R. Müller, R. Yang, F. Fellhauer, M. Kim and S. Sasaki, "Channel models for IEEE 802.11ay", IEEE 802.11-09/0334r8, Sept. 2015. https://mentor.ieee.org/802.11/dcn/15/11-15-1150-03-00ay-channel-models-for-ieee-802-11ay.docx.

[132] M. Peter, W. Keusgen and R. J. Weiler, "On path loss measurement and modeling for millimeter-wave 5G," *2015 9th European Conference on Antennas and Propagation (EuCAP)*, May 2015, pp. 1–5.

[133] R. J. Weiler, M. Peter, W. Keusgen, A. Kortke and M. Wisotzki, "Millimeter-wave channel sounding of outdoor ground reflections," *IEEE Radio and Wireless Symposium (RWS)*, 2015.

[134] R. J. Weiler, M. Peter, W. Keusgen, H. Shimodaira, K. T. Gia and K. Sakaguchi, "Outdoor millimeter-wave access for heterogeneous networks: Path loss and system performance," *2014 IEEE 25th Annual International Symposium on Personal, Indoor, and Mobile Radio Communication (PIMRC)*, Sept. 2014, pp. 2189–2193.

[135] A. Alexiou, ed., *5G Wireless Technologies*, IET: London, 2017.

[136] A. Maltsev, A. Pudeyev, I. Karls, I. Bolotin, G. Morozov, R. J. Weiler, M. Peter, W. Keusgen, M. Danchenko and A. Kuznetsov, "Quasi-deterministic approach to mmwave channel modelling in the FP7 MIWEBA project," *WWRF'33*, 2014.

[137] R. Weiler, M. Peter, W. Keusgen, A. Maltsev, I. Karls, A. Pudeyev, I. Bolotin, I. Siaud and A.-M. Ulmer-Moll, "Quasi-deterministic millimeter-wave channel models in MIWEBA," *EURASIP Journal on Wireless Communications and Networking*, Art. 84, pp. 1–16, Mar. 2016.

[138] J. Kunisch and J. Pamp, "An ultra-wideband space-variant multipath indoor radio channel model," *2003 IEEE Conference on Ultra Wideband Systems and Technologies*, IEEE, 2003, pp. 290–294.

[139] A. F. Molisch, M. Steinbauer and H. Asplund, "Virtual cell deployment areas and cluster tracing: New methods for directional channel modeling in microcells," *Vehicular Technology Conference, 2002. VTC Spring 2002*, vol. 3, IEEE, 2002, pp. 1279–1283.

[140] A. Maltsev, A. Pudeyev, I. Karls, I. Bolotin, G. Morozov, R. Weiler, M. Peter and W. Keusgen, "Quasi-deterministic approach to mmwave channel modeling in a nonstationary environment," *2014 IEEE Globecom Workshops (GC Wkshps)*, Dec. 2014, pp. 966–971.

[141] A. Hammoudeh, M. Sanchez and E. Grindrod, "Modelling of propagation in outdoor microcells at 62.4 GHz," *Microwave Conference*, vol. 1, 1997, pp. 199–123.

[142] K. Sarabandi, E. Li and A. Nashashibi, "Modelling and measurements of scattering from road surfaces at millimeter-wave frequencies," *IEEE Transactions on Antennas and Propagation*, vol. 45, no. 11, pp. 1679–1688, 1997.

[143] ITU, "Reflection from the surface of the Earth," report 1008-1, I. R.

[144] A. Maltsev, I. Bolotin, A. Lomayev, A. Pudeyev and M. Danchenko, "User mobility impact on millimeter-wave system performance," *2016 10th European Conference on Antennas and Propagation (EuCAP)*, Mar. 2016, pp. 1–5.

[145] A. Maltsev, E. Perahia, R. Maslennikov, A. Sevastyanov, A. Lomayev and A. Khoryaev, "Impact of polarization characteristics on 60 GHz indoor radio communication systems," *IEEE Letters in Antennas and Wireless Propagation Letters*, vol. 9, pp. 413–416, 2010.

[146] J. F. Monserrat and M. Fallgren, "Simulation guidelines, Deliverable 6.1 v1," technical report, ICT-317669 METIS project, 2013.

[147] J. Medbo, K. Borner, K. Haneda, V. Hovinen, T. Imai, J. Jarvelainen, T. Jamsa, A. Karttunen, K. Kusume, J. Kyrolainen, P. Kyosti, J. Meinila, V. Nurmela, L. Raschkowski, A. Roivainen and J. Ylitalo, "Channel modelling for the fifth generation mobile communications," *8th European Conference on Antennas and Propagation (EuCAP 2014)*, Apr. 2014, pp. 219–223.

[148] L. Rachowski, P. Kyösti, K. Kusume and T. Jämsä, "METIS channel models, deliverable 1.4 v.1.3," technical report ICT-317669, METIS project, 2015.

[149] P. Kyösti, J. Lehtomäki, J. Medbo and M. Latva-aho, "Map-based channel model for evaluation of 5G wireless communication systems," *IEEE Transactions on Antennas and Propagation*, vol. 65, pp. 6491–6504, Dec. 2017.

[150] A. Hekkala, P. Kyösti, J. Dou, L. Tian, N. Zhang, W. Zhang and B. Gao, "Map-based channel model for 5G wireless communications," submitted invited talk, 32nd URSI GASS, Aug. 2017.

[151] J. Medbo, P. Kyösti, K. Kusume, L. Raschkowski, K. Haneda, T. Jämsä, V. Nurmela, A. Roivainen and J. Meinilä, "Radio propagation modeling for 5G mobile and wireless communications," *IEEE Communications Magazine*, vol. 54, pp. 144–151, June 2016.

[152] J. E. Berg, "A recursive method for street microcell path loss calculations," *Sixth IEEE International Symposium on Personal, Indoor and Mobile Radio Communications (PIMRC'95)*, vol. 1, Sept. 1995, pp. 140–143.

[153] S. Hussain and C. Brennan, "An intra-visibility matrix based environment pre-processing for efficient ray tracing," *2017 11th European Conference on Antennas and Propagation (EUCAP)*, Mar. 2017, pp. 3520–3523.

[154] Z. Zhang, Z. Yun and M. F. Iskander, "New computationally efficient 2.5D and 3D ray tracing algorithms for modeling propagation environments," *IEEE Antennas and Propagation Society International Symposium. 2001 Digest*, vol. 1, July 2001, pp. 460–463.

[155] T. Imai, "Novel ray-tracing acceleration technique employing genetic algorithm for radio propagation prediction," *2006 First European Conference on Antennas and Propagation*, Nov. 2006, pp. 1–6.

[156] A. Escobar, L. Lozano, H. Cadavid and M. F. Catedra, "A new ray-tracing acceleration technique for radio propagation," *2010 IEEE Antennas and Propagation Society International Symposium*, July 2010, pp. 1–4.

[157] T. Pedersen and B. H. Fleury, "A realistic radio channel model based on stochastic propagation graphs," *Proceedings of the 5th MATHMOD Vienna – 5th Vienna Symposium on Mathematical Modelling*, vols. 1, 2, Feb. 2006, pp. 324–331.

[158] T. Pedersen and B. H. Fleury, "Radio channel modelling using stochastic propagation graphs," *IEEE International Conference on Communications. ICC'07*, 2007, pp. 2733–2738.

[159] T. Pedersen, G. Steinböck and B. H. Fleury, "Modeling of reverberant radio channels using propagation graphs," *IEEE Transactions on Antennas and Propagation*, vol. 60, no. 12, pp. 5978–5988, 2012.

[160] G. Steinböck, A. Karstensen, P. Kyösti and A. Hekkala, "A 5G hybrid channel model considering rays and geometric stochastic propagation graph," *IEEE 27th Annual International Symposium on Personal, Indoor and Mobile Radio Communications (PIMRC'16)*, Sept. 2016.

[161] S. Y. Seidel and T. S. Rappaport, "A ray tracing technique to predict path loss and delay spread inside buildings," *IEEE GLOBECOM*, Dec. 7, 1992, pp. 649–653.

[162] K. Schaubach, N. J. Davis and T. S. Rappaport, "A ray tracing method for predicting path loss and delay spread in microcellular environments," *IEEE Vehicular Technology Conference*, May 1992, pp. 932–935.

9 Peer-to-Peer Networking

Andreas F. Molisch, Anmol Bhardwaj, George MacCartney, Jr.,
Yunchou Xing, Shu Sun, Theodore S. Rappaport, Camillo Gentile,
Kate A. Remley and Alenka Zajić

9.1 Background

Device-to-device (D2D) radio channels have fundamentally different properties compared to those of conventional cellular (device-to-infrastructure, D2I) channels. The main reason for this is that most often both the receive antenna and the transmit antenna are located at low heights, and hence there is more interaction with objects in the close neighborhood of the devices. The difference is especially pronounced for outdoor links, where a base station (BS) would be high above ground (typically 10 m for microcells, and up to 100 m for macrocells), while all devices are at street level. Consequently, in D2D, over-the-rooftop propagation is not a viable mechanism, and street-canyon propagation is more strongly affected by shadowing objects such as cars and trucks. One important class of outdoor D2D systems is vehicle-to-vehicle (V2V) communications, which implies strong mobility. A different class are people located more or less stationary in outdoor cafes, plazas, etc. In indoor situations, the difference between D2I and D2D propagation mechanisms is less pronounced, and the range of validity for many indoor channel models includes the D2D case.

In both D2D and D2I networks, user equipment mobility, human presence and finite multipath persistence are the principal factors that degrade link availability. Because the reflecting surfaces in most manmade environments can be modeled as horizontal and vertical planes, ray-based channel models used in D2D simulations in which the user equipment is located at comparable heights are effectively 2D with limited components above or below the horizontal plane. Ray-based channel models used in D2I simulations in which the user equipment and base transceiver station (BTS) are located at different heights are effectively 3D with significant elevation angle components, and are therefore much more complex.

9.2 Path Loss

A first impact of the different propagation conditions is the path-loss model. When the two devices are on the same street, then a conventional path-loss model, as discussed in Chapter 5, is appropriate. If the devices are on orthogonal streets, the model of [1], based on measurements (at 6 GHz) in and around Munich, Germany, proposes

$$PL(d_r, d_t, w_r, x_t, i_s) = C + i_s L_{SU}$$

$$+ \begin{cases} 10 \log_{10} \left(\left(\dfrac{d_t^{ET}}{(x_t w_r)^{ES}} \dfrac{4\pi d_r}{\lambda} \right)^{EL} \right), & \text{if } d_r \leq d_b \\[2em] 10 \log_{10} \left(\left(\dfrac{d_t^{ET}}{(x_t w_r)^{ES}} \dfrac{4\pi d_r^2}{\lambda d_b} \right)^{EL} \right), & \text{if } d_r > d_b \end{cases}, \qquad (9.1)$$

where d_t and d_r denote the distance of the transmitter (TX) and receiver (RX) to the intersection center, respectively, w_r is the width of the RX street, and x_t is the distance of the TX from the wall. The other variables are fitting factors. Note that this type of model might also be applicable for D2I, in particular in a microcellular scenario.

Relatively little previous channel modeling work has addressed the needs of millimeter-wave (mmWave) D2D network simulations. Representative work includes [2], which compares measurements of angle-dependent propagation at 38 and 60 GHz and [3], which uses ray-tracing simulations to assess path-loss exponent and shadowing variance for both line-of-sight (LoS) and indirect paths. The results confirm that LoS channels experienced far less path loss and delay spread than indirect (reflected) channels, and that fewer useful indirect paths are available at 60 GHz than at 38 GHz. Measurement-based models have recently been used to characterize path loss in cellular peer-to-peer outdoor environments and V2V scenarios [3, 4]. Measurements conducted in these scenarios are used to build upon some of the standard path-loss models.

Device-to-device measurement and modeling analysis conducted in [5] demonstrate that the well-known theoretical free space (FS) and Stanford University interim (SUI) empirical path-loss models can be modified to fit the analytical results. Measurements were conducted in outdoor environments (flat and dense vegetation) at 60 GHz. The FS and SUI path-loss models were modified by applying slope correction factors such that they match the measurement based 1 m close-in (CI) FS reference distance path-loss model (eqs. 9.2 and 9.3):

$$PL_{FS,Mod}(d)[dB] = \alpha_{LOS}(PL_{FS}(d) - PL_{FS}(d_0)) + 20 \log_{10}(f_{GHz})$$
$$+ PL(d_0) + X_\sigma \qquad \text{with } d_0 = 1 \text{ meter,}$$
$$\qquad (9.2)$$

$$PL_{SUI,Mod}(d)[dB] = \alpha_{NLOS}(PL_{SUI}(d) - PL_{SUI}(d_0)) + 20 \log_{10}(f_{GHz})$$
$$+ PL(d_0) + X_\sigma.$$
$$\qquad (9.3)$$

The slope correction factors were calculated using the MMSE (minimum mean square error) approach for both the LoS and non-LoS (NLoS) environments. Overall results demonstrate that the modified models resulting from slope correction factors match closely with the trend given by the measurement-based CI model. Similar analysis is also conducted for D2I scenarios at 73 GHz and again the slope correction factors are successfully computed to match SUI and FS models to the CI FS reference distance model [5].

9.3   Shadowing

Shadowing also requires some new and improved modeling. Traditionally, the correlation of the shadowing between the different links originating at the BS has been modeled; this can be described in a rather straightforward way through spatial correlation functions.

For communication between two moving devices only (e.g., two cars), the conventional shadowing model might still hold, in particular under the assumption that the shadowing only depends on the distance between the devices. Under this assumption, it does not matter whether the change in distance is created by the movement of one or two devices. Alternatively, ray-tracing results also show that the autocorrelation function can be modeled as the product of the autocorrelation functions when only the TX and only the RX are moving, respectively [6].

Reference [7] suggests a 2D sum of sinusoids (SOS) method to generate the cellular shadowing landscape. For the P2P case, [8] extends it to a 4D one:

$$s(x, y, u, v) = \sum_{n=1}^{N} c_n \cos[2\pi(f_{x,n}x + f_{y,n}y + f_{u,n}u + f_{v,n}v) + \theta_n], \qquad (9.4)$$

where $s(x, y, u, v)$ is the simulated shadowing value based on a 4D map (x, y, u, v), and where (x, y) and (u, v) give the location of the two mobile stations (MSs) respectively. N is the number of sinusoids we use. The random phase set $\{\theta_n\}$ with uniform distribution between $[0, 2\pi]$ can be generated before the simulation. The spatial frequency set $\{f_n\}$, and amplitude set $\{c_n\}$ are computed by sampling the 4D power spectral density of the shadowing process, which is given by

$$\Phi(f_x, f_y, f_u, f_v) = \mathcal{F}\{(R_{P2P}(\Delta x, \Delta y, \Delta u, \Delta v)\}$$

$$A_n = \int \int \int \int_{I_n} \Phi(f_x, f_y, f_u, f_v)\, df_x df_y df_u df_v \qquad (9.5)$$

$$c_n = 2A_n\sigma(d_0 d_1),$$

where $\mathcal{F}$ indicates the spatial Fourier transform, and $\{I_n\}$ is a sampling frequency set of f_x, f_y, f_u, f_v with unit step size. σ is the shadowing standard deviation, which is a function of $d_0 d_1$. The exact sampling method does not play a significant role here. Note, however, that shadowing itself is not stationary (variances change when moving over large distances). Thus, one can dynamically update the amplitude set when $d_0 d_1$ changes. Note that continuity of the shadowing realizations has to be ensured (similar in spirit to continuous-phase frequency shift keying).

However, when multiple links are investigated, we have to describe the correlation of the shadowing for the different links. Proper implementation of the shadowing in a physically meaningful way can be difficult, and requires careful definition of stationarity regions for shadowing (compare [6]). Measurements at 28 GHz conducted in [9], demonstrate that spatial autocorrelation of individual multipath components (MPCs) reaches 0 after 2 and 5 wavelengths in LoS and NLoS scenarios respectively.

For communication between two pedestrians holding devices, the body shadowing plays an important role; it is noteworthy that a "rotational" shadowing, that is, shadowing depending on the orientation of the person, occurs in addition to a "translational" shadowing as a person is moving on a trajectory (i.e., walking) (see [10, 11]).

Shadowing by building structure and furnishings, finite persistence of MPCs due to interruptions in reflecting surfaces and human presence are the major causes of signal blockage in both D2D and D2I mmWave links. Human presence is particularly difficult to manage and has been extensively studied. Typical results include those reported in [12–17]. These measurements, which together include data collected at 2, 4.7, 26, 27, and 73 GHz, demonstrate the rapid rate and considerable depth of fading encountered when a mmWave path is blocked by a person and the possibility of using beamswitching to mitigate the effect of such fading.

A lot of work has been done toward improving pre-existing human blockage models based on measurement campaigns. A 73 GHz D2D human blockage measurement campaign presented in [12, 15], demonstrates that the double knife-edge diffraction (DKED) approximation underestimates human body shadowing, while uniform theory of diffraction (UTD) models overestimate it. Measurement campaigns reported in [13, 15, 18] have demonstrated that a four-state piecewise linear model can be used to accurately model blockage events in D2D scenarios. According to the model, the shadowing event can be described as:

$$
SE(t)[\text{dB}] = \begin{cases} r_{\text{decay}} \times t, & \text{for } 0 \le t \le \frac{SE_{\text{mean}}}{r_{\text{decay}}} \\ SE_{\text{mean}}, & \text{for } \frac{SE_{\text{mean}}}{r_{\text{decay}}} \le t \le t_D - \frac{SE_{\text{mean}}}{r_{\text{rise}}} \\ SE_{\text{mean}} - r_{\text{rise}} \times t, & \text{for } t_D - \frac{SE_{\text{mean}}}{r_{\text{rise}}} \le t \le t_D \\ 0, & \text{otherwise} \end{cases} . \tag{9.6}
$$

Parameters corresponding to the blockage events, such as signal strength, fade duration, rising and decay rates and level crossing rate were extracted from the measurement data and were successfully used to fit the four-state model. A very similar human blockage parametric model based on empirical data from a comprehensive measurement campaign was also developed for the D2I scenario in [4].

9.4　Dispersion

For the modeling of the angular and delay dispersion, the same modeling methods as for D2I can be used (see Chapter 8). Of course, the numerical values for the delay and angular spreads might be different, but generally there is no change in the fundamental modeling method.

9.5   Temporal Variations

There are two distinct groups of D2D channels, depending on the dynamics of the nodes, that is, if the devices themselves are moving or not. In the first case, devices at *both* link ends can move, sometimes very fast. In addition, scatterers and shadowing objects can also move. This is, e.g., the case for V2V channels, for which extensive research has shown that the channel statistics typically change over time and hence the conventional assumption about wide-sense-stationary uncorrelated scattering (WSSUS) is only fulfilled for rather short time intervals, and moderate frequency intervals. Stationarity bandwidths and times have been measured in the centimeter-wave range [19, 20], and for D2I links at mmWaves [21], but not for D2D links at mmWaves.

In order to handle the nonstationarities from a channel modeling perspective, the most straightforward solution is often to use a geometry-based stochastic channel model (GSCM), where the nonstationarities are automatically taken care of and modeled by the randomly placed scatterers in the environment. Another approach is based on tapped delay lines, such that the location (delay) of the taps is either adjusted continuously, or a birth/death process of the taps is implemented [22].

In static or nomadic scenarios, the two nodes do not change with respect to each other. This occurs, for example, in machine-to-machine communications (static nodes) or peer-to-peer (WiFi direct) links between laptops (nomadic scenarios). In that case the Doppler spectrum of the channel is determined by moving objects in the surroundings. Typically the Doppler spread is low as many of the dominant scatterers are static as well, and hence the coherence time of the channel can be quite large.

9.6   Conclusions and Future Work

Ray-based channel models used in D2D simulations in which the user equipment is located at comparable heights are effectively 2D with limited elevation angle components. Ray-based channel models used in D2I simulations in which the user equipment and BTS are located a different heights are effectively 3D with significant elevation angle components, and are therefore much more complex.

If trace data from mobility models that capture the movement of both the user equipment and people within the coverage area can be overlaid onto building outline data, link availability can be assessed using simple ray-tracing methods. The time required to execute such site-specific simulations may be excessive, however. This raises the possibility that statistical simulations based upon site-general models may be faster and more effective for the purposes of fairly comparing alternative networking schemes. If such approaches are used, it may be helpful to develop correlation models that relate path loss to delay spread and K-factor similar those developed for macrocell environments in [23] so that the capacity of individual links can be more efficiently estimated.

References

[1] T. Mangel, O. Klemp and H. Hartenstein, "A validated 5.9 GHz non-line-of-sight path-loss and fading model for inter-vehicle communication," *2011 11th International Conference on ITS Telecommunications*, Aug. 2011, pp. 75–80.

[2] T. S. Rappaport, E. Ben-Dor, J. N. Murdock and Y. Qiao, "38 GHz and 60 GHz angle-dependent propagation for cellular & peer-to-peer wireless communications," *2012 IEEE International Conference on Communications (ICC)*, June 2012, pp. 4568–4573.

[3] A. Al-Hourani, S. Chandrasekharan and S. Kandeepan, "Path loss study for millimeter wave device-to-device communications in urban environment," *2014 IEEE International Conference on Communications Workshops (ICC)*, June 2014, pp. 102–107.

[4] A. Angles-Vazquez, E. Carreno and L. S. Ahumada, "Modeling the effect of pedestrian traffic in 60-GHz wireless links," *IEEE Antennas and Wireless Propagation Letters*, vol. 16, pp. 1927–1931, 2017.

[5] A. I. Sulyman, A. Alwarafy, G. R. MacCartney, T. S. Rappaport and A. Alsanie, "Directional radio propagation path loss models for millimeter-wave wireless networks in the 28-, 60-, and 73-GHz bands," *IEEE Transactions on Wireless Communications*, vol. 15, pp. 6939–6947, Oct. 2016.

[6] Z. Li, R. Wang and A. F. Molisch, "Shadowing in urban environments with microcellular or peer-to-peer links," *2012 6th European Conference on Antennas and Propagation (EUCAP)*, Mar. 2012, pp. 44–48.

[7] X. Cai and G. B. Giannakis, "A two-dimensional channel simulation model for shadowing processes," *IEEE Transactions on Vehicular Technology*, vol. 52, pp. 1558–1567, Nov. 2003.

[8] Z. Wang, E. K. Tameh and A. R. Nix, "Joint shadowing process in urban peer-to-peer radio channels," *IEEE Transactions on Vehicular Technology*, vol. 57, pp. 52–64, Jan. 2008.

[9] M. K. Samimi and T. S. Rappaport, "28 GHz millimeter-wave ultrawideband small-scale fading models in wireless channels," *CoRR*, vol. abs/1511.06938, 2015.

[10] J. Karedal, A. J. Johansson, F. Tufvesson and A. F. Molisch, "A measurement-based fading model for wireless personal area networks," *IEEE Transactions on Wireless Communications*, vol. 7, pp. 4575–4585, Nov. 2008.

[11] S. L. Cotton, "Human body shadowing in cellular device-to-device communications: Channel modeling using the shadowed $\kappa - \mu$ fading model," *IEEE Journal on Selected Areas in Communications*, vol. 33, pp. 111–119, Jan. 2015.

[12] G. R. MacCartney, S. Deng, S. Sun and T. S. Rappaport, "Millimeter-wave human blockage at 73 GHz with a simple double knife-edge diffraction model and extension for directional antennas," *2016 IEEE 84th Vehicular Technology Conference (VTC-Fall)*, Sept. 2016, pp. 1–6.

[13] G. R. MacCartney, T. S. Rappaport and S. Rangan, "Rapid fading due to human blockage in pedestrian crowds at 5G millimeter-wave frequencies," *IEEE Global Communications Conference (GLOBECOM)*, Dec. 2017.

[14] M. Nakamura, M. Sasaki, M. Inomata and T. Onizawa, "The effect of human body blockage to path loss characteristics in crowded areas," *2016 International Symposium on Antennas and Propagation (ISAP)*, Oct. 2016, pp. 218–219.

[15] M. Peter, M. Wisotzki, M. Raceala-Motoc, W. Keusgen, R. Felbecker, M. Jacob, S. Priebe and T. Kurner, "Analyzing human body shadowing at 60 GHz: Systematic wideband

MIMO measurements and modeling approaches," *2012 6th European Conference on Antennas and Propagation (EUCAP)*, Mar. 2012, pp. 468–472.

[16] K. Saito, T. Imai and Y. Okumura, "Fading characteristics in the 26GHz band indoor quasi-static environment," *2014 IEEE International Workshop on Electromagnetics (iWEM)*, Aug. 2014, pp. 135–136.

[17] L. Talbi, "Human disturbance of indoor EHF wireless channel," *Electronics Letters*, vol. 37, pp. 1361–1363, Oct. 2001.

[18] M. Jacob, C. Mbianke and T. Kurner, "A dynamic 60 GHz radio channel model for system level simulations with MAC protocols for IEEE 802.11ad," *IEEE International Symposium on Consumer Electronics (ISCE 2010)*, June 2010, pp. 1–5.

[19] O. Renaudin, *Experimental channel characterization for vehicle-to-vehicle communication systems*. PhD thesis, Catholic University of Louvain, Louvain-la-Neuve, Belgium, 2013.

[20] L. Bernadó, T. Zemen, F. Tufvesson, A. F. Molisch and C. F. Mecklenbräuker, "The (in-)validity of the WSSUS assumption in vehicular radio channels," *2012 IEEE 23rd International Symposium on Personal Indoor and Mobile Radio Communications (PIMRC)*, IEEE, 2012, pp. 1757–1762.

[21] R. Wang, C. Bas, S. Sangodoyin, S. Hur, J. Park, J. Zhang and A. Molisch, "Stationarity region of mm-wave channel based on outdoor microcellular measurements at 28 GHz," *Military Communications Conference (MILCOM), MILCOM 2017-2017 IEEE*, IEEE, 2017, pp. 782–787.

[22] T. Zwick, C. Fischer and W. Wiesbeck, "A stochastic multipath channel model including path directions for indoor environments," *IEEE Journal on Selected Areas in Communications*, vol. 20, no. 6, 2002, pp. 1178–1192.

[23] L. J. Greenstein, V. Erceg, Y. S. Yeh and M. V. Clark, "A new path-gain/delay-spread propagation model for digital cellular channels," *IEEE Transactions on Vehicular Technology*, vol. 46, pp. 477–485, May 1997.

10 Temporal Variance: Literature Review on Human Blockage Models

Katsuyuki Haneda, Usman Virk, George MacCartney, Jr., Yunchou Xing, Shu Sun, Theodore S. Rappaport, Andreas F. Molisch, Alenka Zajić, Camillo Gentile and Kate A. Remley

10.1 Human Blockage Modeling Overview

Human blockage causes temporal variations to radio channels when a mobile device is in motion and some plane waves constituting the radio channels are blocked by a human body. Even when two sides of communications are static, moving human bodies often shadow some plane waves, leading to time-varying radio channel responses. Shadowing of plane waves due to human bodies makes the shapes of the Doppler spectrum significantly different for the stationary and mobile links. It is most straightforward to incorporate human blockage effects into radio channels through their geometric descriptions. This allows us to define dynamic motions of human bodies in relation to the locations of communicating devices, from which it is possible to analyze interactions the between the human bodies and each plane wave. Available models of human blockage therefore are defined by the shapes and materials of blocking objects. Blockage losses are determined by simple mathematical formulas, most of which are motivated by diffraction of plane waves around the blocking objects. The models are therefore physically sound, while their properties such as shapes, dimensions and materials are statistically defined and determined to fit measurements. The main task of modeling human blockage is therefore to choose reasonable properties of the blocking objects. This chapter covers human blockage models with different shapes and material properties of the blocking objects, with mathematical representations to estimate the shadowing losses in addition to free-space losses of a plane wave. A similar survey addressing existing human blockage models is available in the recent literature [1].

10.2 Absorbing Screen Models

10.2.1 Double Knife-Edge Model of an Absorbing Screen

The human body has been popularly modeled as an absorbing screen. The simplest shape of the screen is a vertically infinitesimal strip and is called a double knife-edge model. This is illustrated in Figure 10.1 [3]. It is possible to obtain reasonable estimates of the receive (RX) field behind the body using double knife-edge diffraction (DKED) from the absorbing screen. Diffracted fields from the two sides of the

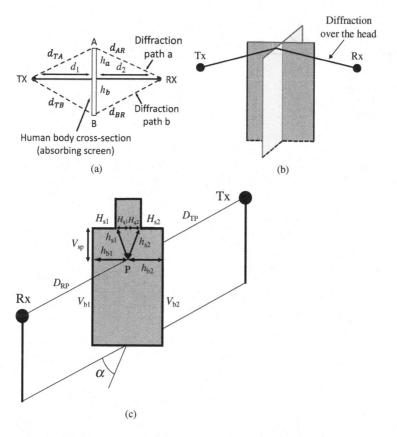

Figure 10.1 Popular absorbing screen models of a human body. (a) Double knife-edge model. © 2020 IEEE. Reprinted, with permission from [2]. (b) Triple knife-edge model. (c) Double knife edge model with a head and shoulders.

absorbing screen are considered. First, let us start from a half-plane absorbing screen with a point transmit (TX) source and RX point, whose geometry is illustrated in Figure 10.2. The RX field is given by

$$E = \frac{1+j}{2} \left\{ \left(\frac{1}{2} - C(v) \right) - j \left(\frac{1}{2} - S(v) \right) \right\} E_0, \tag{10.1}$$

where E_0 is the RX field when there is no absorbing knife edge; $C(v)$ and $S(v)$ are cosine and sine Fresnel integrals given by

$$C(v) + jS(v) = \int_0^v \exp\left(j\frac{\pi}{2}t^2 \right) dt, \tag{10.2}$$

$$v = -h\sqrt{\frac{2}{\lambda}\left(\frac{1}{d_1} + \frac{1}{d_2} \right)}, \tag{10.3}$$

where λ is the wavelength. The formulas apply regardless of polarization of the incident waves. The Fresnel integral eq. (10.2) can easily be solved numerically using

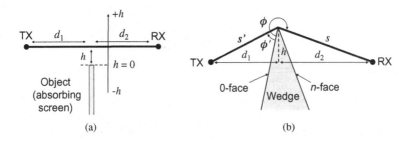

Figure 10.2 (a) A half-plane absorbing screen and (b) a wedge between two points of TX and RX. © 2020 IEEE. Reprinted, with permission from [2].

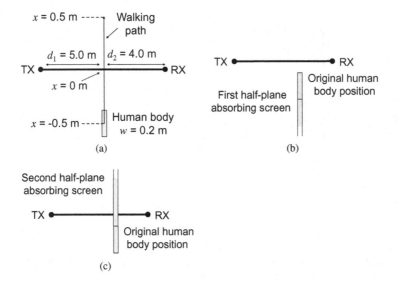

Figure 10.3 (a) A human body blocking the LoS between the TX and RX point sources. (b), (c) Two subproblems for solving the shadowing effects due to a human body.

built-in functions of commonly available computational tools. Equation (10.2) works best under the conditions that $d_1, d_2 \gg h$ and $d_1, d_2 \gg \lambda$.

Let us consider a human body of 0.2 m width walking across 1 m distance on a line perpendicular to the TX–RX line, as shown in Figure 10.3(a). It is possible to calculate the relative field strength behind the body by dividing the original DKED problem in Figure 10.3(a) into two subproblems illustrated in Figures 10.3(b) and 10.3(c). The subproblems consist of half-plane absorbing screens with top edges corresponding to the different sides of the body. The field from each subproblem is solved by eq. (10.1), where the reference line-of-sight (LoS) field is given by

$$E_0 = \frac{\lambda}{4\pi(d_1 + d_2)} \exp\left(-j2\pi f \frac{d_1 + d_2}{c}\right), \tag{10.4}$$

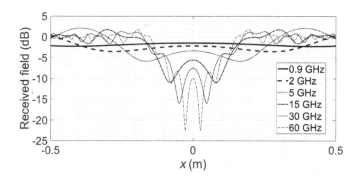

Figure 10.4 The normalized RX field at various radio frequencies due to human shadowing along the walking path. The geometry of the human body and TX and RX locations are defined in Figure 10.3.

where c is the velocity of light. The total field at the RX is expressed by the sum of the fields solved from the two subproblems as

$$E_{\text{DKED}} = E_{\text{a}} \exp\left(-j2\pi f \frac{\Delta d_{\text{a}}}{c}\right) + E_{\text{b}} \exp\left(-j2\pi f \frac{\Delta d_{\text{b}}}{c}\right), \qquad (10.5)$$

where E_{a} and E_{b} are diffracted fields observed at RX, and $\Delta d_{\text{a}} = d_{\text{TA}} + d_{\text{AR}} - d_1 - d_2$, $\Delta d_{\text{b}} = d_{\text{TB}} + d_{\text{BR}} - d_1 - d_2$ are extra propagation distances of the two diffracted paths compared to the LoS, respectively.

The total RX field normalized to the LoS, that is, $|E_{\text{DKED}}/E_0|$, is illustrated in Figure 10.4 for different radio frequencies. It must be noted that the Fresnel integral eq. (10.2) is not necessarily valid at lower radio frequencies since the conditions $d_1, d_2 \gg h$ and $d_1, d_2 \gg \lambda$ are not met. Still, the example illustrates possible influence of the radio frequency on blockage losses. The wave propagation between TX and RX points occurs in a local region defined by the Fresnel zones and is most significant in the first Fresnel zone. As the width of Fresnel zones decreases as the radio frequency of the link increases, the width of the first Fresnel zone becomes smaller at higher frequencies. For the lowest frequency of 0.9 GHz in the given link, the width of the first Fresnel zone is around 0.86 m, which is four times larger than the given width of the human body; hence we observe a lower blockage effect compared to the highest frequency of 60 GHz. It is also noteworthy that the total RX field fluctuates in the shadowed region due to constructive and destructive interference of two diffracted paths from the sides of the absorbing screen. The fluctuation is more apparent in the higher frequencies. It is important to average this small-scale fluctuation out from the RX field when comparing measurements and models. Thanks to its simplicity, the DKED model is also used in estimating link attenuation when multiple human bodies block a propagation path [4, 5]. For evaluating the human blockage attenuation more accurately, [6] modifies the DKED model to account for the TX and RX antenna radiation patterns.

10.2.2   Multiple Knife-Edge Model of an Absorbing Screen

Different from the DKED models, where the human body is modeled as a single infinitesimally long absorbing strip along the z-axis, [7] assumes that the human blockage is characterized by two vertical absorbing strips, as illustrated in Figure 10.1(b), where the top edge height of the strip is defined as that of a human body. The two strips intersect orthogonally and may have different widths, representing the width and thickness of a human body. Depending on the orientation of the two intersecting strips, only one of the two strips with the larger cross-section seen from the TX–RX link is considered for calculating the diffracted paths. A diffracted path from the top edge of the strip is considered in addition to the side diffracted paths, leading to the triple knife-edge diffraction (TKED) model of a human body. As the diffracted field eq. (10.1) does not depend on polarizations of the waves, the field at the RX point is given by

$$E_{\text{TKED}} = E_{\text{DKED}} + E_{\text{h}} \exp\left(-2\pi f \frac{\Delta d_{\text{h}}}{c}\right), \qquad (10.6)$$

where $\Delta d_{\text{h}} = d_{\text{THR}} - d_1 - d_2$, d_{THR} is a 3D distance between the TX source and RX observation point through the top edge of the half-plane vertical absorbing strip.

There are more complex human body blockage models taking into account not only a torso, but also shoulders and a head [7–9], as illustrated in Figure 10.1(c). The models are based on an absorbing screen, making use of eq. (10.1). The multiple knife-edge diffraction (MKED) model considers paths from each edge and estimates the blockage losses by summing their field strength. Furthermore, the orientation of the human body, α, defined in Figure 10.1(c), leads to variation of the human blockage loss. When applying eq. (10.1) to calculate the fields from each edge, the height h, distances d_1 and d_2 in eq. (10.1) can be set by $h = h' \cos \alpha$, $d_1 = D_{\text{TP}} \pm h' \sin \alpha$ and $d_2 = D_{\text{RP}} \pm h' \sin \alpha$ for $h' = h_{\text{b1}}, h_{\text{b2}}, h_{\text{s1}}, h_{\text{s2}}$ for $-\pi/2 \le \alpha \le \pi/2$; when the orientation of the human body is perpendicular to the TX–RX line, that is, $\alpha \sim \pm\pi/2$, the thickness of the human body W_{T} is used instead of h. It is again noted that the diffraction coefficients do not depend on wave polarizations because the human body model is expressed as absorbing screens.

The TKED and MKED models provide better agreement with measurements in general at the expense of the increased complexity of the model and possibly computational load. The increased model complexity will become more apparent when the orientation of a human body is arbitrary and when mobile and base station antenna heights are different so that the Fresnel zones of the LoS path illuminates above the chest of a human body.

10.2.3   Conducting Screen and Wedge Models

The authors of [9] calculate diffraction coefficients from each edge of a finitely conducting human body screen by assuming that each edge is a wedge with a zero wedge angle. The difference from the MKED model is that the screen is now conducting and not absorbing. In this case, the uniform theory of diffraction (UTD) is used for deriving the RX field [10]. Given the wedge geometry depicted in Figure 10.2, the diffracted field is given by

$$E_{wg} = E_0 \frac{e^{-jks'}}{s'} D^{\perp\|} \sqrt{\frac{s'}{s(s'+s)}} e^{-jks}, \tag{10.7}$$

where the polarization-dependent diffraction coefficient $D_\|^\perp$ for a finitely conducting wedge is given by eq. (10.8).

$$
\begin{aligned}
D^{\perp\|} = \frac{-e^{-j\pi/4}}{2n\sqrt{2\pi k}} &\times \left\{ \cot\left(\frac{\pi + (\phi - \phi')}{2n}\right) \cdot F(kLa^+(\phi - \phi')) \right. \\
&+ \cot\left(\frac{\pi - (\phi - \phi')}{2n}\right) \cdot F(kLa^-(\phi - \phi')) \bigg\} \\
&+ \left\{ R_0^{\perp\|} \cot\left(\frac{\pi - (\phi + \phi')}{2n}\right) \cdot F(kLa^-(\phi + \phi')) \right. \\
&+ R_n^{\perp\|} \cot\left(\frac{\pi + (\phi + \phi')}{2n}\right) \cdot F(kLa^+(\phi + \phi')) \bigg\}
\end{aligned}
\tag{10.8}
$$

The function $F(\cdot)$ in eq. (10.8) is the Fresnel integral, given as

$$F(x) = 2j\sqrt{x}e^{jx} \int_{\sqrt{x}}^{\infty} e^{-j\tau^2} d\tau, \tag{10.9}$$

and furthermore, in eq. (10.8),

$$L = \frac{ss'}{s+s'}, \tag{10.10}$$

$$a^\pm(\beta) = 2\cos^2\left(\frac{2n\pi N^\pm - \beta}{2}\right), \tag{10.11}$$

$$\beta = \phi \pm \phi', \tag{10.12}$$

where n defines the exterior wedge angle to be $n\pi$ and $N^\pm$ are the integers that most nearly satisfy the following two equations:

$$2\pi n N^+ - \beta = \pi, \quad 2\pi n N^- - \beta = -\pi. \tag{10.13}$$

Finally, $R_0^{\perp\|}$ and $R_n^{\perp\|}$ are the polarimetric Fresnel reflection coefficients of a plane wave on the 0- and n-faces, where incident and reflecting angles are given by ϕ' and $n\pi - \phi$, respectively. Possible singularity of the cotangent functions in eq. (10.7) around the reflection and shadowing boundaries is mitigated through the approximation

$$\cot\left(\frac{\pi \pm \beta}{2n}\right) \cdot F(kLa^\pm\beta) \approx n\left[\sqrt{2\pi kL}\, \text{sgn}\, \epsilon - 2kLe e^{j\pi/4}\right] e^{j\pi/4}, \tag{10.14}$$

with ϵ defined by

$$\beta = 2\pi n N^\pm \mp (\pi - \epsilon). \tag{10.15}$$

In this manner, it is possible to take into account realistic conductivity and permittivity of a human body. To improve the estimation of diffracted field from right-angle wedges, [11] proposes a different approximation of the diffraction coefficients using inverse problem theory. It is noteworthy that the wedge diffraction becomes a thin screen when $n = 2$, allowing us to calculate blockage losses due to finitely conducting screens using eq. (10.7). The diffraction coefficient from a finitely conducting wedge is also applicable to other types of shadowing objects than human bodies, such as building corners [12].

10.3   Cylinder Models

10.3.1   Circular Cylinder Models

Human blockage models based on cylinders have also been popularly considered in the literature [13, 15–18]. When a cylinder is circular in cross-section and is a perfect electric conductor (PEC), closed-form polarimetric diffracted fields from the cylinder can be derived based on the geometrical theory of diffraction (GTD) as [19]

$$E_z = \sum_{n=1}^{\infty} D_n^e E_i \left[\exp\left\{ -(jk + \Omega_n^e)\tau_1 \right\} + \exp\left\{ -(jk + \Omega_n^e)\tau_2 \right\} \right] \frac{\exp(-jks_d)}{\sqrt{8j\pi k s_d}},$$

(10.16)

where two propagation paths of distances τ_1 and τ_2 deliver energy between points P_1 and P_2 and are attenuated according to a constant Ω_n. For the electric field, this constant is given by

$$\Omega_n^e = \frac{\alpha_n}{a} M e^{j\pi/6},$$

(10.17)

where $-\alpha_n$ denotes the nth root (zero) of the Airy function Ai($\cdot$) [20]. Finally, D_n and M are given by

$$D_n = 2M \text{Ai}'(-\alpha_n)^{-2} e^{j\pi/6},$$

(10.18)

$$M = \left(\frac{ka}{2} \right),$$

(10.19)

where Ai$'$ denotes a derivative of the Airy function and $k = 2\pi/\lambda$ is a wave number. It must be noted that the GTD solutions have singularity around the transition region between the lit and shadowed regions. Analytical UTD solutions for a conducting cylinder are limited, for example, to a thin lossy material coating [21] and to sufficiently far-away RX location from the cylinder [22]. The solutions are for normal incidence of a plane wave to a cylinder, making it possible to analyze the scattering problem in 2D space. When the scattering problem extends to 3D as oblique incidence of a plane wave to consider, for example, scenarios with different heights of TX and RX antennas, closed-form solutions of the scattering field do not exist. It is necessary to rely on a numerical electromagnetic field solver [23] and more extensive numerical integration [24] in this case.

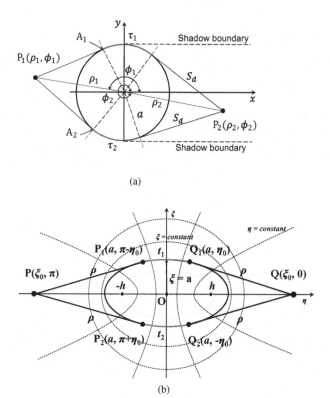

Figure 10.5 (a) Circular [13] and (b) elliptic cylinder models of link blockage. © 2020 IEEE. Reprinted, with permission, from [14].

10.3.2   Elliptic Cylinder Model

Let us consider a cylinder with an elliptic cross-section. We assume that the center of the ellipse coincides with the z-axis of the Cartesian coordinate system [14]. The distance between the two focal points of the ellipse is assumed to be $2h$. The elliptical coordinates on the horizontal plane, defined by the ξ–η domain as depicted in Figure 10.5, have a relationship with the Cartesian x–y coordinate as

$$x = h \cosh \xi \cos \eta,$$
$$y = h \sinh \xi \sin \eta. \tag{10.20}$$

In Figure 10.5, the locations are expressed on the elliptic coordinate system as, for example, $P(\xi, \eta)$. The ellipse representing the human cross-section is given by $\xi = a$. The TX and RX antennas are located at $P(\xi_0, \pi)$ and $Q(\xi_0, 0)$, respectively. Two diffracting rays symmetric to the horizontal axis exist from the TX antenna at P to the RX antenna at Q. The points of tangency where the two rays incident and leave the cylinder are $P_1(a, \pi - \eta_0)$, $Q_1(a, \eta_0)$, $P_2(a, \pi + \eta_0)$, and $Q_2(a, -\eta_0)$. The radial distance from the points of tangency to the TX and RX is $\overline{PP_1} = \overline{QQ_1} = \overline{PP_2} = \overline{QQ_2} = \rho$. The following derives the total field at the RX E_z for vertical polarization. First, the diffracted field at the RX due to the ray $\overline{PP_1Q_1Q}$ is

$$E_{1z} = A_0 \rho^{-1} \exp\{jk(\rho + t_1)\} \cdot \sum_{n=0}^{\infty} B_n(\eta_P) B_n(\eta_Q) \cdot \exp\{jk^{1/3}\tau_n\alpha(t_1)\}$$
$$\left[1 - \exp\{jkT + jk^{1/3}\tau_n\alpha(T)\}\right]^{-1},$$
(10.21)

where $\eta_P = \pi$ and $\eta_Q = 0$ are the η coordinates corresponding to the TX and RX locations P and Q, respectively; k is the wave number; t_1 denotes the arc length from P_1 to Q_1; and T is the total arc length of the ellipse. A_0 is a constant given by

$$A_0 = \frac{e^{j\pi/4}}{2\pi}\sqrt{\lambda}.$$
(10.22)

$B_n(\eta)$ represents the diffraction coefficient

$$B_n(\eta) = \pi^{3/4} 2^{1/4} 6^{-2/3} e^{j\pi/24} k^{-1/12} b^{1/6}(\eta)[\text{Ai}'(q_n)]^{-1},$$
(10.23)

where q_n denotes the nth root (zero) of the Airy function $\text{Ai}(\cdot)$ with Ai' as its derivative [20]. $b(\eta)$ refers to the radius of the ellipse on the major axis as

$$b(\eta) = h(\cosh a \sinh a)^{-1}(\sinh^2 a - \sin^2 \eta + 1)^{3/2}.$$
(10.24)

Furthermore, in eq. (10.21), $\alpha(x)$ is

$$\alpha(x) = \int_0^x b^{-2/3}(\eta)d\eta,$$
(10.25)

and τ_n is

$$\tau_n = e^{j\pi/3} 6^{-1/3} q_n.$$
(10.26)

Then, the diffracted field E_{2z} due to the ray $\overline{PP_2Q_2Q}$ can be evaluated in the same way as E_{1z}. Now the total diffracted field at the RX as a combination of the diffracted fields from the two sides of the elliptic cylinder is given as $E_z = E_{1z} + E_{2z}$.

10.4   Other Heuristic Models

The human blockage models discussed so far estimate the extra attenuation through calculating the diffracted fields behind blocking objects. Analytical formulas are available for the diffracted fields, either based on the GTD or UTD solutions, for simple objects such as absorbing screens, conducting edges and PEC cylinders. The formulas involve the Fresnel integral, which can be straightforwardly calculated thanks to built-in functions available in many computational tools. However, the analytical formulas may still be considered too complex to deserve implementing them in extensive radio network simulations. Heuristic models are therefore devised to simplify the GTD and UTD solutions using further approximations of the formulas, or observations and modeling of measurements.

10.4.1   Measurement-Based Models

Measurement-based models at millimeter waves include, for example, [26] that characterize signal-level attenuation due to human shadowing observed in short-range 60 GHz radio links using a Gaussian distribution. The work in [18] proposes a piecewise linear approximation of time-varying shadowing at 60 GHz due to human blockage. The approximation consists of a decreasing slope, shadowing dip and increasing slope of the received field strength as time goes during a human blockage event. The work in [27] models the transit rapid fading due to human blockage in pedestrian crowds via Markov models based on measurements in a dense urban environment at 73.5 GHz.

10.4.2   mmMAGIC Model

Simplified GTD and UTD solutions of the field behind blocking objects are proposed in a European project, mmMAGIC [25]. The geometry of the blocking object is illustrated in Figure 10.6, where the side and top views of the geometry shown on the right side of the figure are named "projection 1 and 2" hereinafter, respectively. The blocking object is a rectangular screen floating in the air, and is claimed to be comprehensive enough to simulate different physical objects. The shadowing loss is determined by diffracted fields from four edges of the screen as

$$E_{\mathrm{mmMAGIC}} = \left(1 - \prod_{i=1}^{2} \sum_{j=1}^{2} s_{ij} \left[\frac{1}{2} - \frac{ph_{ij}}{Ph} F_{ij} \right] \right) E_0, \tag{10.27}$$

where

$$F_{ij} = \left\{ \frac{1}{2} - \frac{1}{\pi} \tan^{-1} \left(\frac{\nu_{ij}\pi}{2} \right) \right\} \cos \psi_{ij}, \tag{10.28}$$

$$\nu_{ij} = \sqrt{\frac{\pi}{\lambda}(D1_{ij}^{\mathrm{proj}} + D2_{ij}^{\mathrm{proj}} - r_{i}^{\mathrm{proj}})}, \tag{10.29}$$

$$ph_{ij} = \exp\left\{ \frac{-j2\pi}{\lambda}(D1_{ij} + D2_{ij}) \right\}, \tag{10.30}$$

$$Ph = \exp\left(\frac{-j2\pi}{\lambda}r \right). \tag{10.31}$$

Finally, s_{ij} is a sign parameter, which is 1 if the non-LoS (NLoS) condition holds between TX and RX in projection i, while $s_{ij} = \mathrm{sgn}(D1_{ij} + D2_{ij} - D1_{ik} - D2_{ik})$ if the LoS condition is fulfilled in projection $i, k = \mod (j, 2) + 1$. When a multipath which is subject to the present blockage is attributed to specular reflections, either the TX or RX in Figure 10.6 should be replaced by its mirror image with respect to the reflection surfaces. On the other hand, if a multipath which is subject to the present blockage is *not* due to specular reflections, the TX and RX in the geometry should be replaced by the previous and following interacting points of the multipath. The term $\cos \psi_{ij}$ in F_{ij} accounts for increase of diffraction loss in the shadowed zone

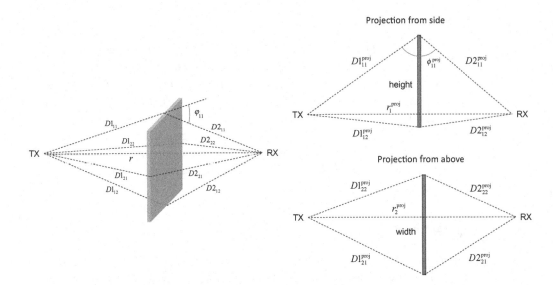

Figure 10.6 Geometry of a link blockage model in [25]. The side and top views of the geometry depicted in the top-right and bottom-right figures are called "projections 1 and 2," respectively in the text. Copyright © 2017 Ericsson AB, all rights reserved.

just behind the screen. When the relative distance to the screen is sufficiently large, this factor may be neglected. The formulations fulfill Babinet's principle, meaning that different shapes of blocking objects such as a truck [25] may be synthesized by combining multiple screens.

10.5   Comparison of Models

Figure 10.7 shows the human blockage loss estimated from different models covered in this section. The comparison is made at three different radio frequencies: 15, 28 and 60 GHz. The loss is estimated for various azimuth orientations of a human body with TX–body and RX–body distances of 2.76 m, and a body width and thickness of 0.5 and 0.2 m, respectively. The body thickness is considered only in the elliptic cylinder model, and otherwise only the width is adopted to define the dimension of absorbing screens and circular cylinders. The azimuth orientation angle is defined so that the cross-section of a human body is the largest at 0° and 180° because the human points either to the TX or RX, while 90° orientation corresponds to the human body pointing perpendicular to the TX–RX link. The human body is 1.0 m high and TX and RX antenna heights are set such that the optical LoS of the TX–RX link hits the center of the blocking object. The curves labeled with "UTD" are derived from eq. (10.7) for a PEC screen of the specified dimension. All the curves are for vertically polarized fields. It is assumed that reflections from a surrounding environment do not exist. Measurement-based heuristic models are not shown here because they do not consider physical dimensions of the human body.

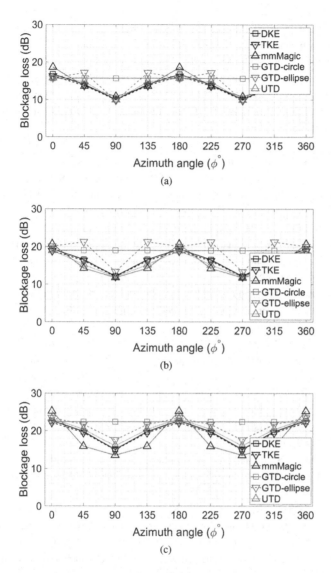

Figure 10.7 Comparison of human blockage models for varying azimuth orientations of a human body: (a) 15, (b) 28 and (c) 60 GHz.

References

[1] W. Qi, J. Huang, J. Sun, Y. Tan, C. X. Wang and X. Ge, "Measurements and modeling of human blockage effects for multiple millimeter wave bands," in *Proceedings of the 13th International Wireless Communications Mobile Computing Conference (IWCMC'17)*, June 2017, pp. 1604–1609.

[2] U. T. Virk and K. Haneda, "Modeling human blockage at 5G millimeter-wave frequencies," *IEEE Transactions on Antennas and Propagation*, vol. 68, no. 3, pp. 2256–2266, Mar. 2020.

[3] J. Kunisch and J. Pamp, "Ultra-wideband double vertical knife-edge model for obstruction of a ray by a person," in *Proceedings of the IEEE International Conference on Ultra-Wideband*, vol. 2, Sept. 2008, pp. 17–20.

[4] M. Jacob, S. Priebe, T. Kurner, M. Peter, M. Wisotzki, R. Felbecker and W. Keusgen, "Extension and validation of the IEEE 802.11ad 60 GHz human blockage model," in *Proceedings of the 7th European Conference on Antennas and Propagation (EuCAP'13)*, Apr. 2013, pp. 2806–2810.

[5] J. S. Lu, D. Steinbach, P. Cabrol and P. Pietraski, "Modeling human blockers in millimeter wave radio links," *ZTE Communications Magazine*, vol. 10, pp. 23–28, Dec. 2012.

[6] T. S. Rappaport, G. R. MacCartney, S. Sun, H. Yan and S. Deng, "Small-scale, local area, and transitional millimeter wave propagation for 5G communications," *IEEE Transactions on Antennas and Propagation*, vol. 65, pp. 6474–6490, Dec. 2017.

[7] M. Jacob, S. Priebe, A. Maltsev, A. Lomayev, V. Erceg and T. Kurner, "A ray tracing based stochastic human blockage model for the IEEE 802.11ad 60 GHz channel model," in *Proceedings of the 5th European Conference on Antennas and Propagation (EUCAP'11)*, Apr. 2011, pp. 3084–3088.

[8] X. Chen, L. Tien, P. Tang and J. Zhang, "Modelling of human body shadowing based on 28 GHz indoor measurement results," in *Proceedings of the IEEE 84th Vehicular Technology Conference (VTC-Fall'16)*, Sept. 2016.

[9] N. Tran, T. Imai and Y. Okumura, "Study on characteristics of human body shadowing in high frequency bands: Radio wave propagation technology for future radio access and mobile optical networks," in *Proceedings of the 80th IEEE Vehicular Technology Conference (VTC-Fall'14)*, Sept. 2014.

[10] R. Luebbers, "Finite conductivity uniform GTD versus knife edge diffraction in prediction of propagation path loss," *IEEE Transactions on Antennas and Propagation*, vol. 32, pp. 760–766, Jan. 1984.

[11] H. Wang and T. S. Rappaport, "A parametric formulation of the UTD diffraction coefficient for real-time propagation prediction modeling," *IEEE Antennas and Wireless Propagation Letters*, vol. 4, pp. 253–257, 2005.

[12] M. Jacob, S. Priebe, R. Dickhoff, T. Kleine-Ostmann, T. Schrader and T. Kurner, "Diffraction in mm and sub-mm wave indoor propagation channels," *IEEE Transactions on Microwave Theory and Technology*, vol. 60, pp. 833–844, Mar. 2012.

[13] C. Gustafson and F. Tufvesson, "Characterization of 60 GHz shadowing by human bodies and simple phantoms," in *Proceedings of the 6th European Conference on Antennas and Propagation (EUCAP)*, Mar. 2012, pp. 473–477.

[14] B. R. Levy, "Diffraction by an elliptic cylinder," *Journal of Mathematics and Mechanics*, vol. 9, pp. 147–165, Nov. 1960.

[15] M. Ghaddar, L. Talbi and T. A. Denidni, "Human body modelling for prediction of effect of people on indoor propagation channel," *Electronics Letters*, vol. 40, pp. 1592–1594, Dec. 2004.

[16] M. Ghaddar, L. Talbi, T. A. Denidni and A. Sebak, "A conducting cylinder for modeling human body presence in indoor propagation channel," *IEEE Transactions on Antennas and Propagation*, vol. 55, pp. 3099–3103, Nov. 2007.

[17] M. Jacob, S. Priebe, T. Kurner, M. Peter, M. Wisotzki, R. Felbecker and W. Keusgen, "Fundamental analyses of 60 GHz human blockage," in *Proceedings of the 7th European Conference on Antennas and Propagation (EuCAP'13)*, Apr. 2013, pp. 117–121.

[18] M. Peter, M. Wisotzki, M. Raceala-Motoc, W. Keusgen and R. Felbecker, "Analyzing human body shadowing at 60 GHz: Systematic wideband MIMO measurements and modeling approaches," in *Proceedings of the 6th European Conference on Antennas and Propagation (EUCAP'12)*, Mar. 2012, pp. 468–472.

[19] G. James, *Geometrical Theory of Diffraction for Electromagnetic Waves*, Peregrinus: Stevenage, 1976.

[20] M. Abramowitz and I. A. Stegun, *Handbook of Mathematical Functions: With Formulas, Graphs, and Mathematical Tables*, Courier Corporation: Chelmsford, MA, 1964.

[21] H.-T. Kim and N. Wang, "UTD solution for electromagnetic scattering by a circular cylinder with thin lossy coatings," *IEEE Transactions on Antennas and Propagation*, vol. 37, pp. 1463–1472, Nov. 1989.

[22] T. B. A. Senior and J. L. Volakis, *Approximate Boundary Conditions in Electromagnetics*, The Institution of Electrical Engineers: London, 1995.

[23] M. Yokota, T. Ikegamai, Y. Ohta and T. Fujii, "Numerical examination of EM wave shadowing by human body," in *Proceedings of the 4th European Conference on Antennas and Propagation (EuCAP'10)*, Apr. 2010.

[24] A. G. Aguilar, P. H. Pathak and M. Sierra-Pzrez, "A canonical UTD solution for electromagnetic scattering by an electrically large impedance circular cylinder illuminated by an obliquely incident plane wave," *IEEE Transactions on Antennas and Propagation*, vol. 61, pp. 5144–5154, Oct. 2013.

[25] M. Peter et al., "Measurement results and final mmMAGIC channel models: Specular wall reflections and diffused scattering measurements," report H2020-ICT-671650-mmMAGIC/D2.2, Millimetre-Wave Based Mobile Radio Access Network for Fifth Generation Integrated Communications (mmMAGIC), May 2017.

[26] P. Karadimas, B. Allen and P. Smith, "Human body shadowing characterization for 60-GHz indoor short-range wireless links," *IEEE Antennas Wireless Propagation Letters*, vol. 12, pp. 1650–1653, Dec. 2013.

[27] G. R. MacCartney, T. S. Rappaport and S. Rangan, "Rapid fading due to human blockage in pedestrian crowds at 5G millimeter-wave frequencies," *IEEE Global Communications Conference (GLOBECOM)*, Dec. 2017.

11 Terahertz Channels

Alenka Zajić, Theodore S. Rappaport, Andreas F. Molisch,
George MacCartney, Jr., Yunchou Xing, Camillo Gentile
and Kate A. Remley

11.1 Introduction

Frequencies from 100 GHz to 3 THz are promising bands for the next generation of wireless communication systems because of the wide swaths of unused and unexplored spectrum [1–4]. These frequencies also offer the potential for revolutionary applications that will be made possible by new thinking, and advances in devices, circuits, software, signal processing, applications and systems. Work in [5] shows that there is no fundamental physical channel impediment (e.g., rain, atmospheric absorption) to utilizing sub-THz and THz bands up to 1 THz for future wireless communications, and propagation on the horizon (e.g., elevation angles $\leq 15°$) may not cause interference (same or adjacent bands) between passive satellite sensors and terrestrial transmitters at frequencies above 100 GHz if the antenna patterns of the transmitters are carefully designed to avoid radiation in space (e.g., adaptive antenna patterns with very low sidelobes).

While 5G, IEEE 802.11ay and IEEE 802.15.3d [6, 7] are being built out for the millimeter-wave (mmWave) spectrum and promise data rates up to 100 Gbps, future 6G networks and wireless applications are probably a decade away from implementation, and are sure to benefit from operation in the 100 GHz to 1 THz frequency bands where even greater data rates will be possible [4, 8, 9]. The short wavelengths at mmWave and THz frequencies will allow massive spatial multiplexing in hub and backhaul communications, as well as incredibly accurate sensing, imaging, spectroscopy and other applications described in [1, 10–13]. The THz band, which we shall describe as being from 100 GHz through 3 THz, can also enable secure communications over highly sensitive links, such as those needed in the military, due to the fact that extremely small wavelengths (orders of microns) enable extremely high-gain antennas to be made in extremely small physical dimensions [14].

The ultrahigh data rates facilitated by mmWave and THz wireless local area and cellular networks will enable super-fast download speeds for computer communications, autonomous vehicles, robotic controls, the information shower [15], high-definition holographic gaming, entertainment, video conferencing and high-speed wireless data distribution in data centers [4, 16]. In addition to the extremely high data rates, there are promising applications for future mmWave and THz systems that are likely to evolve in 6G networks and beyond. These applications can be categorized into main areas such as wireless cognition, sensing, imaging, wireless communication and

Table 11.1 Promising applications as mmWave and THz:PL the potential for 6G [1].

Application	Example use cases
Wireless cognition	Robotic control
	Drone fleet control
	Autonomous vehicles
	Human surrogate
Sensing	Air quality detection
	Personal health monitoring systems
	Gesture detection
	Explosive detection and gas sensing
Imaging	See in the dark (mmWave camera)
	High-definition video resolution radar
	THz security body scan
Communication	Mobile wireless communications
	Wireless fiber for backhaul
	Intra-device radio communication
	Connectivity in data centers
	Information shower (≥ 100 Gbps)
Positioning	Centimeter-level positioning

position location/THz navigation (also called localization or positioning) [17, 18], as summarized in Table 11.1.

There are tremendous challenges ahead for creating commercial transceivers at THz frequencies, but global research is addressing the challenges. For example, the DARPA T-MUSIC program is investigating SiGe HBT, CMOS/SOI and BiCMOS circuit integration, in the hopes of achieving power amplifier threshold frequencies f_t of 500–750 GHz [19]. A survey of power amplifier capabilities since the year 2000 is given in [20]. It should be clear that the semiconductor industry will solve these challenges, although new architectures for highly dense antenna arrays will be needed due to the small wavelengths and physical size of RF transistors in relation to element spacing in THz arrays.

Since there is very high atmospheric attenuation at THz band frequencies, especially at frequencies above 800 GHz, highly directional "pencil beam" antennas (antenna arrays) will be used to compensate for the increased path loss due to the fact that the gain and directivity increase by the square of the frequency for a fixed physical antenna aperture size [21–23]. This feature makes THz signals exceedingly difficult to intercept or eavesdrop [4, 14, 24, 25]. However, a narrow pencil-like beam does not guarantee immunity from eavesdropping, and physical-layer security in THz wireless networks and transceiver designs that incorporate new countermeasures for eavesdropping will be needed [26].

Energy efficiency is always important for communication systems, especially as circuitry moves up to above 100 GHz, and a theoretical framework to quantify energy consumption in the presence of vital device, system and network trade-offs was presented in [27, 28]. The theory, called the *consumption factor theory* (CF, with a metric measured in bps/W), provides a means for enabling quantitative analysis and design approaches for understanding power trade-offs in any communication system. It was shown in [27, 28] that the efficiency of components of a transmitter closest to the output, such as the antenna, have the largest impact on CF [27]. The power efficiency increases with increasing bandwidth when most of the power used by components that are "off," for example, ancillary, to the signal path (e.g., the baseband processor, oscillator or a display) is much greater than the power consumed by the components that are in line with the transmission signal path (e.g., power amplifier, mixer, antenna) [27, 28]. For a very simple radio transmitter, such as one that might be used in low-cost IoT (Internet of Things) or "smart dust" applications where the power required by the ancillary baseband processor and oscillator is small compared to the delivered radiated power, the power efficiency is independent of the bandwidth [27, 29, 30]. Thus, contrary to conventional wisdom, the CF theory proves that for antennas with a fixed physical aperture, it is *more energy efficient* to move up to mmWave and THz frequencies which yield much wider bandwidths and better power efficiency on a bits per second per watt (bps/W) basis, as compared to the current, sub-6 GHz communication networks.

11.2   Challenges in Measuring and Modeling THz Channels

Terahertz wireless communications have two key advantages that can be combined to achieve very high data rates. First, the usable frequency band around each frequency is much larger, so each channel can have a much higher data rate. This alone can increase data rates to several hundreds of Gbit/s, but spatial multiplexing is still needed to reach Tbit/s data rates. Fortunately, THz frequencies allow smaller antennas and antenna spacing, which provides for more communication channels within the same array aperture within a chip package. However, to unlock the potential of THz wireless communications, several challenges in channel measurements and modeling need to be addressed.

11.2.1   Antenna Design

Directional antennas are necessary for THz communications due to the high path loss at these frequencies [31, 32]. The high antenna directivity gives rise to a scattering pattern that is somewhat different from other indoor (GHz or mmWave) channels observed in [33–35]. In addition to scattering mechanisms that are common to all indoor channels, in THz channels signals may reflect off of objects that are behind the receive (RX) antenna, travel back to objects near the transmit (TX) antenna and reflect back to be received by the RX antenna. This essentially produces a second arriving path, even without any scatterers between the TX and RX. This phenomenon has

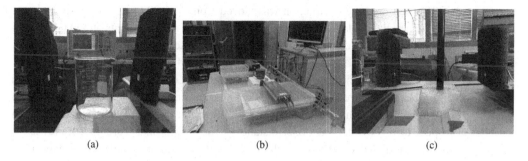

      (a)   (b)   (c)

Figure 11.1 Obstructed LoS measurement setups at (a) 30 GHz, (b) 140 GHz, and (c) 300 GHz. © 2018 IEEE. Reprinted, with permission from [38].

been observed in two independent measurement campaigns [31, 36], and while these reflections can be suppressed in channel sounding experiments by putting absorbers around the TX and RX, in practice either antennas need to be designed to suppress these reflections or channel models and communication systems need to account for them. Furthermore, due to the high directionality of THz antennas, antennas should be designed to be steerable to cover multiple directions a high number of multiple-input, multiple-output (MIMO) antennas needs to be deployed to increase data rates.

11.2.2 Diffraction

Terahertz applications require a good understanding of propagation mechanisms across all THz frequencies in order to determine suitable carrier frequencies and to create reliable systems. Additional challenges arise when these systems are brought to motion and propagation conditions vary over time. At mmWave frequencies, it is very likely that line-of-sight (LoS) paths will be partially (or fully) obstructed by objects in the channel as users change their positions. In contrast to lower-frequency bands, where these objects would create multiple reflections, at mmWave and THz frequencies, in addition to multiple reflections, diffraction can be a prevalent propagation mechanism. Hence, it is very important to study the impact of diffraction on channel propagation and how to properly account for this effect in channel models [37]. To illustrate how diffraction changes across multiple frequencies, Figure 11.1(a)–(c) shows measurement setups that allow us to study diffraction effects at 30, 140 and 300 GHz in obstructed indoor propagation environments [38].

The cylindrical obstruction that initially blocks the LoS path was gradually moved away from the LoS along a trajectory perpendicular to the LoS path. A ceramic mug with a diameter of 8.5 cm was used as the obstruction for the 30 GHz and 140 GHz bands. A metal pipe with diameter of 1.6 cm was used for the 300 GHz band in order to accommodate the narrower beam width of the 300 GHz antennas. The obstructed LoS (OLoS) measurements at 30 GHz had a TX–RX separation distance 40 cm and the obstruction was diagonally moved from 0 to 200 mm off the midpoint with an offset step of 5 mm. The OLoS measurements at 140 GHz had a TX–RX separation distance of 86.36 cm and the obstruction was moved from 0 to 200 mm off the midpoint of LoS with a diagonal offset step of 2 mm. The OLoS measurements at 300 GHz had a TX–RX separation distance 20 cm and the obstruction was moved from 0 to 34 mm off the midpoint of LoS with a diagonal offset step of 1 mm.

To verify the existence of diffraction in the OLoS channel in the 30, 140 and 300 GHz bands, we compare the calculated uniform theory of diffraction (UTD) diffraction gain using the method in [38, 39] to the measured diffraction gain with respect to the cylindrical offset distances. The results in Figure 11.2(a)–(c) show that the predicted UTD models align well with the measured diffraction gain in all frequency bands. Note that in the shadow region of the 30 GHz channel, that is, where the LoS channel is completely blocked by the cylindrical object, the measured diffraction gain deviates 10 dB from the predicted UTD model. On the other hand, this deviation is not observed in the 140 GHz and 300 GHz bands. The reason for this deviation is the fact that cylindrical obstruction is more transparent to waves at 30 GHz than at 140 GHz and 300 GHz. Another interesting observation is that a 10 dB "dip" appears near the shadow–lit boundary in the measured diffraction gain at 30 GHz and 140 GHz. The observed "dip," or abrupt increase in diffraction loss is a result of extra attenuation induced by the thickness of the ceramic mug's wall. As the offset distance increases to 35 mm, where the LoS path is tangent to the ceramic mug, the fields have to propagate the longest distance through the ceramic material, and therefore experience the largest attenuation. On the other hand, the "dip" is not observed at 300 GHz because the increment of the offset step (i.e., 1 mm), is greater than the thickness of the metal pipe, such that the measurement instance does not capture the moment when the "dip" occurs.

The results show that all three frequency bands experience diffraction in obstructed environments, but the diffraction effects become more prominent as frequency increases and this effect needs to be carefully modeled. Furthermore, it was shown that the UTD can be used across several frequency bands to model diffraction in OLoS environments.

11.2.3 Reflection

The lower spectrum of mmWave (i.e., 30 GHz) can cover a longer communication range and penetrate more easily through blockages, while the higher spectrum of THz frequencies provides large available bandwidth along with compact antenna form factors at the cost of a shorter range and higher levels of attenuation. Material properties are expected to impact specular reflections more at THz frequencies than at mmWave frequencies [40]. To illustrate differences in reflection coefficient for different materials and arriving specular angles at 30 GHz and 300 GHz, Figure 11.3 plots the magnitude of the reflection coefficient corresponding to different specular angles for copper and aluminum plates, cardboard and wood. For wood, the magnitude of the reflection coefficient is slightly higher in the 30 GHz channel than in the 300 GHz channel. Metal plates and wood are observed as better reflectors at 30 GHz than at 300 GHz, whereas cardboard is a better reflector at 300 GHz. Furthermore, the

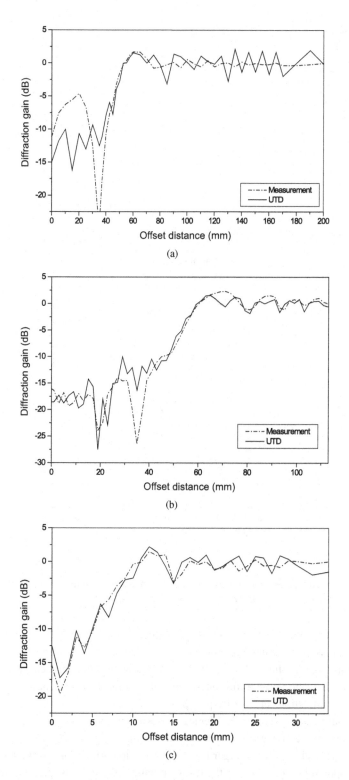

Figure 11.2 Comparison of modeled and measured diffraction gain at (a) 30 GHz, (b) 140 GHz and (c) 300 GHz. © 2018 IEEE. Reprinted, with permission from [38].

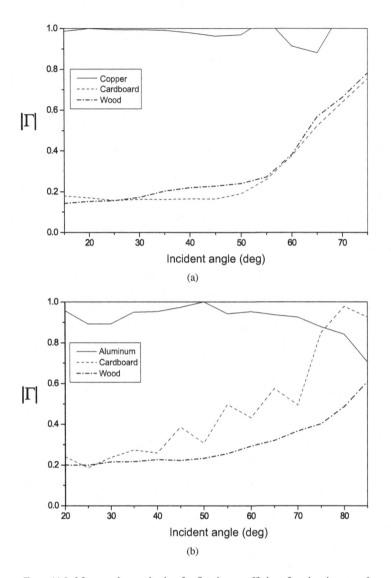

Figure 11.3 Measured magnitude of reflection coefficient for aluminum and copper plates, cardboard, and wood in (a) 30 GHz and (b) 300 GHz bands.

interference from the LoS path causes more fluctuations to the reflection coefficient measurement as the specular angle increases, and this effect is more pronounced in the 30 GHz band than in the 300 GHz band due to the wider antenna beamwidth. All of these results indicate that careful measurements and modeling are needed to capture the impact of reflections at THz frequencies.

Additionally, compared to THz device-to-device propagation, THz propagation in metal enclosures (such as cameras or computer casings) experiences both traveling and resonant waves. This yields a larger number of multiple reflections as well as a larger multipath spread [41, 42]. Due to the resonant nature of the fields, the received power can vary with the transceivers' position. To model propagation in such environments, the casing needs to be carefully modeled.

11.2.4   Scattering

At mmWave and THz frequencies, the wavelength λ becomes small, motivating the use of hybrid beamforming [43, 44] for "practical antenna packaging" [1, 45]. At sub-THz frequencies, λ is comparable to or smaller than the surface roughness of many objects, which suggests that scattering may not be neglected like it was when compared to reflection and diffraction at microwave frequencies (300 MHz to 3 GHz) [47–49].

Measured scattering patterns of different incident angles at 28, 73 and 142 GHz are shown in Figures 11.4(b), 11.4(d), and 11.4(f), respectively. The peak measured power (scattered power plus reflected power) was observed to occur at the specular reflection angle. The peak measured power was greater at larger incident angles than at smaller incident angles (e.g., a 9.4 dB difference between 80° and 10° at 142 GHz), where most of the energy is due to reflection but not scattering [1]. At all angles of incidence, measured power was within 10 dB below the peak power in a ±10° angle range of the specular reflection angle, likely a function of antenna patterns. In addition, backscattered power was observed (e.g., for 10° and 30° incidence at 142 GHz) but was more than 20 dB below the peak received power. This means that the surface of drywall can still be considered to be smooth even at 142 GHz and the specular reflection is the main mechanism for indoor propagation at 142 GHz.

Comparisons between measurements and predictions made by a dual-lobe directive scattering (DS) model (as introduced in [47, 50]) with TX incident angles of $\theta_i = 10°$, 30°, 60° and 80° are shown in Figures 11.4(b), 11.4(d) and 11.4(f). Permittivity $\epsilon_r = 4.7$, 5.2 and 6.4, estimated from the reflection measurements using the Fresnel reflection coefficient equation [51], are used in the dual-lobe DS model at 28, 73 and 142 GHz, respectively. It can be seen that simulations of peak received power (the sum of reflection and scattering) at the specular reflection angle agrees well with measured data (within 3 dB), confirming that scattering can be modeled approximately by a smooth reflector with some loss (see (3)–(5) and (23) in [47]) when material properties are known, while scattering at other scattering angles falls off rapidly.

11.3   Similarities to mmWave Propagation

There are notable differences seen at THz frequencies compared to mmWave systems (e.g., high phase noise and Doppler, limited output power and more directional beams), which makes propagation more challenging [1, 51]. These differences can make a fundamental impact on the channel characteristics for radio frequencies at

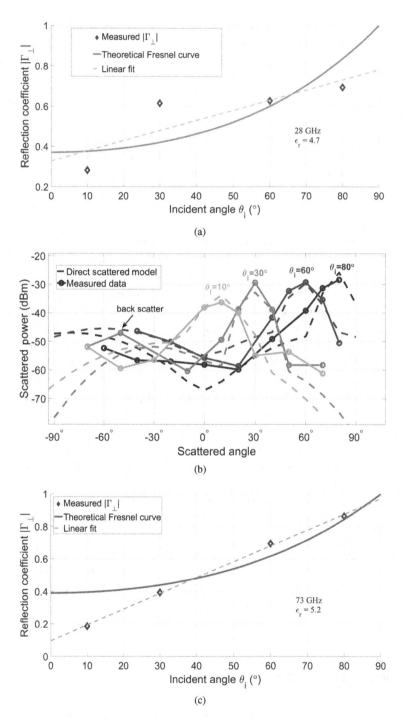

Figure 11.4 Comparison between measurements and dual-lobe directive scattering (DS) model plus reflected power using (3)–(5) and (23) in [47] at incident angles 10°, 30°, 60° and 80° (ϵ_r = 4.7, 5.2 and 6.4 for drywall at 28, 73 and 142 GHz). © 2019 IEEE. Reprinted, with permission, from [47]. (*Cont. next page*)

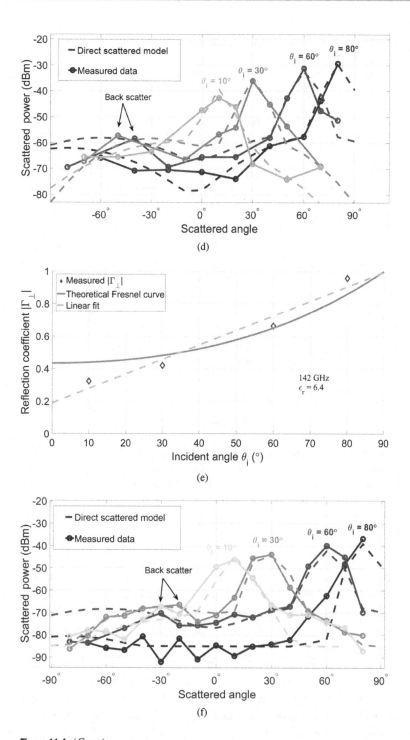

Figure 11.4 (*Cont.*)

Table 11.2 Indoor directional CI path-loss model at 28, 73 and 142 GHz for both LoS and NLoS environments [53, 54].

Env.	28 GHz [53]		73 GHz [53]		142 GHz [51]	
	PLE	σ (dB)	PLE	σ (dB)	PLE	σ (dB)
LoS	1.70	2.50	1.60	3.20	1.99	2.71
NLoS$_{Best}$	3.00	10.80	3.40	11.80	3.03	6.91
NLoS	4.40	11.60	5.30	15.70	4.70	14.10

Early results of 142 GHz directional indoor path loss V-V polarization

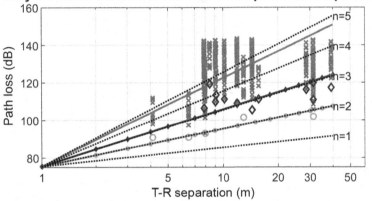

Figure 11.5 142 GHz directional path loss scatterplot and indoor directional CI ($d_0 = 1$ m) path-loss model for both LoS and non-LoS (NLoS) scenarios. Each circle represents an LoS path-loss value, crosses represent NLoS path-loss values measured at arbitrary antenna pointing angles between the TX and RX, and diamonds represent angles with the lowest path loss measured for each NLoS TX–RX location combination. © 2021 IEEE. Reprinted, with permission, from [52].

THz. However, indoor propagation measurements at 28, 73 and 142 GHz show that the path loss after the first meter at THz frequencies (e.g., 142 GHz) is similar to path loss at mmWave frequencies (e.g., 28 and 73 GHz) [51] and the multipath time dispersion (time delay statistics) for mmWave and THz bands are somewhat similar in mobile channels.

Figure 11.5 presents the directional path-loss scatterplot and best-fit CI path-loss model [53, 55] at 142 GHz for both LoS and NLoS environments. The LoS path-loss exponents (PLEs) are 1.7 at 28 GHz, 1.6 at 73 GHz and 2.0 at 142 GHz, as shown in Table 11.2, showing that there is a bit more loss at 142 GHz, likely due to atmospheric attenuation [1]. The NLoS Best PLEs and the NLoS PLEs are similar over all three frequencies, respectively, with NLoS at 142 GHz having slightly less loss than lower frequencies, likely due to greater reflected power as frequency increases (see Figure 11.4).

Table 11.3 Indoor omnidirectional CI path-loss model at 28, 73 and 142-GHz for both LoS and NLoS environments [53, 54]

Env.	28 GHz [53]		73 GHz [53]		142 GHz	
	PLE	σ (dB)	PLE	σ (dB)	PLE	σ (dB)
LoS	1.10	1.80	1.30	2.40	1.75	2.88
NLoS	2.70	9.60	3.20	11.30	2.69	6.59

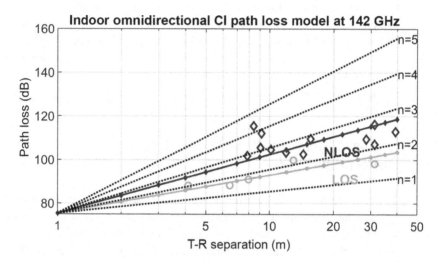

Figure 11.6 NYU best-fit omnidirectional CI path-loss model at 142 GHz for both LoS and NLoS situations. The diamonds represent the measured omnidirectional path loss at 142 GHz in NLoS environments and the circles, conversely, represent the LoS situation. © 2021 IEEE. Reprinted, with permission, from [52].

Figure 11.6 shows the NYU best-fit omnidirectional CI path-loss model at 142 GHz and the scatterplot of synthesized omnidirectional measured path loss at 142 GHz. The LoS omnidirectional PLE at 142 GHz is 1.75, which is less than the directional LoS PLE of 1.99, showing that omnidirectional antennas would capture power from all directions as compared to directional antennas, but they would cover a shorter link range due to the lower antenna gain. The NLoS omnidirectional PLE is 2.69, which is close to the $NLoS_{Best}$ PLE of 3.03 and is less than the NLoS PLE of 4.70 at 142 GHz, indicating that accurate beamforming algorithms are required to maintain indoor NLoS links at 142 GHz.

The omnidirectional PLE and shadowing parameters of the CI path-loss model with 1 m free space reference distance for both LoS and NLoS are summarized in Table 11.3. The indoor LoS PLEs are 1.10, 1.30 and 1.75 at 28, 73 and 142 GHz, respectively, indicating that the partition loss at 142 GHz is higher than at 28 and 73 GHz. In the NLoS case, the PLEs are 2.70, 3.20 and 2.69 at 28, 73 and 142 GHz, respectively.

The indoor office large-scale path-loss results show that there is remarkable similarity in terms of PLEs over 28, 73 and 142 GHz for both LoS and NLoS scenarios, when referenced to the first meter free-space reference distance [52, 56]. The results imply that THz channels are similar to today's mmWave wireless propagation channels except for the path loss in the first meter of propagation, when energy spreads into the far-field.

A 3D spatial statistical channel model was presented in [52] for mmWave and sub-THz frequencies in both LoS and NLoS scenarios based on the extensive measurements at 28 and 140 GHz in an indoor office building. This work showed that mathematical distributions of the number of multipath clusters, RMS delay spread, the number of multipath components (MPCs) or subpaths per cluster can be applied for frequencies above and below 100 GHz, although the statistical means of those distributions decrease with increasing frequency [52, 56]. The similarities in the wireless channels were also observed for outdoor urban microcell environments at 28, 38, 73 and 142 GHz, as presented in [57].

11.4 New Applications and New Channels for New Environments at THz

Terahertz signaling will be applied in a number of emerging applications. Since the applications determine the environments in which channel measurements and models need to be made, we will give here a brief review. In general, the applications fall into two classes: (1) those where the extremely high data rates of THz links are essential, and (2) those where the special propagation properties of THz frequencies allow improved sensing of the environment.

The first type of application includes small-cell systems and Wi-Fi with extremely high throughput – both per-user and in aggregate. The small wavelength of THz radiation allows the construction of massive MIMO arrays (on the order of 10,000 antennas) within a reasonable form factor, which can be exploited not only for improved range, but more importantly for increasing the number of users that can be supplied at the same time. In this context, the angular dispersion, and thus separability, of the MPCs is essential. This holds for both outdoor environments and indoor (Wi-Fi-type) applications. At the same time, some indoor applications, such as virtual reality, require data rates of up to 80 Gbit/s. The transmission of uncoded 8k video requires similarly high data rates; this could be needed both for communication between computers and peripherals, as well as between 8k cameras and infrastructure. This is far beyond the capabilities of current Wi-Fi systems such as 802.11ad (operating in the mmWave band), and constitutes an interesting area of application. Indoor office environments have been measured at THz frequencies for a long time [32, 36, 58–61]. The characterization of outdoor environments is more recent. The first double-directional measurements over a larger (100 m) distance were made in [62]. They showed considerable angular spread due to reflections as well as vegetation scattering (see Figure 11.7).

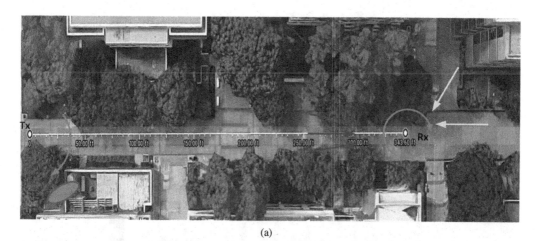

(a)

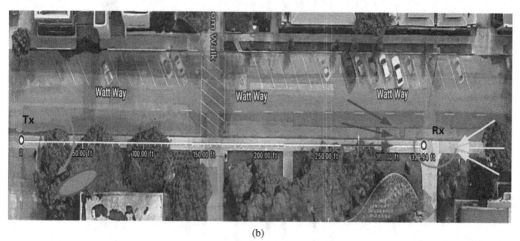

(b)

Figure 11.7 Photo of environment (top) and power angular spectra at 140 GHz carrier frequency in an outdoor environment. © 2019 IEEE. Reprinted, with permission, from [62]. (*Cont. next page*)

Another application for THz communication is the "information kiosk," a TX to which a user would step up. Since TX and RX would be close together, large amounts of information could be downloaded wirelessly in a short time. This concept goes back to the "infostation" proposed by Rutgers University in the late 1990s, though of course the use of THz signaling allows much higher data rates than was envisioned 25 years ago [63]. Infostations have several advantages in the context of THz signals: (1) the short distance allows high data rates even with small TX power, thus eliminating the need for (expensive) high-power transmit amplifiers); (2) the typical arrangement that a person holds an RX to a predetermined location eliminates the need for adaptive antennas and beamtracking; and (3) the short distance between TX and RX reduces the delay spread and thus relaxes requirements for equalizers [64].

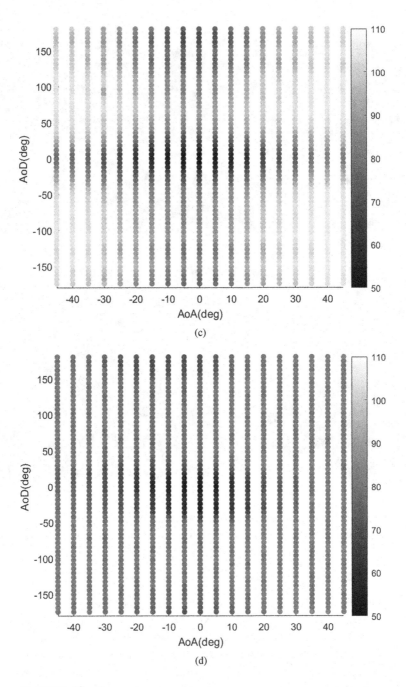

Figure 11.7 (*Cont.*)

In a similar vein, THz signaling is also very suitable for high-speed connections in wireless data centers [16, 65–67]. Traditionally, cables transmit data between different servers. However, this is cumbersome when rearrangements of the server racks are to be done, and may be subject to mechanical damage to the cables. In contrast, wireless connections between servers allow extremely high throughput, and the long time over which a connection remains stationary eliminates concerns about overhead for beam adjustment and tracking. Due to the short distances within data centers, atmospheric attenuation is not a concern. The large number of metallic objects is, however, anticipated to lead to more multipath, so this environment has to be considered differently than traditional indoor environments.

Another "classical" application for extremely high data rates is cellular backhaul [68]. Irrespective of whether the mobile access occurs in the THz, mmWave, or sub-6 GHz bands, the aggregate data rate for a 5G base station (BS) can easily reach tens of Gbit/s. If wired (optical fiber) backhaul is to be avoided, then THz wireless offers the best possibility of meeting those data rate requirements. These THz connections have fixed location and antenna orientation, but due to the longer distances they need to cover, they are more sensitive to atmospheric absorption and disturbances. Many of the outdoor measurements that have been made in the past (see below) are related to this and similar application cases.

Related to the backhaul problem are connections between "mobile base stations" located on high-speed trains and fixed infrastructure nodes [69–72]. For example, an in-train network might accumulate data from all passengers in a high-speed train. Due to the large number of such passengers (several hundred), even at regular Wi-Fi link speed for the individual users, the aggregate data rate is in the tens of Gbit/s. Transmission of the aggregated data to fixed BSs along the train line then constitutes the backhaul problem. Additional difficulties arise from the fact that the target direction of the beamforming is time-variant and appropriate beamtracking needs to be implemented. However, in contrast to standard mobile access, the trajectory of the mobile end is highly predictable, so that only the deviations from the nominal speed need to be compensated. A similar situation occurs in the backhaul from drones that are used as mobile BSs to cover rural areas or temporary hotspots [73].

As the data rates exchanged between boards in a computer, and between chips on a board, increases, the wired connections previously used for such applications become insufficient. In many distributed computing applications, several hundred Gbit/s needs to be exchanged between chips and boards. Wireless THz connections are a promising way to alleviate these bottlenecks. In addition to the requirement of high throughput, there is also a requirement for low latency, since computations on one board might depend on the data fetched from another. Consequently, time-consuming signal processing, such as forward error correction (FEC) decoding for large codeblocks, or complicated equalizers, are not an option for these applications. The links thus must be designed in a manner that enables low error probability with simple detection techniques. Furthermore, the links must be designed to prevent snooping by possible interceptors. For all these purpose, detailed channel models are again required. Extensive measurements and models have been performed in [31, 74, 75] and related considerations on desktops [76]. Channel models for on-chip and on-board devices tend to be stochastic models, though deterministic models can be helpful in determining the optimum placement of the antennas or arrays [74, 77–79].

Moving to yet smaller spatial scales, the field of *nanonetworks*, that is, networks in which TX and RX are on a molecular scale, has excited considerable interest in the literature.[1] Such networks consist not only of communications links, but need to involve sensors, actuators, storage and the capability to communicate with larger networks through suitable gateways. A key application for those networks is healthcare, where the nanodevices could be injected into a human body to sense/help a diagnosis, and also actuate treatments. Other applications include environmental monitoring and agriculture. Challenges for practical implementation are manifold, but from a propagation channel point of view one of the biggest challenges lies in devising experimental setups for channel measurements, and the design of suitable antennas. Note that in this field the main advantage of the use of THz frequencies is not the high bandwidth or data rate, but rather the small wavelength that allows the construction of antennas that fit onto the nanodevices.

Terahertz can also be used for environmental sensing [80–82]. In this case, it is not a high data rate that is important, but rather the presence of characteristic absorption lines and scattering properties of various chemical compounds (including water and oxygen) and pollutants that plays a role. Such a sensing system might set up, for example, a long-distance fixed wireless access link, and measure the absorption/scattering in a particular band to determine the density of pollutants in the air. It has even been shown that (for an indoor scenario) open flames can change the transfer function of the channel in the THz regime, thus enabling fire sensing. Backscatter from particles can also be determined by THz transceivers, again allowing a determination of the density of certain types of scatterers in the environment.

Finally, THz radar offers a number of advantages [83–86]. Compared to LiDAR, it is more robust to dust, clouds, fog and other weather impairments. Compared to traditional microwave radar, it provides significantly improved resolution. Since a channel sounder is essentially a bi-static radar, there is an obvious connection between channel investigations and radar. Applications of THz radar in the civilian space mainly revolve around automotive radars (where they provide better resolution than the current 24 and 73 GHz radars), and have also been investigated for body scanners for both security (airport) and medical applications. Furthermore, air-to-ground radars have been considered for the military space.

From this enumeration of applications, one can see that channel measurements in a large variety of environments are required. As will be outlined next, only a few of those cases have been analyzed in any depth. There thus will be a need in the next years to perform extensive measurement campaigns.

11.5   What Is Known to Date

There are three main channel sounding techniques in the THz range: THz time domain spectroscopy (THz-TDS), vector network analyzer (VNA)-based channel sounding

---

[1] Different papers use the expression "nanonetworks" for different applications. For example, [87] uses it for short-distance (0.1–5 m) local area networks. Others use it for networks of chip-to-chip links. We will follow the terminology first suggested by Jornet and Akyildiz [88].

and correlation-based sounding [22, 23, 60, 89–95]. THz-TDS is based on sending ultrashort pulsed laser light from a common source to the TX and the RX. The TX converts the ultrashort light pulse to the THz range and the detector at the RX transforms the field strength of the received THz impulse into an electrical signal when the optical impulse hits the detector [89, 90]. The short THz-TDS pulses cover a huge bandwidth and are excellent for estimating electrical and scattering parameters of sample materials. However, due to the large size of the spectrometer and the limited output power, this approach is not suitable to be used over a wide range of indoor or outdoor scenarios or for measuring the wireless channels at more than a few meters of distance.

Four-port VNAs are commonly used for THz-range channel sounding, where the two additional ports (compared to the traditional two-port VNAs used at lower frequencies) are used to generate a local oscillator for the mixer in the frequency extenders that are used to increase the VNA stock frequency range to much higher frequencies through heterodyning [90, 91]. A VNA sweeps discrete narrowband frequency tones across the bandwidth of interest to measure the S_{21} parameter of the wireless channel. Due to the long sweep time across a broad spectrum, which can exceed the channel coherence time, VNA-based channel sounders are typically used in a static environment and require a cable that can be a tripping hazard over tens or hundreds of meters [23, 91, 92].

Correlation-based channel sounder systems transmit a known wideband pseudo-random sequence. At the RX, the received signal is cross-correlated with an identical but slightly delayed pseudo-random sequence, providing autocorrelation gain at the expense of a slightly longer acquisition time (on the order of tens of ms) [91, 92]. Sliding correlator chips have recently been produced that offer a 1 Gbps baseband spread spectrum sequence [96], and sliding correlators generally enable cable-free operation over useful mobile communication distances of up to 200 m at sub-THz frequencies, depending on transmit power, bandwidth and antenna gain [23, 91].

Indoor 3D spatial statistical channel models for mmWave and sub-THz bands were derived from indoor radio propagation measurements performed from 2014 to 2019 at 28, 73 and 142 GHz, with link distances up to 40 m, in an indoor office environment on the ninth floor of 2 Metrotech Center, Brooklyn, New York, using a wideband sliding correlation-based channel sounder with steerable horn antennas at both the transmitter and receiver [52]. Over 15,000 power delay profiles (PDPs) were derived from the measurements and were used to extract channel statistics such as the number of time clusters, cluster delays and cluster powers. The resulting channel statistics enable the establishment of a statistical channel model from 28 to 140 GHz for the indoor office scenario. Side-by-side comparisons of propagation characteristics (e.g., large-scale path loss, multipath time dispersion, scattering) across a wide range of frequencies from mmWave to THz were made to study the key similarities and differences in the propagation channels [52].

A correlation-based channel sounder at 300 GHz with 8 GHz bandwidth was presented and evaluated in [90] with the same wired clock source being connected to both the TX and RX, which used a subsampling technique to avoid the expense of high-speed A/D converters. A 12th-order M-sequence was used with a subsampling factor of 128, and the theoretical maximum measurable Doppler frequency was 8.8 kHz, equivalent to a velocity of 31.7 km/h at 300 GHz [90].

Propagation measurements in the 140 GHz band were conducted in a shopping mall [97, 98] using a VNA-based channel sounder with a 19 dBi horn antenna at the RX, and a 2 dBi bicone antenna at the TX. It was shown that the numbers of clusters and MPCs in each cluster in the 140 GHz band, an average of 5.9 clusters and 3.8 MPCs per cluster, were fewer as compared to the 28 GHz band, which had an average of 7.9 clusters and 5.4 MPCs per cluster [97].

Work in [99, 100] presented D-band propagation measurements in a very close-in environment around a personal computer using a VNA-based sounder. Indoor directional path losses at 30, 140 and 300 GHz were compared using different path-loss models in [100]. Although the LoS path loss models predicted a PLE close to 2.0, the multifrequency close-in free-space reference distance model with a frequency-dependent term (CIF) and alpha–beta–gamma (ABG) model had better PLE values and standard deviation stability for these indoor environments than the single-frequency CI and floating interception (FI) models [25, 53, 100–102].

Measurements at 100, 200, 300 and 400 GHz with a 1 GHz RF bandwidth THz-TDS channel sounder showed that both indoor LoS and NLoS (specular reflection from interior building walls) links could provide a data rate of 1 Gbps [103].

Propagation loss measurements for estimating the performance of a communication link in the 350 GHz frequency band were presented in [104], where a VNA-based system was used with 26 dBi gain co-polarized horn antennas at both the TX and RX. The presence of water absorption lines in the spectra at 380 GHz and 448 GHz was very evident. Data rates of 1 Gbps for a 8.5 m link and 100 Gbps for a 1 m TX–RX separation distance were shown to be possible via wireless communication links at 350 GHz [104].

Channel and propagation measurements at 300 GHz were presented in [31, 36, 105], where a VNA-based channel system with 26 dBi gain horn antennas at both TX and RX was used to analyze the channel characteristics at 300–310 GHz with an IF frequency bandwidth of 10 kHz. Maximum transmission rates of several tens of Gbps for LoS and several Gbps for NLoS paths were shown to be achievable [105].

THz-band indoor propagation measurements were conducted in [61] using a VNA-based system covering a frequency range from 260 GHz to 400 GHz with 25 dBi gain horn antennas at both TX and RX within a TX–RX separation range of 0.95 m. Measurement results showed that Tbps throughput was achievable in the THz band. However, robust beamforming algorithms will be required in THz-band communications. Acoustic ceiling panels, which were shown to be good reflectors in the THz band, could be used as low-cost components to support NLoS links [61].

11.6   Summary

Ultra-broadband THz communication systems are expected to help satisfy the ever-growing need for smaller devices that can offer higher-speed wireless communication anywhere and anytime. In the past years, it has become obvious that wireless data rates exceeding 100 Gbit/s will be required several years from now. This large bandwidth paired with higher-speed wireless links can open the door to a large number of novel applications such as ultrahigh-speed picocell cellular links, wireless short-range communications, secure wireless communications for military and defense applications, chip-to-chip communications and on-body communications for health-monitoring systems. To enable future THz-range wireless communications, it is imperative to understand propagation mechanisms that govern communication and to develop channel models that can be used to describe general channel properties needed for system design or for algorithm testing.

The propagation mechanism of electromagnetic (EM) waves at THz frequencies is strongly dependent on the material properties of the constituents in the propagation channel. The high reliability required for high-data-rate and low-latency applications demands the characterization of the impact of blockages from cables, humans and small-scale mobility (e.g., rack vibrations) on THz propagation. It is therefore of utmost importance that the propagation channel environment be characterized before the development of a THz communication system that will operate in this environment. Each of the new applications that will be possible due to THz short-range communications have environments that are unique in their structural layout and scatterer (i.e., reflective objects in the environment) distribution, all of which contribute to path loss and other system-impacting attributes. For example, the data center environment is unique in its peculiar arrangement of server rack stalls with constituent blades perpendicularly assembled in each rack.

The challenges often experienced with propagation channel characterization range from difficulty in constructing channel sounders – equipment used for the wireless channel measurements at a particular frequency – to lack of expertise in conducting the measurement campaign, particularly in unconventional propagation channels such as between chips on a motherboard. There are alternatives to channel measurements, such as *ray-tracing*, which involves simulating the EM properties of the propagation channel using a computer aided design (CAD) model. It is important to note that ray-tracing suffers numerous shortcomings, as the simulation model is never a true reflection of the actual behavior of the propagation environment. Additionally, development of statistical channel models that can be applied to all THz communication-relevant environments are missing.

Although there is ongoing work on a physical layer design for THz communications, achieving super-low latency (<1 ms) for any wireless link implementation remains elusive. Another challenge is how to avoid the use of power-hungry high-resolution baseband devices such as analog-to-digital converters (ADCs) and digital-to-analog converters (DACs). The energy efficiency of even low-resolution ADCs in finer technology nodes is limited by more stringent thermal noise requirements. A physical layer design that addresses the aforementioned challenges is needed for the development of a future wireless THz data centers. Additionally, low latency and low energy consumption channel equalization is needed. Finally, MIMO configurations and designs need to be investigated, especially in LoS environments.

References

[1] T. S. Rappaport, Y. Xing, O. Kanhere, S. Ju, A. Madanayake, S. Mandal, A. Alkhateeb, and G. C. Trichopoulos, "Wireless communications and applications above 100 GHz: Opportunities and challenges for 6G and beyond (invited)," *IEEE Access*, vol. 7, pp. 78729–78757, Feb. 2019.

[2] T. Kurner, "Towards future THz communications systems," *Terahertz Science and Technology*, vol. 5, no. 1, pp. 11–17, 2012.

[3] S. Koenig, D. Lopez-Diaz, J. Antes, F. Boes, R. Henneberger, A. Leuther, A. Tessmann, R. Schmogrow, D. Hillerkuss, R. Palmer, T. Zwick, C. Koos, W. Freude, O. Ambacher, J. Leuthold and I. Kalfass, "Wireless sub-THz communication system with high data rate," *Nature Photonics*, vol. 7, no. 12, pp. 977–981, 2013.

[4] I. F. Akyildiz, J. M. Jornet and C. Han, "Terahertz band: Next frontier for wireless communications," *Physical Communication*, vol. 12, pp. 16–32, 2014.

[5] Y. Xing and T. S. Rappaport, "Terahertz wireless communications: Co-sharing for terrestrial and satellite systems above 100 GHz (invited)," *IEEE Communications Letters*, vol. 25, no. 10, pp. 3156–3160, Oct. 2021.

[6] A. Maltsev, A. Maltsev, A. Pudeyev, Y. Gagiev, A. Lomayev, I. Bolotin, K. Johnsson, T. Sakamoto, H. Motozuka, C. Gentile, P. Papazian, Jae-K. Choi, J. Senic, J. Wang, D. Lai, N. Golmie, K. Remley, J. Luo, Y. Xin, K. Zeng, N. Iqbal, J. He, G. Wang, R. Müller, D. Dupleich, R. S. Thomä, R. Yang, F. Fellhauer, M. Kim, S. Sasaki, K. Umeki, T. Iwata, K. Wangchuk and Jun-ichi Takada, "Channel models for IEEE 802.11ay," IEEE 802.11- 15/1150r9, Mar. 2017.

[7] "IEEE standard for high data rate wireless multi-media networks amendment 2: 100 Gb/s wireless switched point-to-point physical layer," IEEE Std 802.15.3d-2017 (Amendment to IEEE Std 802.15.3-2016 as amended by IEEE Std 802.15.3e- 2017), pp. 1–55, Oct. 2017.

[8] A. A. Boulogeorgos, A. Alexiou, T. Merkle, C. Schubert, R. Elschner, A. Katsiotis, P. Stavrianos, D. Kritharidis, P.-K. Chartsias, J. Kokkoniemi, M. Juntti, J. Lehtomaki, A. Teixeira and F. Rodrigues, "Terahertz technologies to deliver optical network quality of experience in wireless systems beyond 5G," *IEEE Communications Magazine*, vol. 56, no. 6, pp. 144–151, 2018.

[9] T. Kurner and S. Priebe, "Towards THz communications: Status in research, standardization and regulation," *Journal of Infrared, Millimeter, and Terahertz Waves*, vol. 35, no. 1, pp. 53–62, 2014.

[10] M. J. W. Rodwell, Y. Fang, J. Rode, J. Wu, B. Markman, S. T. Suran Brunelli, J. Klamkin and M. Urteaga, "100–340GHz systems: Transistors and applications," *2018 IEEE International Electron Devices Meeting (IEDM)*, Dec. 2018, pp. 14.3.1–14.3.4.

[11] H. Aggrawal, P. Chen, M. M. Assefzadeh, B. Jamali and A. Babakhani, "Gone in a picosecond: Techniques for the generation and detection of picosecond pulses and their applications," *IEEE Microwave Magazine*, vol. 17, no. 12, pp. 24–38, Dec. 2016.

[12] D. M. Mittleman, "Twenty years of terahertz imaging (invited)," *Optics Express*, vol. 26, no. 8, pp. 9417–9431, Apr. 2018.

[13] M. Tonouchi, "Cutting-edge terahertz technology," *Nature Photonics*, vol. 1, no. 2, pp. 97–105, Feb. 2007.

[14] J. Harvey, M. B. Steer and T. S. Rappaport, "Exploiting high millimeter wave bands for military communications, applications, and design," *IEEE Access*, vol. 7, pp. 52350–52359, Apr. 2019.

[15] V. Petrov, D. Moltchanov and Y. Koucheryavy, "Applicability assessment of terahertz information showers for next-generation wireless networks," *2016 IEEE International Conference on Communications (ICC)*, May 2016, pp. 1–7.

[16] C.-L. Cheng, S. Sangodoyin and A. Zajic, "THz cluster-based modeling and propagation characterization in a data center environment," *IEEE Access*, vol. 8, pp. 56544–56558, 2020.

[17] O. Kanhere and T. S. Rappaport, "Position locationing for millimeter wave systems," *2018 IEEE Global Communications Conference (GLOBECOM)*, Dec. 2018, pp. 206–212.

[18] O. Kanhere and T. S. Rappaport, "Position location for futuristic cellular communications: 5G and beyond," *IEEE Communications Magazine*, vol. 59, no. 1, pp. 70–75, Jan. 2021.

[19] W. Chappel, "Briefing prepared for T-MUSIC Proposer's Day," Jan. 2019.

[20] H. Wang, T.-Y. Huang, N. S. Mannem, J. Lee, E. Garay, D. Munzer, E. Liu, Y. Liu, B. Lin, M. Eleraky, H. Jalili, J. Park, S. Li, F. Wang, A. S. Ahmed, C. Snyder, S. Lee, H. T. Nguyen and M. E. D. Smith, "Power amplifiers performance survey 2000–present," online https://gems.ece.gatech.edu/PA survey.html, accessed 2019.

[21] S. Ramo, J. R. Whinnery and T. Van Duzer, *Fields and Waves in Communication Electronics*, Wiley: Chichester, 1994.

[22] T. S. Rappaport, R. W. Heath, R. C. Daniels and J. N. Murdock, *Millimeter Wave Wireless Communications*, Pearson Prentice Hall: Upper Saddle River, NJ, 2015.

[23] Y. Xing and T. S. Rappaport, "Propagation measurement system and approach at 140 GHz moving to 6G and above 100 GHz," *IEEE 2018 Global Communications Conference*, Dec. 2018, pp. 1–6.

[24] R. Piesiewicz, T. Kleine-Ostmann, N. Krumbholz, D. Mittleman, M. Koch, J. Schoebei and T. Kurner, "Short-range ultra-broadband terahertz communications: Concepts and perspectives," *IEEE Antennas and Propagation Magazine*, vol. 49, no. 6, pp. 24–39, Dec. 2007.

[25] T. S. Rappaport, D. Moltchanov and Y. Koucheryavy, "Overview of millimeter wave communications for fifth- generation (5G) wireless networks-with a focus on propagation models," *IEEE Transactions on Antennas and Propagation*, vol. 65, no. 12, pp. 6213–6230, Dec. 2017.

[26] J. Ma, R. Shrestha, J. Adelberg, C.-Y. Yeh, Z. Hossain, E. Knightly, J. M. Jornet and D. M. Mittleman, "Security and eavesdropping in terahertz wireless links," *Nature*, vol. 563, no. 7729, p. 89, Oct. 2018.

[27] J. N. Murdock and T. S. Rappaport, "Consumption factor: A figure of merit for power consumption and energy efficiency in broadband wireless communications," *2011 IEEE GLOBECOM Workshops (GC Wkshps)*, Dec. 2011, pp. 1393–1398.

[28] J. N. Murdock and T. S. Rappaport, "Consumption factor and power-efficiency factor: A theory for evaluating the energy efficiency of cascaded communication systems," *IEEE Journal on Selected Areas in Communications*, vol. 32, no. 2, pp. 221–236, Feb. 2014.

[29] S. Verdu, "Spectral efficiency in the wideband regime," *IEEE Transactions on Information Theory*, vol. 48, no. 6, pp. 1319–1343, 2002.

[30] C. Luo, M. Medard and L. Zheng, "On approaching wideband capacity using multitone FSK," *IEEE Journal on Selected Areas in Communications*, vol. 23, no. 9, pp. 1830–1838, 2005.

[31] S. Kim and A. G. Zajic, "Statistical characterization of 300 GHz propagation on a desktop," *IEEE Transactions on Vehicular Technology*, vol. 64, no. 8, pp. 3330–3338, 2014.

[32] S. Priebe and T. Kurner, "Stochastic modeling of THz indoor radio channels," *IEEE Transactions on Wireless Communications*, vol. 12, no. 9, pp. 4445–4455, 2013.

[33] P. Smulders, "Exploiting the 60 GHz band for local wireless multimedia access: Prospects and future directions," *IEEE Communications Magazine*, vol. 40, no. 1, pp. 140–147, 2002.

[34] M. Fryziel, C. Loyez, L. Clavier, N. Rolland and P. A. Rolland, "Path-loss model of the 60 GHz indoor radio channel," *Microwave and Optical Technology Letters*, vol. 34, no. 3, pp. 158–162, 2002.

[35] H. Xu, V. Kukshya and T. S. Rappaport, "Spatial and temporal characteristics of 60-GHz indoor channels," *IEEE Journal on Selected Areas in Communications*, vol. 20, no. 3, pp. 620–630, Apr. 2002.

[36] S. Priebe, C. Jastrow, M. Jacob, T. Kleine-Ostmann, T. Schrader and T. Kürner, "Channel and propagation measurements at 300 GHz," *IEEE Transactions on Antennas and Propagation*, vol. 59, no. 5, pp. 1688–1698, May 2011.

[37] T. Kleine-Ostmann, M. Jacob, S. Priebe, R. Dickhoff, T. Schrader and T. Kurner, "Diffraction measurements at 60 GHz and 300 GHz for modeling of future THz communication systems," *2012 37th International Conference on Infrared, Millimeter, and Terahertz Waves*, 2012, pp. 1–2.

[38] C.-L. Cheng, S. Kim and A. Zajic, "UTD-based modeling of diffraction loss by dielectric circular cylinders at D-band," *2016 IEEE International Symposium on Antennas and Propagation (APSURSI)*, June 2016, pp. 1365–1366.

[39] D. McNamara, C. Pistorius and J. Malherbe, *Introduction to the Uniform Geometrical Theory of Diffraction*, Artech House: Norwood, MA, 1990.

[40] C. Han, A. O. Bicen and I. F. Akyildiz, "Multi-ray channel modeling and wideband characterization for wireless communications in the terahertz band," *IEEE Transactions on Wireless Communications*, vol. 14, no. 5, pp. 2402–2412, 2014.

[41] J. Fu, P. Juyal and A. Zajic, "THz channel characterization of chip-to-chip communication in desktop size metal enclosure," *IEEE Transactions on Antennas and Propagation*, vol. 67, no. 12, pp. 7550–7560, Dec. 2019.

[42] J. Fu, P. Juyal and A. Zajic, "Modeling of 300 GHz chip-to-chip wireless channels in metal enclosures," *IEEE Transactions on Wireless Communications*, vol. 19, no. 5, pp. 3214–3227, May 2020.

[43] X. Zhang, A. F. Molisch and S.-Y. Kung, "Variable-phase-shift-based RF-baseband codesign for MIMO antenna selection," *IEEE Transactions on Signal Processing*, vol. 53, no. 11, pp. 4091–4103, 2005.

[44] A. F. Molisch, V. V. Ratnam, S. Han, Z. Li, S. L. H. Nguyen, L. Li and K. Haneda, "Hybrid beamforming for massive MIMO: A survey," *IEEE Communications Magazine*, vol. 55, no. 9, pp. 134–141, 2017.

[45] S. Sun, T. S. Rappaport and M. Shafi, "Hybrid beamforming for 5G millimeter-wave multi-cell networks," *IEEE Conference on Computer Communications Workshops (INFOCOM WKSHPS)*, Apr. 2018.

[46] Y. Xing, O. Kanhere, S. Ju, and T. S. Rappaport, "Indoor wireless channel properties at millimeter wave and sub-terahertz frequencies," *IEEE 2019 Global Communications Conference*, Dec. 2019, pp. 1–6.

[47] S. Ju, S. H. A. Shah, M. A. Javed, J. Li, G. Palteru, J. Robin, Y. Xing, O. Kanhere and T. S. Rappaport, "Scattering mechanisms and modeling for terahertz wireless communications," *Proceedings of the IEEE International Conference on Communications*, May 2019, pp. 1–7.

[48] C. Jansen, S. Priebe, C. Moller, M. Jacob, H. Dierke, M. Koch and T. Kurner, "Diffuse scattering from rough surfaces in THz communication channels," *IEEE Transactions on Terahertz Science and Technology*, vol. 1, no. 2, pp. 462–472, 2011.

[49] J. Kokkoniemi, V. Petrov, D. Moltchanov, J. Lehtomaki, Y. Koucheryavy and M. Juntti, "Wideband terahertz band reflection and diffuse scattering measurements for beyond 5G indoor wireless networks," *European Wireless 2016: 22nd European Wireless Conference. VDE*, 2016, pp. 1–6.

[50] V. Degli Esposti, F. Fuschini, E. M. Vitucci and G. Falciasecca, "Measurement and modelling of scattering from buildings," *IEEE Transactions on Antennas and Propagation*, vol. 55, no. 1, pp. 143–153, Jan. 2007.

[51] Y. Xing and T. S. Rappaport, "Indoor wireless channel properties at millimeter wave and sub-terahertz frequencies," *IEEE 2019 Global Communications Conference*, Dec. 2019, pp. 1–6.

[52] S. Ju, Y. Xing, O. Kanhere and T. S. Rappaport, "Millimeter wave and sub-terahertz spatial statistical channel model for an indoor office building," *IEEE Journal on Selected Areas in Communications*, May 2021.

[53] G. R. MacCartney, T. S. Rappaport, S. Sun, S. Deng, "Indoor office wideband millimeter-wave propagation measurements and models at 28 GHz and 73 GHz for ultra-dense 5G wireless networks (invited paper)," *IEEE Access*, vol. 3, pp. 2388–2424, Oct. 2015.

[54] S. Deng, M. K. Samimi and T. S. Rappaport, "28 GHz and 73 GHz millimeter-wave indoor propagation measurements and path loss models," *IEEE International Conference on Communications Workshops (ICCW)*, June 2015, pp. 1244–1250.

[55] T. S. Rappaport, S. Sun, R. Mayzus, H. Zhao, Y. Azar, K. Wang, G. N. Wong, J. K. Schulz, M. Samimi, F. Gutierrez, "Millimeter wave mobile communications for 5G cellular: It will work!" *IEEE Access*, vol. 1, pp. 335–349, May 2013.

[56] Y. Xing, T. S. Rappaport and A. Ghosh, "Millimeter wave and sub-THz indoor radio propagation channel measurements, models, and comparisons in an office environment (invited paper)," *IEEE Communications Letters*, 2021.

[57] Y. Xing and T. S. Rappaport, "Urban microcell radio propagation measurements and channel models for millimeter wave and terahertz bands (invited paper)," *IEEE Communications Letters*, July 2021.

[58] M. Salhi, T. Kleine-Ostmann, T. Schrader, M. Kannicht, S. Priebe and T. Kurner, "Broadband channel measurements in a typical office environment at frequencies between 50 GHz and 325 GHz," *2013 European Microwave Conference*, Oct. 2013, pp. 175–178.

[59] S. Priebe, M. Jacob and T. Kuerner, "Angular and RMS delay spread modeling in view of THz indoor communication systems," *Radio Science*, vol. 49, no. 3, pp. 242–251, Mar. 2014.

[60] S. Priebe, M. Kannicht, M. Jacob and T. Kuerner, "Ultra broadband indoor channel measurements and calibrated ray tracing propagation modeling at THz frequencies," *Journal of Communications and Networks*, vol. 15, no. 6, pp. 547–558, Dec. 2013.

[61] N. Khalid and O. B. Akan, "Wideband THz communication channel measurements for 5G indoor wireless networks," *2016 IEEE International Conference on Communications (ICC)*, 2016, pp. 1–6.

[62] N. A. Abbasi, A. Hariharan, A. M. Nair, A. S. Almaiman, F. B. Rottenberg, A. E. Willner and A. F. Molisch, "Double directional channel measurements for THz communications in an urban environment," *IEEE International Conference on Communications (ICC)*, 2019.

[63] D. J. Goodman, J. Borras, N. B. Mandayam and R. D. Yates, "Infostations: A new system model for data and messaging services," *1997 IEEE 47th Vehicular Technology Conference: Technology in Motion*, vol. 2. IEEE, 1997, pp. 969–973.

[64] D. He, K. Guan, A. Fricke, B. Ai, R. He, Z. Zhong, A. Kasamatsu, I. Hosako and T. Kurner, "Stochastic channel modeling for kiosk applications in the terahertz band," *IEEE Transactions on Terahertz Science and Technology*, vol. 7, no. 5, pp. 502–513, 2017.

[65] B. Peng and T. Kurner, "A stochastic channel model for future wireless THz data centers," *2015 International Symposium on Wireless Communication Systems (ISWCS)*, 2015, pp. 741–745.

[66] S. Mollahasani and E. Onur, "Evaluation of terahertz channel in data centers," *NOMS 2016-2016 IEEE/IFIP Network Operations and Management Symposium*, 2016, pp. 727–730.

[67] N. Boujnah, S. Ghafoor and A. Davy, "Modeling and link quality assessment of THz network within data center," *2019 European Conference on Networks and Communications (EuCNC)*, 2019, pp. 57–62.

[68] T. Narytnyk, "Possibilities of using THz-band radio communication channels for super high-rate backhaul," *Telecommunications and Radio Engineering*, vol. 73, no. 15, 2014.

[69] Y. Chen, K. Guan and C. Han, "Joint network and propagation performance evaluation for high-speed railway communications in the mmwave and THz bands," *Proceedings of the Sixth Annual ACM International Conference on Nanoscale Computing and Communication*, 2019, pp. 1–6.

[70] K. Guan, G. Li, T. Kurner, A. F. Molisch, B. Peng, R. He, B. Hui, J. Kim and Z. Zhong, "On millimeter wave and THz mobile radio channel for smart rail mobility," *IEEE Transactions on Vehicular Technology*, vol. 66, no. 7, pp. 5658–5674, 2016.

[71] K. Guan, B. Ai, B. Peng, D. He, X. Lin, L. Wang, Z. Zhong and T. Kurner, "Scenario modules, ray-tracing simulations and analysis of millimetre wave and terahertz channels for smart rail mobility," *IET Microwaves, Antennas and Propagation*, vol. 12, no. 4, pp. 501–508, 2017.

[72] K. Guan, B. Peng, D. He, J. M. Eckhardt, S. Rey, B. Ai, Z. Zhong and T. Kurner, "Measurement, simulation, and characterization of train-to-infrastructure inside-station channel at the terahertz band," *IEEE Transactions on Terahertz Science and Technology*, vol. 9, no. 3, pp. 291–306, 2019.

[73] A. Saeed, O. Gurbuz and M. A. Akkas, "Terahertz communications at various atmospheric altitudes," *Physical Communication*, vol. 41, no. 101113, pp. 1–15, 2020.

[74] S. Kim and A Zajic, "Characterization of 300-GHz wireless channel on a computer motherboard," *IEEE Transactions on Antennas and Propagation*, vol. 64, no. 12, pp. 5411–5423, 2016.

[75] S. Kim and A Zajic, "Statistical modeling and simulation of short-range device-to-device communication channels at sub-THz frequencies," *IEEE Transactions on Wireless Communications*, vol. 15, no. 9, pp. 6423–6433, 2016.

[76] M. Salhi, T. Kleine-Ostmann, M. Kannicht, S. Priebe, T. Kurner and T. Schrader, "Broadband channel propagation measurements on millimeter and sub-millimeter waves in a desktop download scenario," *2013 Asia-Pacific Microwave Conference (APMC)*, 2013, pp. 1109–1111.

[77] A. Fricke, M. Achir, P. Le Bars and T. Kurner, "Characterization of transmission scenarios for terahertz intra-device communications," *2015 IEEE-APS Topical Conference on Antennas and Propagation in Wireless Communications (APWC)*, 2015, pp. 1137–1140.

[78] R. Geise, A. Fricke, G. Zimmer and B. Neubauer, "Geometrically up-scaled propagation measurements for terahertz intra-device communications," *Antennas and Propagation (EuCAP), 2016 10th European Conference on*, 2016, pp. 1–5.

[79] T. Kurner, A. Fricke, S. Rey, P. Le Bars, A. Mounir and T. Kleine-Ostmann, "Measurements and modeling of basic propagation characteristics for intra-device communications at 60 GHz and 300 GHz," *Journal of Infrared, Millimeter, and Terahertz Waves*, vol. 36, no. 2, pp. 144–158, 2015.

[80] X. He, N. Fujimura, J. M. Lloyd, K. J. Erickson, A. A. Talin, Q. Zhang, W. Gao, Q. Jiang, Y. Kawano, R. H. Hauge, F. Léonard and J. Kono, "Carbon nanotube terahertz detector," *Nano Letters*, vol. 14, no. 7, pp. 3953–3958, 2014.

[81] D. W. Van der Weide, J. Murakowski and F. Keilmann, "Gas-absorption spectroscopy with electronic terahertz techniques," *IEEE Transactions on Microwave Theory and Techniques*, vol. 48, no. 4, pp. 740–743, 2000.

[82] N. Born, D. Behringer, S. Liepelt, S. Beyer, M. Schwerdtfeger, B. Ziegenhagen and M. Koch, "Monitoring plant drought stress response using terahertz time-domain spectroscopy," *Plant Physiology*, vol. 164, no. 4, pp. 1571–1577, 2014.

[83] K. B. Cooper, R. J. Dengler, N. Llombart, B. Thomas, G. Chattopadhyay, and P. H. Siegel, "THz imaging radar for standoff personnel screening," *IEEE Transactions on Terahertz Science and Technology*, vol. 1, no. 1, pp. 169–182, 2011.

[84] K. Iwaszczuk, H. Heiselberg and P. U. Jepsen, "Terahertz radar cross section measurements," *Optics Express*, vol. 18, no. 25, pp. 26399–26408, 2010.

[85] K. B. Cooper and G. Chattopadhyay, "Submillimeter-wave radar: Solid-state system design and applications," *IEEE Microwave Magazine*, vol. 15, no. 7, pp. 51–67, 2014.

[86] B. Zhang, Y. Pi and J. Li, "Terahertz imaging radar with inverse aperture synthe sis techniques: System structure, signal processing, and experiment results," *IEEE Sensors Journal*, vol. 15, no. 1, pp. 290–299, 2014.

[87] H. Elayan, A. Eckford and R. Adve, "Regulating molecular interactions using terahertz communication," *ICC–IEEE International Conference on Communications (ICC)*, 2020, pp. 1–6.

[88] J. M. Jornet and I. F. Akyildiz, "Fundamentals of electromagnetic nanonetworks in the terahertz band," *Foundations and Trends in Networking*, vol. 7, nos. 2–3, pp. 77–233, 2013.

[89] W. L. Chan, J. Deibel and D. M. Mittleman, "Imaging with terahertz radiation," *Reports on Progress in Physics*, vol. 70, no. 8, p. 1325, 2007.

[90] S. Rey, J. M. Eckhardt, B. Peng, K. Guan and T. Kurner, "Channel sounding techniques for applications in THz communications: A first correlation based channel sounder for ultra-wideband dynamic channel measurements at 300 GHz," *2017 9th International Congress on Ultra Modern Telecommunications and Control Systems and Workshops (ICUMT)*, Nov. 2017, pp. 449–453.

[91] G. R. MacCartney and T. S. Rappaport, "A flexible millimeter-wave channel sounder with absolute timing," *IEEE Journal on Selected Areas in Communications*, vol. 35, no. 6, pp. 1402–1418, June 2017.

[92] T. S. Rappaport, *Wireless Communications: Principles and Practice*, 2nd ed., Prentice Hall: Upper Saddle River, NJ, 2002.

[93] R. Wilk, I. Pupeza, R. Cernat and M. Koch, "Highly accurate THz time-domain spectroscopy of multilayer structures," *IEEE Journal of Selected Topics in Quantum Electronics*, vol. 14, no. 2, pp. 392–398, 2008.

[94] C. U. Bas, R. Wang, S. Sangodoyin, D. Psychoudakis, T. Henige, R. Monroe, J. Park, C. J. Zhang and A. F. Molisch, "Real-time millimeter-wave MIMO channel sounder for dynamic directional measurements," *IEEE Transactions on Vehicular Technology*, vol. 68, no. 9, pp. 8775–8789, 2019.

[95] T. W. Crowe, B. Foley, S. Durant, K. Hui, Y. Duan and J. L. Hesler, "VNA frequency extenders to 1.1 THz," *2011 International Conference on Infrared, Millimeter, and Terahertz Waves*, 2011, p. 1.

[96] T. Wu, T. S. Rappaport, M. Knox and D. Shahrjerdi, "A wideband sliding correlator-based channel sounder with synchronization in 65 nm CMOS," *IEEE 2019 International Symposium on Circuits and Systems (ISCAS)*, May 2019, pp. 1–5.

[97] S. L. H. Nguyen, J. Järveläinen, A. Karttunen, K. Haneda and J. Putkonen, "Comparing radio propagation channels between 28 and 140 GHz bands in a shopping mall," *European Conference on Antennas and Propagation*, Apr. 2018, pp. 1–5.

[98] S. L. Nguyen, K. Haneda and J. Putkonen, "Dual-band multipath cluster analysis of small-cell backhaul channels in an urban street environment," *IEEE Global Communications Conference (GLOBECOM)*, Dec. 2016, pp. 1–6.

[99] S. Kim, W. T. Khan, A. Zajic and J. Papapolymerou, "D-band channel measurements and characterization for indoor applications," *IEEE Transactions on Antennas and Propagation*, vol. 63, no. 7, pp. 3198–3207, July 2015.

[100] C. L. Cheng, S. Kim and A. Zajic, "Comparison of path loss models for indoor 30 GHz, 140 GHz, and 300 GHz channels," *2017 11th European Conference on Antennas and Propagation*, Mar. 2017, pp. 716–720.

[101] 3GPP, "Study on channel model for frequencies from 0.5 to 100 GHz," 3rd Generation Partnership Project (3GPP), TR 38.901 V14.0.0, May 2017.

[102] S. Rey and T. Kurner, "Channel characteristics study for future indoor millimeter and submillimeter wireless communications," *2016 10th European Conference on Antennas and Propagation (EuCAP)*, Apr. 2016, pp. 1–5.

[103] J. Ma, R. Shrestha, L. Moeller and D. M. Mittleman, "Channel performance for indoor and outdoor terahertz wireless links," *APL Photonics*, vol. 3, no. 5, pp. 1–13, Feb. 2018.

[104] S. Bhardwaj, N. K. Nahar and J. L. Volakis, "All electronic propagation loss measurement and link budget analysis for 350 GHz communication link," *Microwave and Optical Technology Letters*, vol. 59, no. 2, pp. 415–423, July 2016.

[105] T. Kleine-Ostmann, C. Jastrow, S. Priebe, M. Jacob, T. Kürner and T. Schrader, "Measurement of channel and propagation properties at 300 GHz," *2012 Conference on Precision Electromagnetic Measurements*, July 2012, pp. 258–259.

12 Connection between the Measurements and Models

Akbar Sayeed, Kate A. Remley, Camillo Gentile, Andreas F. Molisch,
Alenka Zajić, Theodore S. Rappaport, Martin Käske and Reiner Thomä

The parameters of the channel models are traditionally determined using the measurements. Thus, it is critical to understand the connection between channel measurements and channel models. This will address two important questions: (1) How do we use the channel measurements in estimating model parameters to "fit" the models to the measurements? And (2) What kind of measurements do we need to make to measure the various model parameters? Finally, understanding this relationship will also help in propagating the uncertainties in channel parameters obtained from the measurements into the models. This chapter starts with a general physical model which is a specialization of the general model introduced in Chapter 2 to the practical case of uniform linear arrays (ULAs). Section 12.2 discusses the sampled representation of the ULA physical model that serves as a starting point for connecting measurements and models, as briefly discussed in Section 12.2.3. Section 12.3 discusses techniques for high-resolution parameter estimation (HRPE) that can improve on the Fourier resolution limits in the sampled representation of Section 12.2. Section 12.4 discusses extensions to 2D uniform planar arrays (UPAs) and Section 12.5 introduces concepts related to extending channel models to incorporate the hardware non-idealities related to the channel measurement hardware. This is an important step in developing uncertainty analyses of channel models..

12.1 Physical Channel Model for ULAs

The general model in Chapter 2 captures the physical propagation parameters (angles, delays, Doppler shifts) more or less independent of the communication system or channel sounder. If an antenna array is used on the transmit or receive side, the continuous model will be sampled in the spatial domain. In this section, we outline the specialization of the general physical model for commonly used ULAs.

Consider a system in which the transmitter (TX) and the receiver (RX) are equipped with ULAs. In this case, the physical model can be expressed as a time-varying transfer function as [1–4]

$$\mathbf{H}(t, f) = \int \mathbf{h}(t, \tau)e^{-j2\pi\tau f}d\tau = \sum_{n=1}^{N_p} \alpha_n \mathbf{a}^{\mathrm{R}}(\theta_n^{\mathrm{R}})\mathbf{a}^{\mathrm{T\dagger}}(\theta_n^{\mathrm{T}})e^{-j2\pi\tau_n f}e^{j2\pi\nu_n t}, \quad (12.1)$$

which represents a model for a multiple-input–multiple-output (MIMO) channel connecting a multi-antenna transmitter with a multi-antenna receiver. The channel is represented by the time-varying frequency response matrix $\mathbf{H}(t, f)$ of dimension $N^{\mathrm{R}} \times N^{\mathrm{T}}$, where N^{T} and N^{R} are the number of (typically half-wavelength spaced) antennas at the TX and the RX. The second equality in eq. (12.1) represents signal propagation over N_p paths, with α_n, θ_n^{T}, θ_n^{R}, τ_n and ν_n denoting the complex amplitude, normalized angle of departure (AoD), normalized angle or arrival (AoA), delay and Doppler shift associated with the nth path. The normalized AoAs and AoDs (spatial frequencies)

$$\text{AoD: } \theta_n^{\mathrm{T}}, \text{ AoA: } \theta_n^{\mathrm{R}},$$

which are related to the physical angles, ϕ_n^{R} ϕ_n^{T}, defined with respect to the broadside direction, via the relationship

$$\theta = \frac{\mathbf{x}^{\mathrm{A}}}{\lambda} \sin(\phi) = \frac{1}{2} \sin(\phi),$$

where $\mathbf{x}^{\mathrm{A}}$ is the antenna spacing, and λ is the operating wavelength (corresponding to the center of the operating frequency band). The second equality corresponds to half-wavelength (critical) antenna spacing. The vector $\mathbf{a}^{\mathrm{T}}(\theta^{\mathrm{T}})$ is an $N^{\mathrm{T}} \times 1$ steering vector at the TX for sending a signal in the direction θ^{T}, and $\mathbf{a}^{\mathrm{R}}(\theta^{\mathrm{R}})$ is an $N^{\mathrm{R}} \times 1$ response vector of RX array for a receiving signal coming from the direction θ^{R}.

For ULAs, the array steering and response vectors take the form of discrete spatial sinusoids with the spatial frequencies $\theta^{\mathrm{T}}, \theta^{\mathrm{R}} \in [-0.5, 0.5]$ [2–4]:

$$\mathbf{a}^{\mathrm{T}}(\theta^{\mathrm{T}}) = \frac{1}{\sqrt{N^{\mathrm{T}}}} \left[1, e^{-j2\pi\theta^{\mathrm{T}}}, \ldots, e^{-j2\pi\theta^{\mathrm{T}}(N^{\mathrm{T}}-1)} \right]^{*\dagger}, \tag{12.2}$$

$$\mathbf{a}^{\mathrm{R}}(\theta^{\mathrm{R}}) = \frac{1}{\sqrt{N^{\mathrm{R}}}} \left[1, e^{-j2\pi\theta^{\mathrm{R}}}, \ldots, e^{-j2\pi\theta^{\mathrm{R}}(N^{\mathrm{R}}-1)} \right]^{*\dagger}, \tag{12.3}$$

where the superscript† denotes the Hermitian (complex conjugate) transpose and $*$ denotes complex conjugation. The quasi-static version of the analytical physical model eq. (12.1) is given by

$$\mathbf{H}(f) = \sum_{n=1}^{N_p} \alpha_n \mathbf{a}^{\mathrm{R}}(\theta_n^{\mathrm{R}}) \mathbf{a}^{\mathrm{T}\dagger}(\theta_n^{\mathrm{T}}) e^{-j2\pi\tau_n f}, \tag{12.4}$$

in which the Doppler shifts are small enough that the temporal channel variations over the duration of measurement can be ignored.

The models of eqs. (12.1) and (12.4) are widely used for simulating wireless channels. However, they assume knowledge of the parameters at perfect (infinite) angle–delay–Doppler resolution. On the other hand, any sounder/system in practice has a finite resolution in the angle–delay–Doppler domains. Furthermore, the statistical characteristics of the estimated channel parameters also depend on the finite angle–delay–Doppler resolution of the channel sounding instrument [2, 5–7]. These challenges are accentuated at millimeter-wave (mmWave) frequencies due to: (1) the lack of sufficient measurements in different operational environments;

and (2) limited capabilities of existing channel sounders, for example, low spatial resolution and/or mechanical beam pointing. Fundamentally, many technical issues need to be addressed for estimating the angle–delay–Doppler channel parameters from measurements collected by sounders in practice, especially for sounders equipped with antenna arrays for directional measurements. The sampled channel representation discussed next provides a good starting point for understanding the issues.

12.2   Sampled (Virtual) Representation of the Physical Model

A fundamental connection between the measurements made in practice and the physical model above (with continuous parameters) is revealed by a sampled representation of the idealized model given in eq. (12.1) induced by four key parameters of the channel sounder: (1) temporal (delay) resolution, (2) frequency (Doppler resolution), (3) spatial resolution at the TX and (4) spatial resolution at the RX:

$$\Delta\tau = \frac{1}{W}, \ \Delta\nu = \frac{1}{T}; \ \Delta\theta^T = \frac{1}{N^T}, \ \Delta\theta^R = \frac{1}{N^R}, \tag{12.5}$$

where W is the (two-sided) bandwidth of the system and T is the duration of the probing signal (and coherent integration time of the system) [2, 7, 10]. The spatial resolutions are based on the assumption of critically (half-wavelength) spaced antennas, which is generally the case for preserving all spatial information. In effect, the continuous physical model in eq. (12.1) is replaced by a sampled (virtual) representation with respect to uniformly spaced angles, delays and Doppler shifts at the above resolutions. The impact of larger antenna spacings is discussed in [3] and smaller spacings in [9].

The sampled representation of the physical model eq. (12.1) is given by

$$\mathbf{H}(t,f) = \sum_{i=1}^{N^R} \sum_{k=1}^{N^T} \sum_{\ell=0}^{L} \sum_{m=-M}^{M} H_v(i,k;\ell,m) \mathbf{a}^R(i\Delta\theta^R) \mathbf{a}^{T\dagger}(k\Delta\theta^T) e^{-j2\pi\ell\Delta\tau f} e^{j2\pi m\Delta\nu t},$$

$$\tag{12.6}$$

where L and M represent the maximum number of resolvable delays and resolvable Doppler shifts within the delay and Doppler spreads, τ_{max} and ν_{max}, respectively, and are given by

$$L = \left\lceil \frac{\tau_{max}}{\Delta\tau} \right\rceil = \lceil \tau_{max} W \rceil \ ; \ M = \left\lceil \frac{\nu_{max}}{\Delta\nu} \right\rceil = \lceil \nu_{max} T \rceil, \tag{12.7}$$

where $\lceil \cdot \rceil$ denotes the "ceiling" operation. The sampled representation is completely characterized by the angle–delay–Doppler (virtual) channel coefficients, $\{H_v(i,k;\ell,m)\}$, which can be computed from the measured $\mathbf{H}(t,f)$ as [2, 3]

$$H_v(i,k;\ell,m) = \frac{1}{TW} \int_0^T \int_{-W/2}^{W/2} \mathbf{a}^{R\dagger}(i\Delta\theta^R) \mathbf{H}(t,f) \mathbf{a}^T(k\Delta\theta^T) e^{j2\pi\ell\Delta\tau f} e^{-j2\pi m\Delta\nu t} \, dt \, df.$$

$$\tag{12.8}$$

In essence, the sampled channel representation eq. (12.6) is a 4D Fourier series expansion of the time-varying spatial frequency response matrix $\mathbf{H}(t,f)$ in terms of temporal, spectral and spatial sinusoids, with the sampled (angle–delay–Doppler) channel coefficients serving as the Fourier series coefficients computed in eq. (12.8). For a given system/configuration, the sampled representation is an equivalent representation of $\mathbf{H}(t, f)$ over the signaling duration T and bandwidth W and contains all information about it. It is also an extension of the delay–Doppler sampled representation of (single-antenna) time-varying frequency response functions, introduced by Bello [10], to include the spatial dimension.

The spatial sampling in eq. (12.6) induces an equivalent *beamspace representation* of $\mathbf{H}(t, f)$ given by

$$\mathbf{H}_b(t, f) = \mathbf{U}^{\mathrm{R}\dagger}\mathbf{H}(t, f)\mathbf{U}^{\mathrm{T}} \iff \mathbf{H}(t, f) = \mathbf{U}^{\mathrm{R}}\mathbf{H}_b(t, f)\mathbf{U}^{\mathrm{T}\dagger}, \qquad (12.9)$$

where $\mathbf{H}_b(t, f)$ is the beamspace representation, and the matrices $\mathbf{U}^{\mathrm{R}}$ and $\mathbf{U}^{\mathrm{T}}$ represent the spatial DFT matrices, corresponding to uniformly sampled directions from eq. (12.5), which map the antenna domain into the angle domain (beamspace). The beamspace channel representation is particularly useful at mmWave frequencies due to the highly directional nature of propagation. It is a natural domain for representing channel measurements made with directional antennas; for example, rotated horn antennas or lens antenna arrays [6, 11, 12]. However, the resolution (beamwidth) of the directional antennas needs to be carefully taken into account via their aperture/size and their far-field patterns, as captured by eq. (12.9) for directional measurements made with ULAs and lens arrays.

The Fourier series of $\mathbf{H}(t, f)$ in eq. (12.6) can also be written in terms of delay–Doppler component matrices as

$$\mathbf{H}(t, f) = \sum_{\ell=0}^{L} \sum_{m=-M}^{M} \mathbf{H}_{DD}(\ell, m)e^{-j2\pi\ell\Delta\tau f}e^{j2\pi m\Delta vt},$$

$$\mathbf{H}_{DD}(\ell, m) = \frac{1}{TW} \int_0^T \int_{-W/2}^{W/2} \mathbf{H}(t, f)e^{j2\pi\ell\Delta\tau f}e^{-j2\pi m\Delta vt}\, dt df. \qquad (12.10)$$

Similar to eq. (12.10), the beamspace time–frequency response matrix $\mathbf{H}_b(t, f)$ can also be decomposed in terms of delay–Doppler component matrices as

$$\mathbf{H}_b(t, f) = \sum_{\ell=0}^{L} \sum_{m=-M}^{M} \mathbf{H}_{b, DD}(\ell, m)e^{-j2\pi\ell\Delta\tau f}e^{j2\pi m\Delta vt}$$

$$\mathbf{H}_{b, DD}(\ell, m) = \frac{1}{TW} \int_0^T \int_{-W/2}^{W/2} \mathbf{H}_b(t, f)e^{j2\pi\ell\Delta\tau f}e^{-j2\pi m\Delta vt}\, dt df. \qquad (12.11)$$

12.2.1   Partitioning of Paths Induced by the Sampled Representation

The sampled representation induces a partitioning of propagation paths that is very useful for relating measurements to the physical channel parameters, as illustrated in Figure 12.1. Essentially: (1) each sampled angle–delay–Doppler channel coefficient in eq. (12.6) is associated with an angle–delay–Doppler resolution bin of size

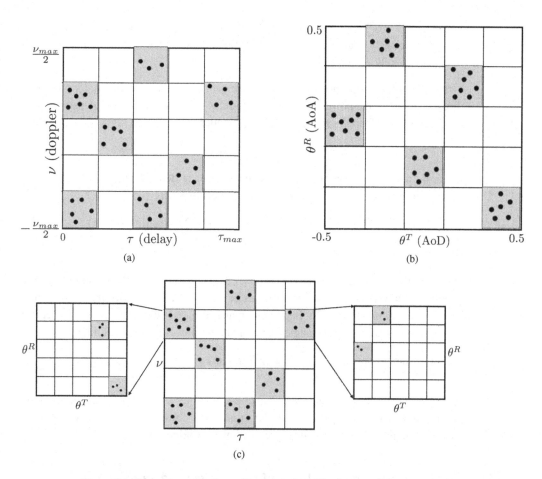

Figure 12.1 Illustration of path partitioning induced by the sampled representation. (a) path partitioning in delay–Doppler. (b) Path partitioning in angle. (c) Path partitioning in angle–delay–Doppler. First the paths are partitioned into different delay–Doppler resolution bins. Then the paths in each delay–Doppler resolution bin can be further resolved in angle. Or we could equivalently start with partitioning in angle and then further partition the paths in each angle resolution bin in terms of delay–Doppler. Each sampled angle–delay–Doppler channel coefficient has a distinct angle–delay–Doppler resolution bin.

defined by the system/sounder resolutions in eq. (12.5); and (2) a particular sampled channel coefficient is approximately the sum of the complex path gains of all paths whose angles (AoAs/AoDs), delays and Doppler shifts lie within the angle–delay–Doppler resolution bin corresponding to the channel coefficient. This follows from explicitly writing the sampled channel coefficients in eq. (12.8) for the physical model eq. (12.1) [2]:

$$H_v(i,k;\ell,m) = \sum_{n=1}^{N_p} \alpha_n f_{N^R}(\theta_n^R - i\Delta\theta^R) f_{N^T}(\theta_n^T - k\Delta\theta^T)$$

$$\text{sinc}(W(\tau_n - \ell\Delta\tau))\text{sinc}(T(v_n - m\Delta v)), \qquad (12.12)$$

where $f_N(x) = \frac{\sin(\pi N x)}{\sin(\pi x)}$ and $\mathrm{sinc}(x) = \frac{\sin(\pi x)}{\pi x}$ denote the Dirichlet sinc and sinc functions, respectively, that are peaky at the origin and have a half-power mainlobe width of approximately $\frac{1}{N}$ and 1, respectively.

Delay–Doppler–angle power profiles: The sampled channel coefficients provide a direct estimate of the power profile in angle–delay–Doppler that is commensurate with the angle–delay–Doppler resolution of the sounder:

$$\Psi_{ADD}(i,k,\ell,m) = |H_v(i,k,\ell,m)|^2, \tag{12.13}$$

and from which the power profile in angle (A) only, or delay–Doppler (DD) only, can be derived by summing over the remaining dimensions as:

$$\Psi_A(i,k) = \sum_{\ell,m} \Psi_{ADD}(i,k,\ell,m) \; ; \; \Psi_{DD}(i,k) = \sum_{i,k} \Psi_{ADD}(i,k,\ell,m). \tag{12.14}$$

Similarly, the power profile in TX angle only or RX angle only can be obtained by summing the 2D angle profile $\Psi_A(i,k)$ over the remaining dimension, and the power profile in delay only or Doppler only can be obtained from $\Psi_{DD}(\ell,m)$ by summing over the remaining dimension. In particular, an estimate of the total channel power can be obtained by summing the angle–delay–Doppler power profile over all dimensions:

$$\text{total channel power} = \sum_{i,k,\ell,m} \Psi_{ADD}(i,k,\ell,m) = \sum_{i,k,\ell,m} |H_v(i,k,\ell,m)|^2. \tag{12.15}$$

The power profile estimates are also constrained by the resolution of the sounder, as also implied by the path partitioning discussed above. Averaging over multiple measurements can be used to improve the power profile estimates.

12.2.2   Matrix Channel Representation of the Physical and Sampled Models

We can also sample $\mathbf{H}(t, f)$ in time and frequency at the delay (time) and Doppler (frequency) resolutions, defined in eq. (12.5), to facilitate digital processing. The result is two equivalent matrix channel representations, one in the time–frequency–aperture domain, say $\mathbf{H}_{t,f}$, obtained from $\mathbf{H}(t, f)$ in eq. (12.6), and one in the Doppler–delay–angle domain, say $\mathbf{G}_{\tau,v}$, obtained from the virtual channel coefficients defined in eq. (12.8) [2]. Let $N = TW$ denote the (approximate) dimension of the temporal signal space (time–bandwidth product). The matrices $\mathbf{H}_{t,f}$ and $\mathbf{G}_{\tau,v}$ are both of dimension $NN^R \times NN^T$, representing the dimensions of the spatiotemporal signal spaces at the TX and the RX. Furthermore, the two equivalent matrix channel representations are related through a linear transform:

$$\mathbf{H}_{t,f} = \mathbf{U}_{NN^R} \mathbf{G}_{\tau,v} \mathbf{V}_{NN^T}^{\dagger} \iff \mathbf{G}_{\tau,v} = \mathbf{U}_{NN^R}^{\dagger} \mathbf{H}_{t,f} \mathbf{V}_{NN^T}. \tag{12.16}$$

The first equation is a matrix version of eq. (12.6), and the second equation is the matrix version of eq. (12.8). For sounders with ULAs or UPAs, the linear transformations are unitary; that is, the matrices $\mathbf{U}_{NN^R}$ and $\mathbf{V}_{NN^T}$ are unitary [2]; in fact, they are DFT matrices induced by eqs. (12.6) and (12.8). However, the equivalent channel representations $\mathbf{H}_{t,f}$ and $\mathbf{G}_{\tau,\nu}$ have very different structures. In particular, the matrix $\mathbf{G}_{\tau,\nu}$ has a limited support and sparse structure due to the limited angular, delay and Doppler spreads, and the sparse nature of propagation, mainly consisting of line of sight (LoS) and strong (single-bounce) non-LoS (NLoS) paths, expected at mmWave frequencies [13]. The concept of path partitioning is very useful for relating the sparse structure of $\mathbf{G}_{\tau,\nu}$ to the parameters of the physical model. Furthermore, if we assume that the complex path amplitudes α_n in the physical model (12.1) are uncorrelated, then the elements of $\mathbf{G}_{\tau,\nu}$ (the angle–delay–Doppler coefficients in eq. (12.8)) are approximately uncorrelated and the elements of $\mathbf{H}_{t,f}$ are samples of a multidimensional wide sense stationary (WSS) process in time, frequency and aperture [2]. This is an extension of the WSS uncorrelated scattering (WSSUS) model in time and frequency introduced by Bello in [10].

12.2.3 Extracting Channel Model Parameters from Measurements

The basic idea is to exploit the fundamental multidimensional Fourier relationship between the time–frequency–aperture channel representation $\mathbf{H}_{t,f}$ and the angle–delay–Doppler representation $\mathbf{G}_{\tau,\nu}$. Consider the quasi-static case eq. (12.4) for simplicity. For channel modeling, the parameters of the physical model in eq. (12.4) – the number of paths, and the delays, angles and complex amplitude for each path – need to be estimated from channel measurements. At a basic level, the measurements are limited by the Fourier resolution of the sounder as described above and captured by the sampled representation.

As a starting point we can collect the measurements and estimate the model parameters using the matrix version eq. (12.16). For example, the measurements may be in the aperture (antenna) and frequency domains to obtain $\mathbf{H}_f$. We can then use eq. (12.16) to estimate the matrix $\mathbf{G}_\tau$ through a 2D DFT of $\mathbf{H}_f$ – one DFT for mapping the frequency domain into the delay domain, and one for mapping the antenna domain to the angle domain (beamspace). This estimate of $\mathbf{G}_\tau$ provides a baseline estimate of the sampled coefficients eq. (12.8) in the delay–angle domain commensurate with the spatiotemporal resolutions of the sounder defined in eq. (12.5). In particular, depending on the noise floor and dynamic range, the entries of $\mathbf{G}_\tau$ can be appropriately thresholded to determine the "significant" angle–delay channel coefficients. This directly leads to a baseline estimate of the physical model parameters in eq. (12.4): the number of the "significant" thresholded coefficients is an estimate of the number of resolvable MPCs, the coefficient values represent the complex gains of the resolvable paths, and the delay–angle indices associated with each coefficient define the (sampled) delay and angles associated with the corresponding MPC.

In many situations, this baseline estimate of the channel model parameters, based on the critically sampled representation, is adequate. Oversampling in the angle–delay (and Doppler) domains is also directly possible in the baseline estimate via eq. (12.8) by sampling at a finer resolution than the critical resolutions in eq. (12.5), as discussed more concretely in the single-antenna special case in Section 12.2.4. In some cases we may use HPRE methods, as discussed in Section 12.3, to improve the resolution of the angle–delay–Doppler parameters even further. Based on the discussion above, we see that it is critical to specify parameters such as the signaling duration, bandwidth, antenna characteristics, signal-to-noise ration (SNR)-dependent threshold, sampling resolutions and sounder dynamic range when reporting channel coefficients or extracting model parameters.

We note that for some sounders the measurements may directly be in the beamspace domain, as in the case of beamspace MIMO sounders using lens arrays from which the angle–delay channel parameters can be estimated [6, 12, 14]. It is also worth noting that directional measurements using mechanically rotated directional antennas (such as horn antennas) also directly measure in the beamspace domain although typically at a lower resolution. Such measurements can be used for channel parameter estimation by using appropriate models for the far-field beampatterns for the horn.

12.2.4   Special Cases

12.2.4.1   Single Antenna Static Channel

Let us consider the simplest channel – a single-antenna multipath channel with no time variation (no relative motion) – to illustrate the essential ideas behind estimating channel parameters from channel measurements and the role of sampled representation. In this case, the physical model for the channel impulse response (CIR) is given by

$$h(\tau) = \sum_{n=1}^{N_p} \alpha_n \delta(\tau - \tau_n), \qquad (12.17)$$

and the corresponding frequency response, defined in eq. (12.4), simplifies to

$$H(f) = \sum_{n=1}^{N_p} \alpha_n e^{-j2\pi\tau_n f} \approx \sum_{\ell=0}^{L} H_v(\ell)e^{-j2\pi\frac{\ell}{W}f}, \qquad (12.18)$$

where the first equality represents the physical model, the physical model and the approximate captures most of the channel power within the channel delay spread by choosing $L = \tau_{max}W$. The sampled channel coefficients can be computing from $H(f)$ using eq. (12.8) as

$$H_v(\ell) = \tilde{h}(\ell\Delta\tau) = \frac{1}{W} \int_{-W/2}^{W/2} H(f)e^{j2\pi\ell\Delta\tau f}df = \sum_{n=1}^{N_p} \alpha_n \text{sinc}\left(W\left(\tau_n - \ell\Delta\tau\right)\right),$$

$$(12.19)$$

where the second equality represents the sampled representation in (12.6) and the third equality corresponds to the physical model. The relation (12.19) states that due to the finite sounder bandwidth, the "infinite resolution" physical model in eqs. (12.17) and (12.18) is replaced by a "smoothed version" in which the delta functions in the idealized physical model are replaced by sinc(Wx) functions in eq. (12.19) centered at the physical delays for different MPCs. This is illustrated in Figure 12.2, in which the top plot shows the idealized physical MPCs. The bandlimited impulse responses for different MPCs are illustrated in the middle plot of Figure 12.2 for a bandwidth of $W = 1$ GHz. Finally, the bottom plot in Figure 12.2 shows the aggregate bandlimited CIR (the sum of the individual bandlimited responses for different MPCs shown in Figure 12.2 middle). The sampled (virtual) delay channel coefficients are also shown, which are uniformly spaced samples of the "smoothed" (aggregate) impulse response $\tilde{h}(\tau)$ (see eq. (12.19)).

Estimating delay parameters from measurements. Consider channel parameter estimation in the context of Figure 12.2. The goal is to estimate the number of paths, and the delay and amplitude for each path for a total of $2N_p + 1$ parameters. The sampled representation provides an approximate estimate of the number of paths by considering only the "dominant" channel coefficients – the samples whose magnitude is above a certain threshold that depends on the SNR. For example, in the bottom plot of Figure 12.2 five sampled coefficients (marked as "x") are dominant. For this case, each physical path would lead to about two dominant coefficients since the actual physical path delays do not exactly align with the sample locations. However, in Figure 12.2 two paths are very closely spaced – closer than the temporal resolution, $\Delta\tau = \frac{1}{W} = 1$ ns – and thus cannot be distinguished in the smoothed aggregate CIR in the bottom plot. By choosing a smaller threshold for "dominant" coefficients, the sampled representation can include more coefficients/samples, and can approximate the true bandlimited impulse response arbitrarily closely.

Oversampling in the sampled representation is also an attractive option. If the sampling interval in eq. (12.19) is chosen smaller than the critical value (see eq. (12.5)), then the peaks of the smoothed impulse response can be identified more accurately, as illustrated in Figure 12.2 (bottom). These peaks correspond to the true underlying delays for well-separated paths but one still cannot resolve paths whose delays are within the critical delay sampling resolution of $\Delta\tau = \frac{1}{W}$, as is the case for the middle two paths in Figure 12.2. The oversampling would identify three peaks in Figure 12.2: two representing the true underlying delays of two well-separated paths, and one representing a pair of unresolvable paths. We note that HRPE methods, discussed in Section 12.3, are also impacted by this limitation; in particular, the maximum likelihood estimation approach in Section 12.3.2 essentially corresponds to oversampling and finding the peaks.

In essence, the oversampled representation captures the bandlimited CIR in terms of the smallest number of *nonuniformly spaced* sinc(Wx) basis functions that are *not orthogonal*. On the hand, the *critically sampled* virtual channel representation captures the CIR in terms of a larger number of *uniformly spaced* sinc(Wx) basis functions that *are orthogonal*. The two approaches represent different trade-offs from signal processing, channel estimation and communication perspectives.

Path partitioning. The ℓth sampled channel coefficient in eq. (12.19) is associated with a virtual propagation delay of $\ell\Delta\tau = \frac{\ell}{W}$. Since sinc(Wx) is concentrated around

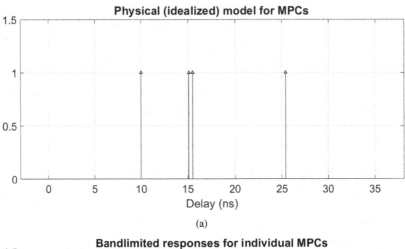

(a)

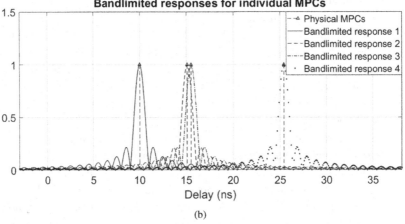

(b)

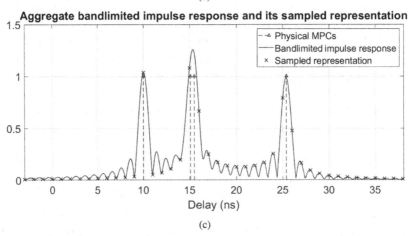

(c)

Figure 12.2 Top: Physical model represented by idealized (infinite bandwidth) MPCs. Middle: The bandlimited impulse responses for individual MPCs. Bottom: The aggregate bandlimited CIR and its sampled representation.

$x = 0$ (with a half-power mainlobe width of approximately $\Delta\tau = \frac{1}{W}$), the critically sampled virtual channel representation induces an intuitively appealing partitioning of paths that follows from eq. (12.18) and is illustrated in Figure 12.2 (bottom plot): The ℓth sampled channel coefficient represents (approximately) the sum of the responses of all (unresolvable) MPCs whose delays lie within a resolution bin of size $\Delta\tau = \frac{1}{W}$ centered around the ℓth virtual delay. Thus, $\Psi_D(\ell) = |H_v(\ell)|^2$ (or an oversampled version) is also a direct estimate of the channel power delay profile (PDP).

Matrix representation. The sampled channel representation in eqs. (12.17)–(12.19) can be further sampled in frequency with resolution $\Delta\nu = \frac{1}{T}$ to yield a matrix representation of the channel that is very useful for digital processing of measurements:

$$h_f[k] = H(k\Delta\nu) = \sum_{\ell=0}^{L} H_v(\ell)e^{-j2\pi\frac{\ell k}{WT}} = \sum_{\ell=0}^{L} g_\tau[\ell]e^{-j\frac{2\pi\ell k}{N}} \iff \mathbf{h}_f = \mathbf{U}_N \mathbf{g}_\tau,$$

(12.20)

where $N = TW$ is the (approximate) dimension of the space of signals of duration T and bandwidth W, $\mathbf{h}_f$ and $\mathbf{g}_\tau$ are N-dimensional sampled representations of the channel frequency response and the CIR, respectively, and $\mathbf{U}_N$ is an $N \times N$ DFT matrix. As is evident from eq. (12.20), the first $L + 1$ entries of $\mathbf{g}_\tau$, corresponding to the channel delay spread, carry most of the channel energy; the rest of the entries are relatively small and can be set to zero. Note that for the matrix representation to hold, oversampling in delay must be accompanied with a corresponding oversampling in frequency satisfying $\frac{\Delta\tau}{\Delta\nu} = \frac{T}{W}$. This simply increases the dimension of the matrix representation to $N' = \frac{T}{\Delta\tau} = \frac{W}{\Delta\nu}$. From this matrix representation, it is clear that the number of samples (in time or frequency or time–frequency), $N = TW$, must be greater than the number of channel parameters to be estimated, $2N_p + 1$. Using a larger N (by increasing T and/or W) and/or by oversampling, the accuracy of the estimates can be improved.

12.2.4.2 Single Antenna Delay–Doppler Channel

A time-varying single-antenna channel can be represented in terms of the delay–Doppler spreading function $h(\tau, \nu) = \sum_{n=1}^{N_p} \alpha_n \delta(\tau - \tau_n)\delta(\nu - \tau_n)$ or equivalently by its 2D Fourier transform, the time-varying frequency response function:

$$H(t, f) = \sum_{n=1}^{N_p} \alpha_n e^{-j2\pi\tau_n f} e^{j2\pi\nu_n t} \approx \sum_{\ell=0}^{L}\sum_{m=-M}^{M} H_v(\ell, m)e^{-j2\pi\frac{\ell}{W}f}e^{2\pi\frac{m}{T}t}, \quad (12.21)$$

where the first equality is the physical model and the second approximation is the sampled representation, in terms of uniformly spaced virtual delays and Doppler shifts, whose coefficients can be computed as

$$H_v(\ell, m) = \tilde{h}(\ell \Delta \tau, m \Delta v) = \frac{1}{TW} \int_0^T \int_{-W/2}^{W/2} H(t, f) e^{j2\pi \ell \Delta \tau f} e^{-j2\pi m \Delta v t} \, dt \, df$$

$$= \sum_{n=1}^{N_p} \alpha_n \, \mathrm{sinc} \left(W \left(\tau_n - \ell \Delta \tau \right) \right) \mathrm{sinc} \left(T ((v_n - m \Delta v)) \right). \tag{12.22}$$

As in the case of the static channel, the virtual delay–Doppler channel coefficients induce a partitioning of propagation paths in delay–Doppler as illustrated in Figure 12.1(a). Analogous to eq. (12.20), a matrix representation for eq. (12.21) can be developed by sampling $H(t, f)$ as $h_{t,f}[i, k] = H(i \Delta t, k \Delta f)$, resulting in $\mathbf{H}_{t,f} = \mathbf{U}_N \mathbf{G}_{\tau,v} \mathbf{U}_N^\dagger$ where $\mathbf{H}_{t,f}$ and $\mathbf{G}_{\tau,v}$ are $N \times N$ matrices representing $h_{t,f}[i, k]$ and $g_{\tau,v}[\ell, m] = H_v(\ell, m)$, respectively, and $\mathbf{U}_N$ is an $N \times N$ DFT matrix. Again, the dominant nonzero entries are limited to $2M + 1$ Doppler indices and $L + 1$ delay indices, as reflected in eq. (12.21).

12.2.4.3   Multi-antenna Channel

Another important special case of the general model in eqs. (12.6) and (12.8) is a multi-antenna channel that is nonselective in time and frequency; that is, there is negligible variation in time and frequency over the duration and bandwidth of interest, resulting in $\mathbf{H}(t, f) = \mathbf{H}$ and $L = M = 0$ in eq. (12.8). In this case, using eq. (12.8), the spatial channel matrix $\mathbf{H}$ can be expressed as

$$\mathbf{H} = \sum_{n=1}^{N_p} \alpha_n \mathbf{a}_R(\theta_{R,n}) \mathbf{a}_T^H(\theta_{T,n})$$

$$= \sum_{i=1}^{N_R} \sum_{k=1}^{N_T} H_v(i, k) \mathbf{a}_R(i \Delta \theta_R) \mathbf{a}_T^H(k \Delta \theta_T) = \mathbf{U}_{N_R} \mathbf{H}_b \mathbf{U}_{N_T}^\dagger, \tag{12.23}$$

where the first equality represents the physical model and the second equality is the virtual (beamspace) channel representation in terms of uniformly spaced AoAs and AoDs at resolutions defined in eq. (12.5). The last equality is a matrix representation of the $N \times N$ antenna-domain matrix $\mathbf{H}$ in terms of a 2D DFT of the $N \times N$ beamspace channel matrix $\mathbf{H}_b$ whose entries are the virtual (beamspace) channel coefficients that can be computed as

$$H_b(i, k) = H_v(i, k) = \mathbf{a}_R^\dagger(i \Delta \theta_R) \mathbf{H} \mathbf{a}_T(k \Delta \theta_T)$$

$$= \sum_{n=1}^{N_p} \alpha_n f_{N_R}(\theta_{R,n} - i \Delta \theta_R) f_{N_T}(\theta_{T,n} - k \Delta \theta_T). \tag{12.24}$$

The last equality in eq. (12.24) relates the beamspace channel coefficients to the physical paths in terms of the Dirichlet sinc function $f_N(\theta) = \frac{\sin(\pi N \theta)}{\sin(\pi \theta)}$, which is peaky around the origin with a half-power mainlobe width of approximately the spatial resolution of the array of size N. The matrices $\mathbf{U}_{N_R}$ and $\mathbf{U}_{N_T}$ in eq. (12.23) represent the (unitary) spatial DFT matrices that map the antenna domain into the angle domain (beamspace). The columns of these matrices are given by steering/response vectors, defined in eq. (12.3), corresponding to uniformly sampled directions with spacings defined in eq. (12.5). The beamspace channel coefficients partition the propagation paths in terms of angular resolution bins, as illustrated in Figure 12.1(b).

12.3   High-Resolution Parameter Extraction and the Physical Model

The resolutions in the sampled representation, defined in eq. (12.5), are the Fourier resolution limits, imposed by the finite bandwidth, temporal duration and array apertures of actual channel sounders or systems. Thus the sampled representation serves as a natural bridge between the infinite resolution physical model and the finite resolution sounder measurements. If the channel is (as is common) represented as a finite sum of MPCs or plane waves (plus, possibly, a low-parameter-dimension diffuse component), and the spatiotemporal signal space dimension is significantly larger than the number of paths, then the resulting channel sparsity can be exploited to obtain estimates of the physical parameters (angles, delays) at a higher resolution [15, 16]. This situation usually applies in mmWave channels, since the large bandwidth and large array sizes (in units of wavelength) imply that the number of MPCs is much smaller than the signal space dimension, even if the number of MPCs is quite large in absolute terms. A number of methods may be used for such HRPE, including maximum likelihood (ML), MUSIC and ESPRIT [8, 17, 18], and compressed sensing/sparse recovery methods for sparse channel estimation [13]. Since the first application of an HRPE to mmWave measurements in [19], a number of papers have used such techniques to provide improved accuracy for MPC parameter estimation.

For a given channel sounder, HRPE techniques have the potential to increase the resolution and accuracy of channel data analysis beyond the Fourier limits by exploiting a-priori knowledge about the form of channel measurements for different sets of parameters. One example is frequency estimation of two sinusoids whose separation in frequency is less than the Fourier resolution $\Delta \nu = \frac{1}{T}$. In Fourier analysis, resolving two frequencies would require two distinct peaks in the spectrum to be visible. However, the result of Fourier analysis is a spectrum with not just two distinct frequency values. In the case of two sinusoids it would be apparent from the spectrum that there is more than one component since the spectrum would not look like one of a single sinusoid. But there is no direct way to determine both frequencies. In HRPE on the other hand, the estimator exploits a-priori information about the number of sinusoids contributing to the spectrum and the analytical form of the resulting Fourier transform – in effect a parametric representation of the Fourier spectrum in terms of the hypothesized frequencies of the sinusoids. The values of the parameters that result in the "best match" (e.g., in an ML sense) to the measured signal form an estimate of the frequencies.

A channel model usually provides a function to compute the impulse response (or frequency) of the channel (for a given configuration, such as bandwidth and array structure) from the physical parameters as in eq. (12.1). For example, the channel impulse function h translates from the parameter domain to the measurement domain, where the measurement domain is defined as the domain in which the sounder measures the channel:

$$h: \left\{\alpha_n, \tau_n, \nu_n, \phi_n^R, \phi_n^T\right\}_n \rightarrow \left\{\tau, t, \mathbf{x}_i^R, \mathbf{x}_k^T\right\}_{ik}, \tag{12.25}$$

where $\left\{\alpha_n, \tau_n, \nu_n, \phi_n^R, \phi_n^T\right\}_n$ denote the propagation path parameters that are mapped by h to the time-varying impulse response that is a function of the time t, the delay lag τ and depends on the position $\mathbf{x}_k^T$ of the kth antenna at the transmitter and the position $\mathbf{x}_i^R$ of the ith antenna at the receiver of the channel sounder. For the ULA model described in eq. (12.1) this model for the CIR corresponds to

$$h_{i,k}(t, \tau) = \sum_{n=1}^{N_p} \alpha_n e^{-j2\pi\theta_n^R i} e^{j2\pi\theta_n^T k} \delta(\tau - \tau_n) e^{j2\pi\nu_n t}, \tag{12.26}$$

where the time variation is captured by the Doppler shift ν_n for constant-velocity relative motion for the nth path, or it could be captured by temporal variation in the path amplitudes $\alpha_n \rightarrow \alpha_n(t)$ for more general relative motion for the nth path.

HRPE can be viewed as the inverse problem: Given a set of observations in the measurement domain, the estimator tries to recover the underlying parameters of the channel model. This corresponds to "parametric model identification" where the impulse (or frequency) response of the linear time-invariant (LTI) system is described by a set of parameters through a specific function for the impulse (or frequency) response, as in eq. (12.26) (or eq. (12.1)). The inverse mapping f underlying an HRPE technique, corresponding to the impulse response model in eq. (12.25), can be formally described as

$$f: \left\{\tau, t, \mathbf{x}_i^R, \mathbf{x}_k^T\right\}_{ik} \rightarrow \left\{\alpha_n, \tau_n, \nu_n, \phi_n^R, \phi_n^T\right\}_n. \tag{12.27}$$

That is, the inputs to the HRPE methods are the channel measurements made by the sounder, for example, $H_{i,k}(t, f)$ in eq. (12.1) or $h_{i,k}(t, \tau)$ in eq. (12.26). These inputs are then processed by the HRPE method to extract the physical channel model parameters by exploiting the a-priori information about the functional relationship between the measurements and the physical model parameters, as described in eqs. (12.1) or (12.26).

The observations required for the estimator are provided by a channel sounder. While a mapping h from the physical parameters to the measured impulse response in general always exists, the measured data need to satisfy certain requirements in order for an estimator f to exist. The measurement domain has to be chosen such that there is no ambiguity in the parameters; that is, there should be only a single set of parameters that lead to a certain set of observations. If this is true, then the estimator can uniquely determine the parameters from the observations. This also provides a direct approach for exploring the parameter space and identifying the parameters that result in the closest fit to the observed channel impulse response/transfer function. The sampled channel representation described in Section 12.2 quantifies these requirements on the sounder.

For estimation of the delay parameters in a temporal impulse response, a common requirement is that the sounding signal is long enough in time to capture the whole impulse response. Otherwise, a component with a delay larger than length of the sounding signal can appear as an earlier delay in subsequent consecutive measurements. One way to avoid this problem is to put sufficient spacing (longer than the channel delay spread) between successive sounding signals.

For the case of angle estimation using a ULA, there exist an infinite number of azimuth and elevation pairs $\left(\left\{\phi^A, \phi^E\right\}\right)$ that result in the same observation (steering vector); see, for example, [20]. Such ambiguities can be readily avoided by using a 2D UPA. If using a 2D array (e.g., a UPA) is not feasible, sometimes the ambiguities can be accounted for by fixing one of the parameters (e.g., assuming a certain elevation in the ULA example) but only if fixing one parameter does not affect the other parameters. Another example is angular estimation using a circular array when the estimator is designed not to estimate elevation; this can lead to false estimates of the azimuth. Finally, the set of observations have to provide sufficient information to enable estimation of all parameters, that is, the measurements have to be linearly independent and their number larger than the number of parameters to be estimated. Again, for a given channel sounder, the sampled representation helps quantify these requirements.

12.3.1   Measurement Data Model: Measurement Equation

Any estimator requires knowledge about a function, such as in eqs. (12.25) or (12.26), in order to solve the inverse problem stated in eq. (12.27) since it needs to know how a given set of parameters would affect a measurement. The possibility of estimating continuous parameters from sampled measurement data in space and time arises from the observation that different values of parameters result in different measurements. This observation is formalized by the sampling theorem for bandlimited signals, which is at the heart of the sampled channel representation in angle–delay–Doppler. Furthermore, while the sounder samples at discrete points in space and time that are theoretically continuous-valued in amplitude and phase, in practice they are limited by the finite-bit resolution of the sounder. As a result, the ability to detect small changes in measurements is limited by measurement and/or quantization noise, and how accurately the mapping between parameters and the measurement domain – the channel model – is known a-priori.

While the function (12.25) is the general definition of how a channel model would map its parameters to the measurement domain, the function/model for the estimator is specific to the channel sounder used for making measurements, and accounts for all relevant characteristics of the sounder components. This specific function that maps parameters of the channel model to the sounder's measurement domain will be called the "measured impulse response," h_{meas}. The function $f_{i,k}(\tau)$ (see eq. (12.27)) corresponds to the sounder equipment in the measurement equation, representing the composite impulse response of the filters for the kth transmitter antenna and the ith receiver antenna. For ULAs, the ideal sounder-independent CIR is given by eq. (12.26), and it can be convolved with $f_{i,k}(\tau)$ to yield the measured impulse response

$$h_{\text{meas},i,k}(t,\tau) = h_{i,k}(t,\tau) * f_{i,k}(\tau) = h_{i,k}(t,\tau) * f(\tau), \qquad (12.28)$$

where the last equality corresponds to the situation, also discussed in Section 12.1, in which all antenna elements at the transmitter have the same impulse response and all antenna elements at the receiver have the same impulse response to yield a common composite impulse response $f(\tau) = f_{i,k}(\tau)$. In the case of a sliding-correlator estimator, this correction also has to account for the autocorrelation function of the sounding signal. If back-to-back-calibration is done to estimate $f_{i,k}(\tau)$ with subsequent deconvolution applied to h_{meas}, then the sounder-independent CIR would reduce to sinc functions in delay, or rectangular functions with constant amplitude in frequency, reflecting the finite bandwidth of the sounder; this is explicitly stated in eq. (12.12) in the sampled channel representation and forms the basis of path partitioning.

The complex patterns $\mathbf{g}_k^T(\phi^T)$ and $\mathbf{g}_i^R(\phi^R)$, defined in Section 12.1, represent the characteristics of the antennas (arrays) used during measurements. If the bandwidth is large enough to reveal a frequency dependence of the antennas, the complex patterns also become frequency-dependent.

The performance and accuracy of an estimator is limited by how accurately one can model the measurement equation with respect to the actual response of the sounder. The measurement data model or the measurement equation has, therefore, to be properly chosen to fit to the sounder used. This is the objective of sounder calibration and verification, as discussed in Chapters 3 and 4 (see Table 3.3). For the delay- or frequency-domain this is relatively easy if it is possible to measure the frequency response of the RF components of the sounder (e.g., in a back-to-back calibration). The measurement of the sounder given an MPC with a certain delay would result in a shift by the same delay of the sounder's impulse response. For the spatial domain this is more challenging since the response of the antennas to an MPC with a certain angle is in general not a shifted version of the response of another angle. For ideal ULAs, this property holds, except for extreme angles near $\mp 90°$. Therefore, it is usually necessary to measure the complex radiation pattern of the antenna arrays, in an anechoic chamber, for example, for a number of source angles. Since the measurement equation needs the complex radiation pattern for an arbitrary angle, but one can only measure it at discrete angles, some form of interpolation is needed. A good method of interpolation is based on the "effective aperture distribution function" (EADF) introduced by [17]. The EADF is effectively a 2D discrete Fourier transform of the measured radiation pattern for a single antenna element. Therefore, it represents a (Fourier) series expansion of the underlying continuous complex radiation pattern obtained from sampled data and can be used to obtain the pattern at arbitrary angles. Obviously the sampling during measurements has to be dense enough in angle such that the sampling theorem holds, and the interpolation accuracy is also limited by the

SNR of the measurements. Another approach of optimal interpolation would be the use of vector spherical harmonics [21].

The sampled representation in Section 12.2 generalizes this concept of Fourier series expansion of radiation patterns to antenna arrays, especially ULAs, and helps quantify the sampling requirements in angle for both single-antenna and antenna-array radiation patterns. For individual antenna elements, the radiation pattern is relatively smooth and thus relatively few measurements over the angular spread are needed – essentially equal to the length of the antenna normalized by half-wavelength in each dimension. Once the radiation pattern for each antenna element is known, the far-field radiation patterns for an array can be computed using the radiation pattern for each element and the relative locations of the elements in the array.

Some estimators, or better implementations thereof, do not make use of measurement equations obtained from measurements but assume an analytical model of the frequency response and complex radiation patterns. This is the essence of the sampled representation for the ULAs, which assumes idealized omnidirectional radiation patterns for each antenna element. It is quite common to assume constant frequency response, omnidirectional (constant gain in, e.g., azimuth) or even isotropic radiation (constant gain of the antenna independent of direction), no coupling between array elements and perfect knowledge about the relative position of each element in the array (in order to compute the phase offset due to out-of-phase-center positioning). Those models have to be used very carefully because the estimator might behave unpredictably if the analytical model does not match the properties of the physical antenna array. The benefit of using measured patterns (with, e.g., the EADF as an interpolator) is that any imperfection of the antennas is automatically taken into account. However, the relative locations and orientations of the antenna elements are still needed with sufficient accuracy. Estimators like ESPRIT and Root-MUSIC require the antenna array to have certain properties (such as being a perfect ULA), making it impossible to use them if the requirements cannot be fulfilled. We note that while sounder calibration in the time or frequency domain is relatively straightforward or well understood, the calibration of sounders from a spatial or angular perspective requires additional research, which is one of the goals of the 5G Alliance moving forward, especially in terms of interactions between measurement and modeling.

12.3.2   Maximum-Likelihood Estimation

The class of ML estimators is attractive for HRPE as they generally pose no limitations on the sounder/arrays used for measurements, other than the requirement of providing ambiguity-free measurements of the propagation channel. The type of antenna array used can also be quite general (nonuniform elements, irregular spacing, etc.), as long as the array produces a unique output for all possible angles. However, there obviously exist better or worse configurations with regard to the estimator performance and accuracy; for example, elements spaced too closely together can introduce ill-conditioning in the estimator. But there is no general limitation imposed by the method (such as shift-invariance in ESPRIT). On the other hand, ML estimators do rely on the measurement equation to be as accurate as possible. Any model mismatch in the measurement equation limits the accuracy and reliability of the estimates. Finally, as in any estimator, the accuracy of the estimates is limited by the SNR of the measurements.

Maximum-likelihood estimators are based on the principle of maximizing the likelihood or probability that a given channel measurement is generated by certain values of underlying model parameters. A channel measurement can be modeled as

$$\tilde{\mathbf{x}} = \mathbf{h}_{\mathrm{meas}}\left(\Theta\right) + \mathbf{n}, \tag{12.29}$$

where Θ denotes all the parameters of $\mathbf{h}_{\mathrm{meas}}$, the measurement model for the channel as discussed in Section 12.3.1, which can be further decomposed into a deterministic discrete component and a stochastic diffuse component:

$$\mathbf{h}_{\mathrm{meas}} = \mathbf{h}_s\left(\Theta_s\right) + \mathbf{h}_d\left(\Theta_d\right), \tag{12.30}$$

where Θ_s are the parameters of the discrete deterministic model, such as the delays, angles and complex amplitudes of the different paths in the ULA physical model in eq. (12.4), and Θ_d are the parameters of the stochastic diffuse channel component. Assuming a Gaussian distributed and possibly correlated stochastic component $\mathbf{h}_d$, the channel measurements are distributed as

$$\tilde{\mathbf{x}} \sim \mathcal{N}_C\left(\mathbf{h}_s\left(\Theta_s\right), \mathbf{R}\left(\Theta_{\mathrm{dan}}\right)\right), \tag{12.31}$$

where Θ_{dan} denotes the parameters of both the diffuse channel component and the Gaussian noise $\mathbf{n}$. The corresponding likelihood function can be written as

$$\ell\left(\tilde{\mathbf{x}}|\Theta_s, \mathbf{R}\left(\Theta_{\mathrm{dan}}\right)\right) = \frac{1}{\pi^M \det\left(\mathbf{R}\left(\Theta_{\mathrm{dan}}\right)\right)} e^{-(\tilde{\mathbf{x}}-\mathbf{h}_s(\Theta_s))^\dagger \mathbf{R}^{-1}(\Theta_{\mathrm{dan}})(\tilde{\mathbf{x}}-\mathbf{h}_s(\Theta_s))}, \tag{12.32}$$

which needs to be maximized for ML estimation. Equivalently, ML estimates of the channel model parameters can be obtained by minimizing the negative log-likelihood function

$$\mathcal{L}\left(\tilde{\mathbf{x}}|\Theta_s, \mathbf{R}\left(\Theta_{\mathrm{dan}}\right)\right) = \ln\left(\det\left(\mathbf{R}\left(\Theta_{\mathrm{dan}}\right)\right)\right) + \left(\tilde{\mathbf{x}} - \mathbf{h}_s\left(\Theta_s\right)\right)^\dagger \mathbf{R}^{-1}\left(\Theta_{\mathrm{dan}}\right)\left(\tilde{\mathbf{x}} - \mathbf{h}_s\left(\Theta_s\right)\right) \tag{12.33}$$

as

$$\left(\hat{\Theta}_s, \hat{\Theta}_{\mathrm{dan}}\right) = \arg \min_{\Theta_s, \Theta_{\mathrm{dan}}} \mathcal{L}\left(\tilde{\mathbf{x}}|\Theta_s, \mathbf{R}\left(\Theta_{\mathrm{dan}}\right)\right). \tag{12.34}$$

This poses a nonlinear optimization problem where estimating the parameters of the discrete channel model is a weighted nonlinear least-squares problem.

The problem can be simplified by making assumptions about or guaranteeing (by designing the sounder) certain properties of the measurement equation. However, as stated above, making assumptions can be dangerous if they are not met in reality.

In general the number of MPCs and thus the number of parameters to estimate is large (>100 components) and therefore the dimensionality of the problem is usually large and there is no analytical solution available. There are three main challenges that an estimator has to address: (1) How many model parameters need to be estimated from measurements, that is, the model order? (2) Can the dimensionality of the problem be reduced? And (3) How do we find the minimum of a nonlinear function? The model order can be estimated prior to starting the estimator or iteratively while estimating (as it is done in, e.g., RiMAX [17]). The dimensionality of the search procedure can be reduced by assuming that certain parameters do not influence each other, as is done in EM (expectation maximization) or SAGE (space alternating generalized expectation maximization) algorithms by treating the problem of estimating P parameters as estimating P individual components independently. However, this assumption is not valid in sparse multipath channels as encountered at mmWave frequencies, in particular. However, sparsity offers other approaches to effectively reducing the dimensionality (or the number of measurements needed) by employing sparse estimation approaches that may be combined with ML estimation by introducing an ℓ_1 regularization term to the least-squares optimization problem in eq. (12.34) (see, e.g., [13]).

Finding the global minimum is generally challenging if the optimization function eq. (12.33) is not convex in the model parameters. A brute force approach may be applied using a grid-based search with further refinement of the grid (commonly done in SAGE). However, this can be prohibitively expensive for a high-dimensional parameter space. One approach to reducing complexity is found in RiMAX, where components are grouped based on their influence on each other. Typically iterative gradient-based algorithms are used, such a Gauss–Newton or Levenberg–Marquardt (done in RiMAX) to reach local minima. Gradient-based methods have the benefit of faster convergence, especially in the case of correlated parameters. However, they require the first- and sometimes also second-order partial derivatives of the negative log-likelihood function. The gradient-based methods usually converge to a local minimum, while one is interested in finding the global minimum. This problem can be solved by initializing the algorithm in the vicinity of the global minimum. The initialization could be done, for example, using the sampled representation introduced in eq. (12.2) as a starting point.

12.4 Extension to 2D Antenna Arrays

The physical model in eq. (12.1) for 1D ULAs can be extended to 2D arrays [5, 6, 22] consistent with the general model in Section 12.1. Consider a system with UPAs at the TX and RX. Essentially the physical model in eq. (12.1) and the sampled representation in eq. (12.6) can be used by replacing the definitions of the steering/response vectors and the AoDs/AoAs to include both azimuth and elevation. Specifically, the AoDs and AoAs can be defined as

$$\theta^{\mathrm{T}} = (\theta^{\mathrm{T,A}}, \theta^{\mathrm{T,E}}), \theta^{\mathrm{R}} = (\theta^{\mathrm{R,A}}, \theta^{\mathrm{R,E}});$$

$$-\frac{\pi}{2} \leq \theta^{\mathrm{T,A}}, \theta^{\mathrm{T,E}} \leq \frac{\pi}{2}, \quad -\frac{\pi}{2} \leq \theta^{\mathrm{T,E}}, \theta^{\mathrm{R,E}} \leq \frac{\pi}{2} \qquad (12.35)$$

$$\theta^{\mathrm{T}} = (\theta^{\mathrm{T,A}}, \theta^{\mathrm{T,E}}), \theta^{\mathrm{R}} = (\theta^{\mathrm{R,A}}, \theta^{\mathrm{R,E}}) ;$$

$$\theta^{\mathrm{T,A}} = \frac{d^{\mathrm{A}}}{\lambda} \sin(\phi^{\mathrm{T,A}}), \ \theta^{\mathrm{T,E}} = \frac{d^{\mathrm{E}}}{\lambda} \sin(\phi^{\mathrm{T,E}}), \quad\quad (12.36)$$

and the array steering and response vectors are given by

$$\mathbf{a}^{\mathrm{T}}(\theta^{\mathrm{T}}) = \mathbf{a}^{\mathrm{T,A}}(\theta^{\mathrm{T,A}}) \otimes \mathbf{a}^{\mathrm{T,E}}(\theta^{\mathrm{T,E}}), \ \mathbf{a}^{\mathrm{R}}(\theta^{\mathrm{R}}) = \mathbf{a}^{\mathrm{R,A}}(\theta^{\mathrm{R,A}}) \otimes \mathbf{a}^{\mathrm{R,E}}(\theta^{\mathrm{R,E}}) \quad (12.37)$$

in terms of Kronecker products of 1D steering/response vectors in azimuth and elevation. The dimensions of the TX and RX UPAs can be factored as

$$N^{\mathrm{T}} = N^{\mathrm{T,A}} \times N^{\mathrm{T,E}} ; \ N^{\mathrm{R}} = N^{\mathrm{R,A}} \times N^{\mathrm{R,E}},$$

representing the number of antennas in the azimuth and elevation directions. Similarly, the spatial resolutions in azimuth and elevation are given by

$$\Delta\theta^{\mathrm{T,A}} = \frac{1}{N^{\mathrm{T,A}}}, \ \Delta\theta^{\mathrm{T,E}} = \frac{1}{N^{\mathrm{T,E}}} ; \ \Delta\theta^{\mathrm{R,A}} = \frac{1}{N^{\mathrm{R,A}}}, \ \Delta\theta^{\mathrm{R,E}} = \frac{1}{N^{\mathrm{R,E}}}. \quad (12.38)$$

12.5   The Extended Sampled Channel Model: Hardware Nonidealities

Future work will include extending the sampled channel model to include hardware nonidealities. At mmWave frequencies, the nonideal characteristics of the hardware, including the antennas, filters and mixers, are significantly more pronounced due to the high angle–delay resolution afforded by the electrically large arrays and large bandwidths [5, 6, 11]. Thus, the ideal sampling described in eq. (12.2) needs to be extended to account for the nonideal hardware characteristics. In effect, the measured beam-frequency channel matrix $\mathbf{H}$ (in the quasi-static case) can be decomposed into three (linear) components:

$$\mathbf{H} = \mathbf{H}^{\mathrm{T}} \mathbf{H}_P \mathbf{H}^{\mathrm{R}}, \quad\quad (12.39)$$

where $\mathbf{H}_P$ denotes the propagation channel and $\mathbf{H}^{\mathrm{T}}$ and $\mathbf{H}^{\mathrm{R}}$ denote the nonideal beam-frequency responses of the sounder hardware at the TX and the RX, respectively. In fitting measurements to models, it is important to isolate the nonideal characteristics of the hardware from the underlying propagation characteristics. Three main research tasks are needed to develop this extended model:

Task 1: Develop and validate the models for $\mathbf{H}^{\mathrm{T}}$ and $\mathbf{H}^{\mathrm{R}}$ that account for the nonideal beam-frequency characteristics induced by the antenna array and the wideband mmWave hardware. These models will combine the nonideal characteristics of individual hardware components.

Task 2: Develop an approach for jointly estimating $\mathbf{H}_P$, $\mathbf{H}^{\mathrm{T}}$ and $\mathbf{H}^{\mathrm{R}}$ from measurements of $\mathbf{H}$. We can exploit the rich structure of $\mathbf{H}_P$ induced by the multi-dimensional Fourier basis waveforms underlying the angle–delay–Doppler sampling [2, 7] as well as the structure of $\mathbf{H}^{\mathrm{T}}$ and $\mathbf{H}^{\mathrm{R}}$ induced by individual components. Some key issues to address include the design of beam-frequency probing signals at the TX and measurements at the RX and understanding the fundamental limits to decomposing $\mathbf{H}$ into $\mathbf{H}_P$, $\mathbf{H}^T$ and $\mathbf{H}^R$.

Task 3: Refine and validate the modeling framework with actual channel measurements. In particular, the calibration methods outlined in Chapter 3 address Tasks 2 and 3 for single-antenna systems and will be extended to multi-antenna systems. Specifically, new methods for calibration of phased array and lens array antennas are needed. The diversity of MIMO channel measurements, using horn antennas, phased arrays and lens arrays, will play an important role in developing the channel modeling framework.

References

[1] L. Liu, C. Oestges, J. Poutanen, K. Haneda, P. Vainikainen, F. Quitin, F. Tufvesson and P. Doncker, "The COST 2100 MIMO channel model," *IEEE Transactions on Wireless Communications*, vol. 19, no. 6, pp. 92–99, 2012.

[2] A. Sayeed and T. Sivanadyan, "Wireless communication and sensing in multipath environments using multi-antenna transceivers," in *Handbook on Array Processing and Sensor Networks* (K. J. R. Liu and S. Haykin, Eds.), IEEE-Wiley: Piscataway, NJ, 2010.

[3] A. M. Sayeed, "Deconstructing multi-antenna fading channels," *IEEE Transactions on Signal Processing*, vol. 50, pp. 2563–2579, Oct. 2002.

[4] M. Steinbauer, A. F. Molisch and E. Bonek, "The double-directional radio channel," *IEEE Antennas and propagation Magazine*, vol. 43, no. 4, pp. 51–63, 2001.

[5] J. Brady, N. Behdad and A. Sayeed, "Beamspace MIMO for millimeter-wave communications: System architecture, modeling, analysis and measurements," *IEEE Transactions on Antenna and Propagation*, vol. 61, no. 7, pp. 3814–3827, July 2013.

[6] A. Sayeed and J. Brady, "Beamspace MIMO channel modeling and measurement: Methodology and results at 28 GHz," *IEEE Globecom Workshop on Millimeter-Wave Channel Modeling*, Dec. 2016.

[7] A. M. Sayeed, "A virtual representation for time- and frequency-selective correlated MIMO channels," *Proceedings of the 2003 International Conference on Acoustics, Speech and Signal Processing*, vol. 4, 2003, pp. 648–651.

[8] R. Roy and T. Kailath, "ESPRIT-estimation of signal parameters via rotational invariance techniques," *IEEE Transactions on Acoustics, Speech and Signal Processing*, vol. 37, no. 7, pp. 984–995, July 1989.

[9] A. M. Sayeed and V. Raghavan, "Maximizing MIMO capacity in sparse multipath with reconfigurable antenna arrays," *IEEE Journal of Selected Topics in Signal Processing*, vol. 1, no. 1, pp. 156–166, June 2007.

[10] P. Bello, "Characterization of randomly time-variant linear channels," *IEEE Transactions on Communications Systems*, vol. 11, no. 4, pp. 360–393, 1963.

[11] J. Brady, J. Hogan and A. Sayeed, "Multi-beam MIMO prototype for real-time multiuser communication at 28 GHz," *IEEE Globecom Workshop on Emerging Technologies for 5G*, Dec. 2016.

[12] A. Sayeed, J. Brady, P. Cheng and U. Tayyab, "Indoor channel measurements using a 28 GHz multi-beam MIMO prototype," *IEEE Vehicular Technology Conference Workshop on Millimeter-Wave Channel Models*, Sept. 2016.

[13] W. Bajwa, J. Haupt, A. Sayeed and R. Nowak, "Compressed channel sensing: A new approach to estimating sparse multipath channels," *Proceedings of the IEEE*, vol. 98, no. 6, June 2010.

[14] A. Sayeed, C. Hall and Y. Zhu, "A lens array multi-beam MIMO testbed for real-time mmWave communication and sensing," *First ACM Workshop on Millimeter-Wave Networks and Sensing Systems (Mobicom)*, Oct. 2017.

[15] J. Fuhl, J. Rossi and E. Bonek, "High-resolution 3-D direction-of-arrival determination for urban mobile radio," *IEEE Transactions on Antennas and Propagation*, vol. 45, no. 4, pp. 672–682, 1997.

[16] T. Santos, J. Karedal, P. Almers, F. Tufvesson and A. F. Molisch, "Modeling the ultra-wideband outdoor channel: Measurements and parameter extraction method," *IEEE Transactions on Wireless Communications*, vol. 9, no. 1, pp. 282–290, 1989.

[17] Richter, A., "Estimation of radio channel parameters: Models and algorithms," PhD Thesis, Vienna University of Technology, 2005.

[18] R. O. Schmidt, "Multiple emitter location and signal parameter estimation," *IEEE Transactions on Antennas and Propagation*, vol. 34, no. 3, pp. 276–280, Mar. 1986.

[19] C. Gustafson, F. Tufvesson, S. Wyne, K. Haneda and A. F. Molisch, "Directional analysis of measured 60 GHz indoor radio channels using SAGE," *IEEE Vehicular Technology Conference (VTC Spring)*, 2011, pp. 1–5.

[20] M. Landmann, M. Kaske and R. Thoma, "Impact of incomplete and inaccurate data models on high resolution parameter estimation in multidimensional channel sounding," *IEEE Transactions on Antennas and Propagation*, vol. 60, pp. 557–573, Feb. 2012.

[21] G. Del Galdo, J. Lotze, M. Landmann and M. Haardt, "Modelling and manipulation of polarimetric antenna beam patterns via spherical harmonics," *European Signal Processing Conference*, 2006.

[22] A. Sayeed and J. Brady, "Millimeter-wave MIMO transceivers: Theory, design and implementation," in *Signal Processing for 5G: Algorithms and Implementations* (F.-L. Luo and J. Zhang, Eds.), IEEE-Wiley: Piscataway, NJ, 2016.

13 Conclusions

Theodore S. Rappaport, Kate A. Remley, Camillo Gentile,
Andreas F. Molisch and Alenka Zajić

With this book, researchers from around the globe under the common umbrella of the *5G mmWave Channel Model Alliance* have intended to provide practical guidance on both verification of channel sounding measurements and on channel models that must capture a wide range of channel characteristics. Throughout, our goal has been to illustrate the important link between channel measurements and channel models at mmWave, sub-THz, and THz frequencies. As the frequency of operation increases, wireless device design must become more agile, dynamic and powerful. Such stringent design criteria require channel models describing the many effects to which the device may be subjected. Yet, the nonidealities in the channel measurement hardware can cause issues unless care is taken to understand the metrics and verify that the hardware is performing as desired. It is the hope of the many contributors to this book that the measurement and modeling techniques presented here will be useful to readers from many disciplines, from beginner to advanced practitioners and academia to industry. We all have the common goal of advancing the state of the art in channel measurement and modeling at mmWave and sub-THz frequencies.

Index

Printed in the United States
by Baker & Taylor Publisher Services